THE PHYSICS OF SYNCHROTRON RADIATION

This book explains the underlying physics of synchrotron radiation and derives its main properties. It is divided into four parts. The first covers the general case of the electromagnetic fields created by an accelerated relativistic charge. The second part concentrates on the radiation emitted by a charge moving on a circular trajectory, deriving its distribution in angle, frequency, and polarization modes. The third part looks at undulator radiation. Starting from the simple case of a plane weak undulator with a spatially periodic field that emits quasi-monochromatic radiation, the author then discusses strong undulators, emitting more complicated radiation and containing higher harmonics. More general undulators are also considered, with a non-planar (helical) electron trajectory or non-harmonic field. The final part deals with applications and investigates the optics of synchrotron radiation dominated by diffraction due to the small opening angle. It also includes a description of electron-storage rings as radiation sources and the effect of the emitted radiation on the electron beam.

This book provides a valuable reference for scientists and engineers in the field of accelerators, and for all users of synchrotron radiation.

ALBERT HOFMANN received his doctorate in physics from the ETH (Swiss Federal Institute of Technology) in Zürich in 1964. From 1966 to 1972 he was a Research Fellow at the Cambridge Electron Accelerator, a joint laboratory of Harvard University and MIT. He then spent the next ten years working as Senior Physicist at CERN, Geneva. In 1983 he became a professor at Stanford University, working on the Stanford Linear Collider (SLC) and on optimizing the storage rings SPEAR and PEP for synchrotron-radiation use. He spent two years as head of the SLAC beam-dynamics group. He then returned to CERN, in 1987, and was jointly responsible for the commissioning of the Large Electron–Positron ring (LEP). After its completion, he worked on accelerator-physics problems with this machine until his retirement from CERN in 1998.

Over the years Albert Hofmann has done consulting work for other machines, such as the European Synchrotron Radiation Facility (ESRF), the Synchrotron Radiation Research Center (SRRC) in Taiwan, and the Swiss Light Source (SLS). He has taught in over 25 short-term schools on accelerator physics and synchrotron radiation, and has published numerous papers. In 1992 he was elected to become a fellow of the American Physical Society, and in 1996 he received the Robert Wilson Prize from this Society. In 2001 he obtained the degree Doctor honoris causa from the University of Geneva.

T0186148

CAMBRIDGE MONOGRAPHS ON
PARTICLE PHYSICS
NUCLEAR PHYSICS AND COSMOLOGY
20

General Editors: T. Ericson, P. V. Landshoff

THE PHYSICS OF SYNCHROTRON RADIATION

ALBERT HOFMANN

Formerly CERN, Geneva

CAMBRIDGE
UNIVERSITY PRESS

CAMBRIDGE UNIVERSITY PRESS
Cambridge, New York, Melbourne, Madrid, Cape Town, Singapore, São Paulo

Cambridge University Press
The Edinburgh Building, Cambridge CB2 8RU, UK

Published in the United States of America by Cambridge University Press, New York

www.cambridge.org
Information on this title: www.cambridge.org/9780521308267

First published 2004
This digitally printed version 2007

A catalogue record for this publication is available from the British Library

Library of Congress Cataloguing in Publication data

Hofmann, Albert, 1933–
The physics of synchrotron radiation / Albert Hofmann.
p.   cm. – (Cambridge monographs on particle physics, nuclear physics, and cosmology; 20)
Includes bibliographical references and index.
ISBN 0 521 30826 7
1. Synchrotron radiation. I. Title. II. Series.
QC793.5.E627H64 2003
539.7′35 – dc21   2003043939

ISBN 978-0-521-30826-7 hardback
ISBN 978-0-521-03753-2 paperback

To my wife Elisabeth
for her support

Contents

Preface

Under the rubric of synchrotron radiation we understand the electromagnetic waves emitted by a charge moving with relativistic velocity and undergoing a transverse acceleration. It is characterized by a small opening angle and a high frequency caused by the velocity of the charge being close to that of light. Owing to the relatively simple motion of the charge, the radiation has clear polarization properties. Ordinary synchrotron radiation is emitted by a charge moving on a circular arc determined by a deflecting magnetic field. It has a broad spectrum, a typical frequency being γ^3 times higher than the Larmor frequency of the charge. This spectrum can be modified by varying the curvature of the trajectory $1/\rho$ within a distance smaller than the formation length of the radiation, as is realized in undulators.

Synchrotron radiation has been investigated theoretically for over a century and experimentally for about half this time. Thanks to its unique properties, this radiation has become a research tool for many fields of science and electron-storage rings serving as radiation sources are spread over the whole globe.

This book tries to explain synchrotron radiation from basic principles and to derive its main properties. It is divided into four parts. First the general case of the electromagnetic fields created by an accelerated relativistic charge is investigated. This gives the angular distribution with the small opening angle of the emitted radiation and distinguishes between the 'near' (Coulomb) and the 'far' (radiation) field. The second part concentrates on the radiation emitted by a charge moving on a circular trajectory, which we usually call synchrotron radiation. Its distributions in angle, frequency, and polarization modes are derived. Undulator radiation is treated in the next part. We start with the simple case of a plane weak undulator with a spatially periodic field that emits quasi-monochromatic radiation. A strong undulator emits radiation that is more complicated and contains higher harmonics. There are more general undulators having a non-planar (helical) electron trajectory or a non-harmonic field. The last part deals with applications and investigates first the optics of synchrotron radiation, which is dominated by diffraction due to the small opening angle. This is followed by a description of electron-storage rings serving as radiation sources and the effect of the emitted radiation on the electron beam.

There are some technical remarks to be made. Throughout the book MKSA units are used. With very few exceptions the radiation field refers to a single positive elementary charge e as a source. For convenience sometimes the radiation emitted by a current I is also given and, in the last chapter, the temporal coherence of the radiation from different particles is considered. As a basis for the properties of the radiation we give first the total emitted power or energy. In the case of ordinary synchrotron radiation we denote by P_s the power radiated by the electron *while* it is going through the magnet and by U_s the energy radiated during one revolution. For undulators we denote by P_u the power radiated in the undulator but averaged over one period and by U_s the energy emitted during one traversal through the undulator. These powers and energies can also be expressed in terms of the photon number or photon flux. Distributions in terms of angle and frequency are then given with these total values as a factor that makes it easy to express them in terms of power, energy, photon-number or photon-flux distributions or in other units. Vectors are printed in bold. They are also written as an array with three components between square brackets, like $\mathbf{E} = [E_x, E_y, E_z]$. For radiation fields the z-component can often be neglected. The remaining two-component vector is written as $E_\perp = [E_x, E_y]$. These field components give the polarization of the radiated power. To mark the contributions of the horizontal or vertical polarization to the power, which is of course a scalar, we write it as a sum $P = P_\sigma + P_\pi$. The calculation of synchrotron radiation leads to some integrals that can be expressed in terms of modified Bessel functions or Airy functions. Here the second type is chosen, but the important results are given in both. Some properties, integrals, and sums of Airy and Bessel functions are given in the appendices, partly for convenience and partly because they are not so easy to find. However, this is not meant to provide rigorous mathematical derivations but rather to provide some insight into how some results are obtained.

There are lots of publications on synchrotron radiation and related topics. Apart from well-known books and journals they appear often in laboratory reports and proceedings of workshops. The bibliography to this volume is by no means complete and refers mostly to the topics covered and the methods used to investigate them.

Acknowledgments

I received much help from many people while writing this book. I owe many thanks to my colleague and friend Bruno Zotter from CERN. He not only answered many questions concerning the mathematics I had to use, but also read the whole book and made many suggestions, corrections, and significant improvements. Jim Murphy from Brookhaven National Laboratory read and corrected part of the book and clarified questions concerning mainly coherent radiation. On many occasions I sought advice from my friend and colleague Hermann Winick from Stanford University. Thanks to his insight and experience, he could answer many questions on synchrotron radiation and explained difficult topics to me. I also profited from discussions with many experts in the field: R. Coisson, K. J. Kim, B. M. Kincaid, S. Krinsky, F. Méot, M. Sands, and H. Wiedemann. I also thank the staffs of the laboratories where I had the opportunity to work on and learn about synchrotron radiation: CEA, Cambridge Electron Accelerator, Harvard University – MIT, Cambridge, MA, U.S.A.; CERN, European Laboratory of Particle Physics, Geneva, Switzerland; SLAC, Stanford Linear Accelerator Center, SSRL division, Stanford, CA, U.S.A.; and LNLS, Laboratório Nacional de Luz Sínchrotron, Campinas, Brazil.

Notation

$\mathbf{A}$	vector potential
$\mathrm{Ai}(x),\ \mathrm{Ai}'(x)$	Airy function and its derivative
$\mathbf{B}$	magnetic-field vector
$\tilde{\mathbf{B}}(\omega)$	Fourier transformed B-field of radiation
B_0	amplitude of magnetic undulator field
$\tilde{B}_y(k_\mathrm{g})$	weak-magnet Fourier component at k_g
c	speed of light
C_q	quantum excitation factor
$D_x,\ D_x'$	particle-beam-optics dispersion
e	elementary charge
$\mathbf{E}$	electric-field vector
$\tilde{\mathbf{E}}(\omega)$	Fourier-transformed E-field of radiation
$E_\mathrm{e} = m_0 c^2 \gamma$	particle energy
$E_\gamma = \hbar\omega$	photon energy
$F_\mathrm{s}(\psi, \omega/\omega_c)$	normalized angular spectral density of SR
$F_\mathrm{u}(\theta, \phi)$	normalized angular power density of UR
$h,\ \hbar = h/(2\pi)$	Planck's constant
$\mathcal{H}$	emittance function
$I_{\mathrm{s}2},\ I_{\mathrm{s}3},\ I_{\mathrm{s}4},\ I_{\mathrm{s}5}$	synchrotron-radiation integrals
$J_n(x)$	Bessel function of order n
$J_\epsilon,\ J_x,\ J_y$	longitudinal and transverse damping partitions
$K_{1/3},\ K_{2/3}$	modified Bessel function of order $1/3, 2/3$
K_f	quadrupole focusing parameter
k_g	wave number of general weak magnet
$k_\mathrm{u} = 2\pi/\lambda_\mathrm{u}$	undulator period wave number
$K_\mathrm{u} = e B_0/(m_0 c k_\mathrm{u})$	undulator parameter
K_u^*	reduced plane undulator parameter
K_uh^*	reduced helical undulator parameter
$L_\mathrm{u} = N_\mathrm{u}\lambda_\mathrm{u}$	undulator length

m_0	rest mass of a particle
$\mathbf{n} = \mathbf{r}/r$	unit vector in $\mathbf{r}$-direction
$n_B = -\rho^2 K_f$	field index
$n_s, \dot{n}_s$	photons per revolution, photon flux
$n_u, \dot{n}_u$	photon number per traversal, photon flux
N_u	undulator period number
P_s	instantaneous radiated power of SR
P_u	period-averaged total power of plane UR
$\mathbf{r}(t')$	distance from source to observer
$\mathbf{R}(t')$	vector from origin to particle
$\mathcal{R}, \Phi$	polar coordinates in image plane
$r_0 = e^2/(4\pi\epsilon_0 m_0 c^2)$	classical electron radius
$\mathbf{r}_p$	vector from origin to observer
$\mathbf{S} = [\mathbf{E} \times \mathbf{B}]/\mu_0$	Poynting vector
$S_{hm}(\omega_m)$	normalized spectral power of helical UR
$S_s(\omega/\omega_c)$	normalized spectral power densities of SR
$t = t' + r/c$	observation time
$t_p = t - r_p/c$	reduced observation time
$T_0 = 2\pi/\omega_0$	revolution time without straight sections
$T_{rev} = 2\pi/\omega_{rev}$	revolution time with straight sections
t'	emission time
U_s	energy radiated per turn of SR
$\mathbf{v}(t') = d\mathbf{R}/dt'$	particle velocity
V	scalar potential
$\hat{V}$	peak voltage of RF system
w	transverse coordinate, x or y
X, Y	rectangular coordinates in image plane
α_c	momentum compaction
$\alpha_\epsilon, \alpha_h, \alpha_v$	longitudinal and transverse damping rates
$\alpha_f = e^2/(2\epsilon_0 ch)$	fine structure constant
$\alpha_w = -\beta'_w/2$	particle-optics functions
$\beta = v/c$	normalized velocity
$\boldsymbol{\beta} = \mathbf{v}/c$	normalized velocity vector
β_w	particle-optics functions
β^*	normalized drift velocity in plane undulator
β_h^*	normalized drift velocity in helical undulators
$\gamma = 1/\sqrt{1-\beta^2}$	Lorentz factor
γ_w	particle-optics functions
γ^*	Lorentz factor of drift velocity
γ_h^*	drift Lorentz factor in helical undulators
ϵ_0	vacuum permittivity
ϵ_x, ϵ_y	horizontal and vertical particle-beam emittance

$\epsilon_{\gamma x}$, $\epsilon_{\gamma y}$	horizontal and vertical photon-beam emittance
$\boldsymbol{\eta}_x$, $\boldsymbol{\eta}_y$	unit vectors in x- and y-directions
$\lambda_{Comp} = h/(m_0 c)$	Compton wavelength
λ_u	undulator period length
μ_0	vacuum permeability
ρ	bending radius
σ_x, σ_x'	RMS electron-beam size and angular spread
φ_B	bending angle in a dipole magnet
$\varphi_w(s)$	betatron phase within one turn
φ_s	synchrotron phase angle in RF acceleration
ψ	angle between median plane and $\mathbf{r}_p$
$\omega_0 = \beta c/\rho = 2\pi/T_0$	angular velocity, Larmor frequency
ω_1	fundamental UR frequency off axis
ω_{10}	fundamental UR frequency on axis
$\omega_c = 3\omega_0 \gamma^3/2$	critical frequency
ω_m	mth harmonic UR frequency off axis
ω_{m0}	mth harmonic UR frequency on axis
$\omega_{rev} = 2\pi/T_{rev}$	revolution frequency with straight sections
$\Omega_u = \beta c k_u$	particle-motion frequency in undulator
$dP/d\Omega$	power radiated per unit solid angle
$d^2 P/(d\Omega\, d\omega)$	angular spectral radiated power density
$(\ )_\sigma$, $(\ )_\pi$	horizontal, vertical linear polarization
$(\ )_+$, $(\ )_-$	positive, negative helicity circular polarization
$\{\ \ \}_{ret}$	parenthesis evaluated at emission time t'

Part I
Introduction

1

A qualitative treatment of synchrotron radiation

1.1 Introduction

We consider the radiation emitted by a charged particle moving with constant, relativistic velocity on a circular arc. It is called *synchrotron radiation*, or sometimes also *ordinary synchrotron radiation*, abbreviated as SR, to distinguish it from the related case of undulator radiation, abbreviated as UR. We start with a qualitative discussion of synchrotron radiation in order to obtain some insight into its physical properties such as the opening angle, spectrum, and polarization. This will also help us to judge the validity of some approximations used in later calculations.

The physical properties of synchrotron radiation have their basis in the fact that the charge moves with relativistic velocity towards the observer. The charge and the emitted radiation travel with comparable velocities in about the same direction. The fields created by the charge over a relatively long time are received by the observer within a much shorter time interval. This time compression determines the spectrum of synchrotron radiation.

1.2 The opening angle

We consider a charge moving in the laboratory frame F on a circular trajectory with radius of curvature ρ, Fig. 1.1. We go into a frame F$'$ that moves with a constant velocity that is the same as that of the charge at the instant it traverses the origin. The particle trajectory has in this frame the form of a cycloid with a cusp at the origin. At this location the particle is momentarily at rest, but undergoes an acceleration in the $-x'$-direction. Like any accelerated charge, it emits radiation having an approximately uniform distribution in this frame F$'$.

We go back to the laboratory frame F by applying a Lorentz transformation. The emitted radiation is now peaked in the forward direction. A photon emitted along the x'-axis in the moving frame F$'$ appears in the laboratory frame at an angle θ given by

$$\sin \theta = \frac{1}{\gamma} \quad \text{or} \quad \theta \approx \frac{1}{\gamma}, \tag{1.1}$$

where $\gamma = 1/\sqrt{1 - \beta^2}$ is the Lorentz factor and $\beta = v/c$ is the normalized velocity. The typical opening angle of the emitted synchrotron radiation is therefore expected to be of

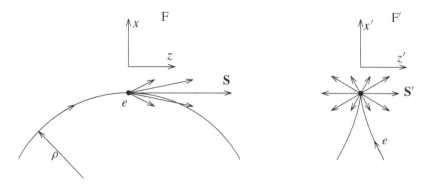

Fig. 1.1. The opening angle of synchrotron radiation.

order $1/\gamma$. For ultra-relativistic particles, $\gamma \gg 1$, the radiation is confined to very small opening angles around the direction of the particle velocity.

1.3 The spectrum emitted in a long magnet

Next we estimate the typical frequency of the emitted radiation. We consider a charge moving on a circular trajectory through a long magnet as shown in Fig. 1.2. We try to estimate the length Δt of the radiation pulse received by the observer P. Owing to the small natural opening angle the observer receives only radiation that was emitted along an arc of approximate angle $\pm 1/\gamma$. Therefore, the radiation observed first was emitted at point A, where the trajectory has an angle $1/\gamma$ with respect to this direction pointing towards the observer, whereas the radiation observed last was emitted at point A', where the trajectory has a corresponding angle $-1/\gamma$. The length of the radiation pulse seen by the observer is therefore just the difference in travel time between the charge and the radiation for going from point A to point A':

$$\Delta t = t_{\mathrm{e}} - t_{\gamma} = \frac{2\rho}{\beta\gamma c} - \frac{2\rho\sin(1/\gamma)}{c}.$$

For the ultra-relativistic velocities considered here we have $1/\gamma \ll 1$ and the trigonometric function can be expanded to give

$$\Delta t \approx \frac{2\rho}{\beta\gamma c}\left(1 - \beta + \frac{\beta}{6\gamma^2}\right) \approx \frac{\rho}{\gamma c}\left(\frac{1}{\gamma^2} + \frac{1}{3\gamma^2}\right) = \frac{4\rho}{3c\gamma^3}.$$

Here we use the ultra-relativistic approximation

$$1 - \beta = \frac{1 - \beta^2}{1 + \beta} \approx \frac{1}{2\gamma^2}. \tag{1.2}$$

From the length Δt of the radiation pulse we get the typical frequency of the spectrum,

$$\omega_{\mathrm{typ}} \approx \frac{1}{\Delta t} \approx \frac{3c\gamma^3}{4\rho}. \tag{1.3}$$

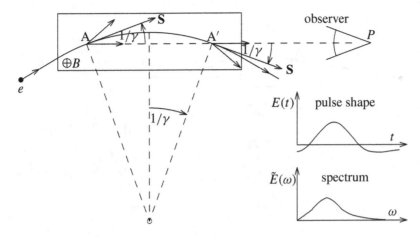

Fig. 1.2. The typical frequency of the synchrotron-radiation spectrum.

Later, on the basis of a quantitative treatment, we will introduce the critical frequency, which is twice as large, $\omega_c = 2\omega_{typ}$. For a large value of the Lorentz factor γ the radiation pulse can become very short and the resulting typical frequency very high.

The above derivation of the typical frequency is quite simple but illustrates some of the most important physical principles of synchrotron radiation. The length of the radiation pulse received is given by the difference in travel time between the particle and the photon for going from point A to point A′. The observed radiation originates from a trajectory arc of approximate length $\ell_r \approx 2\rho/\gamma$. The length L of the magnet has to be larger than this for the above treatment to be valid.

1.4 The spectrum emitted in a short weak magnet

We consider a short weak magnet as shown in Fig. 1.3 with length $L < \rho/\gamma$. It deflects the particle by an angle

$$\Delta\phi = 2\arcsin\left(\frac{L}{2\rho}\right) \approx \frac{L}{\rho} < \frac{1}{\gamma},$$

which is less then the natural opening angle of the radiation. The length Δt_{sm} of the radiation pulse now becomes

$$\Delta t_{sm} = t_e - t_\gamma = \frac{2\rho}{\beta c}\arcsin\left(\frac{L}{2\rho}\right) - \frac{L}{c} \approx \frac{L}{\beta c}(1-\beta) \approx \frac{L}{2c\gamma^2},$$

assuming again that we have the ultra-relativistic case $\beta \approx 1$. The length of the radiation pulse is now proportional to the magnet length L. Reducing it will therefore lead to shorter wavelengths.

The spectrum of the emitted radiation is also modified if the magnetic field changes within the length L, which is the case for undulators. In order for synchrotron radiation to

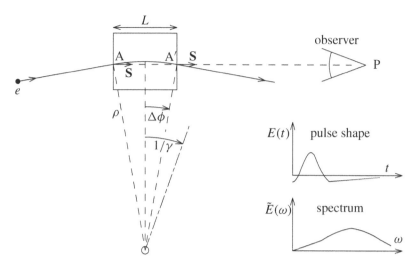

Fig. 1.3. The spectrum radiated in a short magnet.

have a spectrum described by (1.3) it has to be emitted from a magnet with a field that is homogeneous over at least a length of $L > 2\rho/\gamma$. By 'synchrotron radiation' we usually mean the radiation from a long magnet. Sometimes it is also called 'ordinary' synchrotron radiation or 'long-magnet' radiation and will sometimes be abbreviated here to 'SR'. This distinguishes it from undulator or 'short-magnet' radiation. This term 'short magnet' is now commonly used but describes a magnet that is short and weak such that the trajectory angle is everywhere smaller than $1/\gamma$ with respect to the main direction.

1.5 The wave front of synchrotron radiation

In estimating the typical frequency of synchrotron radiation we found that the field which is received by the observer P at the time t within a very short time interval Δt has been emitted by the particle at a different location and at a time t' over a longer time interval $\Delta t'$. Let us consider a particle moving in the general direction towards an observer with a speed close to that of light, emitting pulses of radiation at regular intervals along its trajectory. These pulses are received by the observer at time intervals that are much shorter. The compression of the time sequences Δt of reception compared with the time sequences $\Delta t'$ of emission is stronger the closer the particle velocity is to that of light and the closer its direction to that pointing towards the observer. This is well known from the Doppler effect.

We illustrate this situation in Fig. 1.4 for a charged particle moving with a constant speed $v = \beta c$ ($\beta = 0.8$) anti-clockwise on a circle of radius ρ and emitting a pulse of radiation at regular intervals indicated by small full circles (bullets). These pulses of radiation propagate at the speed of light on circular wave fronts around the sources in their centers. At a certain time t they have reached the situation shown in Fig. 1.4. The pulse emitted first at the time

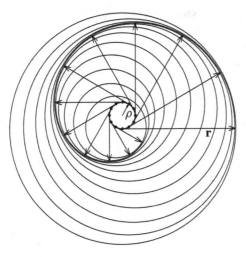

Fig. 1.4. Global propagation of synchrotron radiation for $\beta = 0.9$.

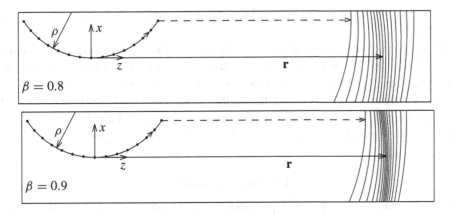

Fig. 1.5. Forward propagation of synchrotron radiation for $\beta = 0.8$ and 0.9.

$t' = 0$ originates from the bottom point and has reached the largest circle. The particle takes some time $\Delta t'$ to reach the next point of emission. Since it moves slower than light the wave emitted at this second point can never catch up with the first one but lags behind only by a small amount in the forward direction indicated by the arrow. Figure 1.4 shows the wave fronts emitted during one revolution of the particle executed at an earlier time. At a certain distance in the forward direction these wave fronts are concentrated in the radial direction. As a consequence an observer at this location receives within a short time interval Δt the radiation emitted during a large interval $\Delta t'$ of the particle motion. In Fig. 1.5 this is illustrated in more detail for the radiation emitted from a finite arc of the trajectory for two velocities $v = \beta c$ of the particle. Clearly the higher velocity ($\beta = 0.9$) leads to a stronger concentration of the wave front than does the lower one ($\beta = 0.8$).

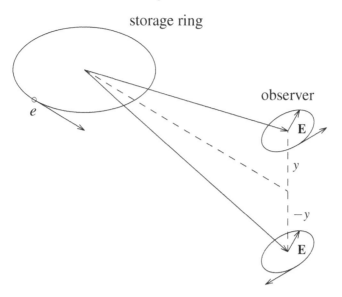

Fig. 1.6. Linear and elliptical polarization of synchrotron radiation.

The emission of short pulses is an artificial picture that we can use in order to obtain a simple illustration. In reality the charge radiates continuously, which is more difficult to draw. Very nice displays of the actual emission of radiation are presented in [1, 2].

We saw at the beginning of this chapter that the radiation is emitted mainly in the forward direction. Therefore, from the wave-front circles drawn in Figs. 1.4 and 1.5 only a limited arc around the forward direction contributes to the field received by the observer.

1.6 The polarization

Since the acceleration of the charge is radial and lies in the plane of the trajectory, we expect that the emitted radiation is mostly linearly polarized, with the electric-field vector also lying in this plane. The radiation observed at a finite angle above or below this plane has some elliptic polarization of opposite helicities, as illustrated in Fig. 1.6.

2

Fields of a moving charge

2.1 Introduction

In the previous chapter we used some qualitative arguments to estimate the basic nature of synchrotron radiation. The results of this exercise are very useful for understanding the underlying physics, estimating the quantities involved, and judging the validity of certain approximations we will make. Now synchrotron radiation is treated in a quantitative manner. We will distinguish between the time t at which the radiation is observed and t' when it was created by the moving charge at a distance r. Since the relation between the two is in general rather complicated, some of the derivations are lengthy. As final results we obtain expressions for the radiation field and the emitted power, which will be applied to calculate synchrotron and undulator radiation in the next two parts. Treatments of synchrotron radiation can be found in many books, journal publications, articles, proceedings of conferences and workshops, and laboratory reports. The first book on the topic of synchrotron radiation [3] was published in 1912. Complete coverage of the topic is presented in [4–8], some of which give also a quantum-mechanical treatment. Many books on electrodynamics treat the radiation from relativistic particles and cover also theoretical aspects of synchrotron radiation [9–13]. On the other hand, many publications on particle accelerators have chapters on synchrotron radiation, giving details of its properties and effects on the electron beam. Among those are the books [14–17]. There are many proceedings from conferences, workshops, and schools and laboratory reports concerned mainly with accelerators but containing also articles on synchrotron radiation [18–20]. Furthermore, there are several handbooks and proceedings concerned mainly with the science done with synchrotron radiation [21–24], which describe the properties and technical possibilities of this source [25]. There are overviews on the history of synchrotron radiation, such as [26], which gives mainly the early development, and [27], which concentrates on the work done in the U.S.S.R.

2.2 The particle motion relevant to the retarded potentials

To relate the observed radiation to the motion of charge and vice versa we invoke so-called retarded potentials and fields, which have their basis in the finite propagation velocity c

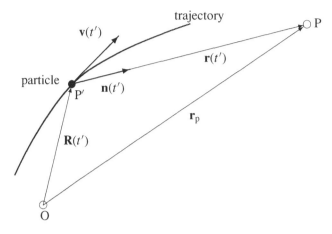

Fig. 2.1. The particle trajectory and radiation geometry.

of the electromagnetic radiation. To calculate the fields measured at time t by a stationary observer we have to know the position and motion of the charge at this earlier time t' of emission.

We discuss now the motion relevant for these two time scales and consider an *elementary positive charge e* moving on a trajectory given by the vector $\mathbf{R}(t')$ and creating an electric field $\mathbf{E}$ and a magnetic field $\mathbf{B}$. These fields are measured at time t by the observer located at P as illustrated in Fig. 2.1. We introduce a vector $\mathbf{r}$ with absolute value r, pointing from the location P′ of emission to the observer P. Owing to the finite propagation velocity c, the field received at time t by the observer P had to have been emitted by the source P′ at the earlier time t' given by the relation

$$t = t' + \frac{r(t')}{c}.$$
(2.1)

Therefore, we have to know the position $\mathbf{R}(t')$ and velocity $\mathbf{v}(t') = d\mathbf{R}/dt'$ of the charged particle at this earlier time t'. We have for the vectors $\mathbf{R}$ (pointing from the origin to the radiating charge), $\mathbf{r}_\mathrm{p}$ (pointing from the origin to the observer), and $\mathbf{r}$ (pointing from the charge to the observer) the relation

$$\mathbf{R}(t') + \mathbf{r}(t') = \mathbf{r}_\mathrm{p} = \text{constant.}$$
(2.2)

Differentiating this with respect to t' gives the change of the vector $\mathbf{r}$,

$$\frac{d\mathbf{r}(t')}{dt'} = -\frac{d\mathbf{R}}{dt'} = -\mathbf{v}(t') = -c\boldsymbol{\beta}(t'),$$
(2.3)

from which we obtain the corresponding change of its absolute value $r = |\mathbf{r}|$:

$$\mathbf{r}\frac{d\mathbf{r}}{dt'} = \frac{1}{2}\frac{d(\mathbf{r}^2)}{dt'} = r\frac{dr}{dt'} = -(\mathbf{r} \cdot \mathbf{v}).$$

Introducing the unit vector

$$\mathbf{n} = \mathbf{r}/r \tag{2.4}$$

pointing from the charge in the direction towards the observer gives for the change of the distance between the source and the observer at time t'

$$\frac{dr}{dt'} = -(\mathbf{n} \cdot \mathbf{v}) = -c(\mathbf{n} \cdot \boldsymbol{\beta}), \tag{2.5}$$

which is just the negative particle-velocity component of the particle in the direction towards the observer as shown in Fig. 2.1. The differential relation between the two time scales t' and t is obtained from (2.1):

$$\boxed{dt = \left(1 + \frac{1}{c}\frac{dr}{dt'}\right)dt' = (1 - \mathbf{n} \cdot \boldsymbol{\beta})\,dt'.} \tag{2.6}$$

2.3 The retarded electromagnetic potentials

In this section we derive expressions for the electromagnetic potentials $\mathbf{A}(t)$ and $V(t)$ observed at P and created by a charge moving along a trajectory given by the vector $\mathbf{R}(t')$. This result will be used in the next section to obtain the electric and magnetic fields $\mathbf{E}$ and $\mathbf{B}$ which are related to the scalar and vector potentials V and $\mathbf{A}$ through Maxwell's equations, which can be found in standard textbooks on electrodynamics listed earlier:

$$\mathbf{E} = -\nabla V - \frac{\partial \mathbf{A}}{\partial t} = \operatorname{grad} V - \frac{\partial \mathbf{A}}{\partial t}$$
$$\mathbf{B} = [\nabla \times \mathbf{A}] = \operatorname{curl} \mathbf{A} \tag{2.7}$$

with the Lorentz convention $\nabla \cdot \mathbf{A} = -\dot{V}/c^2$. The vector potential $\mathbf{A}$ is measured in units of V s m^{-1}.

The potentials created by time-dependent charge $\eta(t')$ and current density $\mathbf{J}(t')$ are given by the expressions

$$V(t) = \frac{1}{4\pi\epsilon_0} \int \frac{\eta(t')}{r(t')}\,dx'\,dy'\,dz'$$

$$\mathbf{A}(t) = \frac{\mu_0}{4\pi} \int \frac{\mathbf{J}(t')}{r(t')}\,dx'\,dy'\,dz'.$$

The above expressions are very similar to those used to calculate the potentials of static charge and stationary current distributions. However, here the charges move and the local charge and current densities change. Since the potentials created propagate at the speed of light, the signals received by the observer P depend on the positions of the charges at the earlier time t'. The integration is carried out over the coordinates x', y', and z' of the earlier distribution.

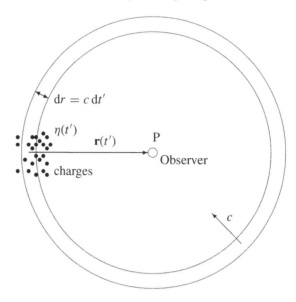

Fig. 2.2. Integration over charges represented by a collapsing sphere.

Since this is important for understanding the potentials and fields created by moving charges, we would like to illustrate it using Fig. 2.2 and follow the discussion given in [10]. It shows charges with a time-dependent density η and an observer P, at whose location we would like to know the potential V at time t. For this it is necessary to integrate over the charge density $\eta(t')$ at time $t' = t - r(t')/c$, which will contribute to the potential at time t and location P. This process represents a thin sphere that is converging with velocity c towards the point P while integrating over all charges that contribute to the potential $V(t)$. Charges moving towards P while this integration is carried out are 'counted' for a longer time and contribute more to the potential $V(t)$. On the other hand, charges moving away from P will contribute less to the potential.

We use this picture of a collapsing, integrating sphere to calculate the potential $V(t)$ at time t at P, created by a single elementary charge e having a finite radius b as shown in Fig. 2.3. First we take the charge to be at rest and find for the time $\Delta t'_0$ during which it contributes to the integration

$$\Delta t'_0 = 2b/c.$$

The potential due to this stationary charge at rest is

$$V(t) = \frac{e}{4\pi \epsilon_0 r} = \text{constant.}$$

Next we assume that the charge moves with velocity $\mathbf{v} = \boldsymbol{\beta} c$ having the component $v_r = \mathbf{n} \cdot \mathbf{v}$ in the direction towards P, Fig. 2.3. The time $\Delta t'_v$ during which this charge

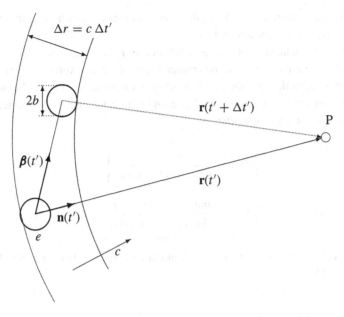

Fig. 2.3. The contribution of a moving charge to the potential.

contributes is

$$\Delta t'_v = \frac{2b}{c - v_r} = \frac{2b}{c(1 - \mathbf{n} \cdot \boldsymbol{\beta})}.$$

Since the ratio

$$\frac{\Delta t'_v}{\Delta t'_0} = \frac{1}{1 - \mathbf{n} \cdot \boldsymbol{\beta}}$$

is independent of the size of the charge we can let b go to zero. This gives the scalar potential (and, in a similar way, also the vector potential) of a moving charge:

$$V(t) = \frac{e}{4\pi \epsilon_0} \frac{1}{r(t')(1 - \mathbf{n}(t') \cdot \boldsymbol{\beta}(t'))}$$

$$\mathbf{A}(t) = \frac{\mu_0 e}{4\pi} \frac{\mathbf{v}(t')}{r(t')(1 - \mathbf{n}(t') \cdot \boldsymbol{\beta}(t'))}.$$

To obtain the potentials at the observation time t we have to evaluate r, $\mathbf{n}$, and $\boldsymbol{\beta}$ at the time t' of emission. This is indicated explicitly above by expressing these quantities as functions of t'. Since we observe the potentials at time t, we would like to express the final result in terms of it and have to know the relation between the two times

$$t' = t'(t),$$

which can be a complicated expression for a general particle motion and hence could make the calculation of the potentials difficult. In most cases the inverse function $t = t(t') =$

$t' + r(t')/c$ is found more easily. It can then serve as a basis on which to find the desired relation $t' = t'(t)$ using some approximations.

Expressing each quantity explicitly as a function of t' often leads to rather complicated expressions that are difficult to read. It is preferable to place the expression which has to be evaluated at the earlier time t' within curly brackets with the subscript 'ret'. We will use this convention for all important expressions but might omit it for intermediate calculations.

The two potentials are now written as

$$
\begin{aligned}
V(t) &= \frac{e}{4\pi\epsilon_0}\left\{\frac{1}{r(1-\mathbf{n}\cdot\boldsymbol{\beta})}\right\}_{\text{ret}} \\
\mathbf{A}(t) &= \frac{\mu_0 e}{4\pi}\left\{\frac{\boldsymbol{\beta}c}{r(1-\mathbf{n}\cdot\boldsymbol{\beta})}\right\}_{\text{ret}}.
\end{aligned}
\tag{2.8}
$$

They are called the retarded potentials of a moving charge or also the Liénard–Wiechert potentials [28, 29].

2.4 The fields of a moving charge

In the previous section we derived the scalar and vector potentials of a moving charge. For most practical applications we need the electric and magnetic fields at the observer. They can be obtained from the equations (2.7) relating fields and potentials. These relations involve a differentiation with respect to the time t and the location $\mathbf{r}_p$ since the observer measures potentials and fields in terms of these coordinates. However, for the evaluation of the potentials the position and velocity of the moving charge have to be evaluated at the earlier time t'. Any change Δt or $\Delta\mathbf{r}_p$ of the observation time and location will change the time t' of emission as well as the parameters $\mathbf{r}(t')$, $\mathbf{R}(t')$, and $\mathbf{n}(t')$ which depend on it.

We evaluate first the derivative with respect to the time t and consider two photons A and B observed with a time difference Δt at the *same location* P. These two photons had to be emitted with different values of the time interval $\Delta t'$ and, owing to the motion of the charge, also at different locations A' and B' as illustrated in Fig. 2.4. For the time differences $\Delta t'$ between the emissions and Δt between the observations of the two photons we use the differential relation (2.6):

$$
dt = (1 - \mathbf{n}\cdot\boldsymbol{\beta})\,dt'.
$$

This gives for the derivative of the vector potential $\mathbf{A}$ with respect to t

$$
\begin{aligned}
\frac{\partial\mathbf{A}}{\partial t} &= \frac{dt'}{dt}\frac{\partial\mathbf{A}}{\partial t'} = \frac{1}{1-\mathbf{n}\cdot\boldsymbol{\beta}}\frac{\partial\mathbf{A}}{\partial t'} \\
&= \frac{\mu_0 ec}{4\pi}\frac{1}{1-\mathbf{n}\cdot\boldsymbol{\beta}}\left(\frac{\dot{\boldsymbol{\beta}}}{r(1-\mathbf{n}\cdot\boldsymbol{\beta})} - \frac{\boldsymbol{\beta}\,\partial(r-\mathbf{r}\cdot\boldsymbol{\beta})/\partial t'}{r^2(1-\mathbf{n}\cdot\boldsymbol{\beta})^2}\right).
\end{aligned}
$$

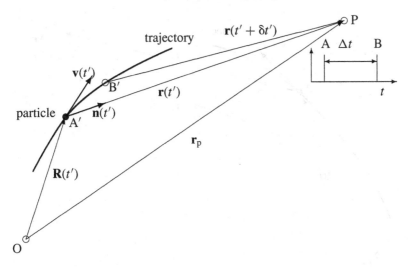

Fig. 2.4. Calculation of the derivative with respect to t.

The derivative of $r - \mathbf{r} \cdot \boldsymbol{\beta}$ is

$$\frac{\partial}{\partial t'}(r - \mathbf{r} \cdot \boldsymbol{\beta}) = \frac{dr}{dt'} - \frac{d\mathbf{r}}{dt'} \cdot \boldsymbol{\beta} - \mathbf{r} \cdot \dot{\boldsymbol{\beta}} = -c(\mathbf{n} \cdot \boldsymbol{\beta}) + c\beta^2 - \mathbf{r} \cdot \dot{\boldsymbol{\beta}},$$

where we used the relations (2.5) and (2.3). The derivative of the vector potential $\mathbf{A}$ with respect to the time t becomes

$$\frac{\partial \mathbf{A}}{\partial t} = \frac{\mu_0 e c}{4\pi} \left\{ \left(\frac{\dot{\boldsymbol{\beta}}}{r(1 - \mathbf{n} \cdot \boldsymbol{\beta})^2} + \frac{c(\mathbf{n} \cdot \boldsymbol{\beta}) - c\beta^2 + (\mathbf{r} \cdot \dot{\boldsymbol{\beta}})}{r^2(1 - \mathbf{n} \cdot \boldsymbol{\beta})^3} \boldsymbol{\beta} \right) \right\}_{\text{ret}}.$$

Sorting this expression according to powers of r leads to

$$\frac{\partial \mathbf{A}}{\partial t} = \frac{\mu_0 e c}{4\pi} \left\{ \frac{(\mathbf{n} \cdot \boldsymbol{\beta} - \beta^2) c \boldsymbol{\beta}}{r^2(1 - \mathbf{n} \cdot \boldsymbol{\beta})^3} + \frac{(1 - \mathbf{n} \cdot \boldsymbol{\beta}) \dot{\boldsymbol{\beta}} + (\mathbf{n} \cdot \dot{\boldsymbol{\beta}}) \boldsymbol{\beta}}{r(1 - \mathbf{n} \cdot \boldsymbol{\beta})^3} \right\}_{\text{ret}}. \tag{2.9}$$

Next we have to calculate the gradient of the scalar potential (2.8),

$$\nabla V = \frac{e}{4\pi \epsilon_0} \nabla \left(\frac{1}{r(1 - \mathbf{n} \cdot \boldsymbol{\beta})} \right) = -\frac{e}{4\pi \epsilon_0} \frac{\nabla(r - \mathbf{r} \cdot \boldsymbol{\beta})}{r^2(1 - \mathbf{n} \cdot \boldsymbol{\beta})^2},$$

which is the change of the expression for V for a small change in the vector $\mathbf{r}_p$, i.e. of the observation position, such that $\Delta V = \nabla V \cdot \Delta \mathbf{r}_p$. Again the situation is made complicated by the fact that changing the position of the observer will impose a change in the time t' of emission and in all parameters that depend on it. This is illustrated in Fig. 2.5, where we have two observers A and B separated in position by a small vector $\Delta \mathbf{r}_p$ and consider two photons arriving at the observers A and B, respectively, at the *same time* t. On moving from observer A to observer B the vector $\mathbf{r}$ undergoes a change consisting of two contributions. The first one, $\Delta \mathbf{r}_1 = \Delta \mathbf{r}_p$, is directly caused by the change of $\mathbf{r}_p$. The second contribution

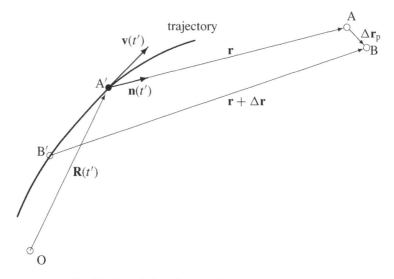

Fig. 2.5. Calculation of the gradient with respect to $\mathbf{r}_p$.

$\Delta \mathbf{r}_2 = (\partial \mathbf{r}/\partial t') \, \Delta t'$ is caused by the difference $\Delta t'$ in time of emission imposed by the condition that the two photons arrive at the same time t at A and B. Using $\partial \mathbf{r}/\partial t' = -c\boldsymbol{\beta}$ we obtain

$$\Delta \mathbf{r} = \Delta \mathbf{r}_p - c\boldsymbol{\beta} \, \Delta t' \quad \text{and} \quad \Delta r = \mathbf{n} \cdot \Delta \mathbf{r} = \mathbf{n} \cdot \Delta \mathbf{r}_p - c(\mathbf{n} \cdot \boldsymbol{\beta}) \, \Delta t'.$$

Since the two photons arrive at the same time t at their respective observers, the difference Δr between the distances traveled is, according to (2.1), given by the difference $\Delta t'$ between the emission times multiplied by the velocity of light,

$$\Delta r = -c \, \Delta t' = \mathbf{n} \cdot \Delta \mathbf{r}_p - c(\mathbf{n} \cdot \boldsymbol{\beta}) \, \Delta t',$$

which gives the relation

$$\Delta t' = -\frac{\mathbf{n} \cdot \Delta \mathbf{r}_p}{c(1 - \mathbf{n} \cdot \boldsymbol{\beta})}.$$

On comparing this with $\Delta t' = \nabla t' \, \Delta \mathbf{r}_p$, we can write for the gradient of t'

$$\nabla t' = -\frac{\mathbf{n}}{c(1 - \mathbf{n} \cdot \boldsymbol{\beta})}. \tag{2.10}$$

For the change of the expression $r - \mathbf{r} \cdot \boldsymbol{\beta}$ appearing in the denominator of equation (2.8) for the potentials V and $\mathbf{A}$ we obtain

$$\begin{aligned}
\Delta(r - \mathbf{r} \cdot \boldsymbol{\beta}) &= \Delta r - (\Delta \mathbf{r} \cdot \boldsymbol{\beta}) - (\mathbf{r} \cdot \Delta \boldsymbol{\beta}) \\
&= -c \, \Delta t' - (\Delta \mathbf{r}_p \cdot \boldsymbol{\beta}) + c\beta^2 \, \Delta t' - (\mathbf{r} \cdot \dot{\boldsymbol{\beta}}) \, \Delta t'.
\end{aligned}$$

Using the expression for $\Delta t'$ obtained above leads to

$$\Delta(r - \mathbf{r} \cdot \boldsymbol{\beta}) = \left(-\boldsymbol{\beta} + \frac{(1 - \beta^2)c + (\mathbf{r} \cdot \dot{\boldsymbol{\beta}})}{c(1 - \mathbf{n} \cdot \boldsymbol{\beta})}\mathbf{n}\right)\Delta \mathbf{r}_p.$$

On making again a comparison with $\Delta(r - \mathbf{r} \cdot \boldsymbol{\beta}) = \nabla(r - \mathbf{r} \cdot \boldsymbol{\beta}) \cdot \Delta \mathbf{r}_p$ we obtain for the gradient of the expression $r - \mathbf{r} \cdot \boldsymbol{\beta}$

$$\nabla(r - \mathbf{r} \cdot \boldsymbol{\beta}) = \left\{\frac{(c(1 - \beta^2) + r(\mathbf{n} \cdot \dot{\boldsymbol{\beta}}))\mathbf{n} - c(1 - \mathbf{n} \cdot \boldsymbol{\beta})\boldsymbol{\beta}}{c(1 - \mathbf{n} \cdot \boldsymbol{\beta})}\right\}_{\text{ret}}. \qquad (2.11)$$

The gradient of the scalar potential V now becomes

$$\begin{aligned}
\nabla V &= \frac{e}{4\pi \epsilon_0} \nabla\left(\frac{1}{r(1 - \mathbf{n} \cdot \boldsymbol{\beta})}\right) = -\frac{e}{4\pi \epsilon_0}\left(\frac{\nabla(r - \mathbf{r} \cdot \boldsymbol{\beta})}{r^2(1 - \mathbf{n} \cdot \boldsymbol{\beta})^2}\right) \\
&= -\frac{e}{4\pi \epsilon_0}\left(\frac{(1 - \beta^2)\mathbf{n} - (1 - \mathbf{n} \cdot \boldsymbol{\beta})\boldsymbol{\beta}}{r^2(1 - \mathbf{n} \cdot \boldsymbol{\beta})^3} + \frac{(\mathbf{n} \cdot \dot{\boldsymbol{\beta}})\mathbf{n}}{cr(1 - \mathbf{n} \cdot \boldsymbol{\beta})^3}\right).
\end{aligned}$$

To obtain the electric field (2.7) we combine the above gradient with the time derivative (2.9) of the vector potential and use $\epsilon_0 \mu_0 = 1/c^2$:

$$\mathbf{E} = \frac{e}{4\pi \epsilon_0}\left\{\frac{(1 - \beta^2)(\mathbf{n} - \boldsymbol{\beta})}{r^2(1 - \mathbf{n} \cdot \boldsymbol{\beta})^3} + \frac{(\mathbf{n} \cdot \dot{\boldsymbol{\beta}})(\mathbf{n} - \boldsymbol{\beta}) - (1 - \mathbf{n} \cdot \boldsymbol{\beta})\dot{\boldsymbol{\beta}}}{cr(1 - \mathbf{n} \cdot \boldsymbol{\beta})^3}\right\}_{\text{ret}}.$$

The expression for a triple vector product

$$[\mathbf{A} \times [\mathbf{B} \times \mathbf{C}]] = (\mathbf{A} \cdot \mathbf{C})\mathbf{B} - (\mathbf{A} \cdot \mathbf{B})\mathbf{C} \qquad (2.12)$$

simplifies the second term of the above equation:

$$\mathbf{E}(t) = \frac{e}{4\pi \epsilon_0}\left\{\frac{(1 - \beta^2)(\mathbf{n} - \boldsymbol{\beta})}{r^2(1 - \mathbf{n} \cdot \boldsymbol{\beta})^3} + \frac{[\mathbf{n} \times [(\mathbf{n} - \boldsymbol{\beta}) \times \dot{\boldsymbol{\beta}}]]}{cr(1 - \mathbf{n} \cdot \boldsymbol{\beta})^3}\right\}_{\text{ret}}. \qquad (2.13)$$

To obtain the magnetic field $\mathbf{B} = [\nabla \times \mathbf{A}]$ we have to execute the curl operation on the vector potential (2.8):

$$\mathbf{A}(t) = \frac{\mu_0 ec}{4\pi}\left\{\frac{\boldsymbol{\beta}}{r(1 - \mathbf{n} \cdot \boldsymbol{\beta})}\right\}_{\text{ret}}.$$

Applying the relation

$$[\nabla \times (u\mathbf{C})] = u[\nabla \times \mathbf{C}] + [\nabla u \times \mathbf{C}] \qquad (2.14)$$

gives

$$[\nabla \times \mathbf{A}] = \frac{\mu_0 ec}{4\pi}\left(\frac{[\nabla \times \boldsymbol{\beta}]}{r(1 - \mathbf{n} \cdot \boldsymbol{\beta})} - \frac{[\nabla(r - \mathbf{r} \cdot \boldsymbol{\beta}) \times \boldsymbol{\beta}]}{r^2(1 - \mathbf{n} \cdot \boldsymbol{\beta})^2}\right). \qquad (2.15)$$

The first term contains the curl operation of the vector $\boldsymbol{\beta}$, which involves differentiation with respect to the coordinates of the observer. The normalized velocity $\boldsymbol{\beta}$ is a function of

t', which itself depends on the observation position as illustrated in Fig. 2.5 and discussed for the calculation of the gradient ∇V. We can therefore write $\Delta \boldsymbol{\beta} = \dot{\boldsymbol{\beta}} \, \Delta t'$ and obtain from (2.14)

$$[\nabla \times \boldsymbol{\beta}] = [\nabla \times (\Delta t' \, \dot{\boldsymbol{\beta}})] = \Delta t' \, [\nabla \times \dot{\boldsymbol{\beta}}] + [\nabla t' \times \dot{\boldsymbol{\beta}}].$$

The first term on the right-hand side can be neglected since it is of second order in $\Delta t'$. We obtain from (2.10)

$$[\nabla \times \boldsymbol{\beta}] = -\frac{[\mathbf{n} \times \dot{\boldsymbol{\beta}}]}{c(1 - \mathbf{n} \cdot \boldsymbol{\beta})}.$$

With the expression for $\nabla(r - \mathbf{r} \cdot \boldsymbol{\beta})$ derived earlier, (2.11), we find from (2.15)

$$[\nabla \times \mathbf{A}] = -\frac{\mu_0 e c}{4\pi} \left\{ \frac{(c(1 - \beta^2) + r(\mathbf{n} \cdot \dot{\boldsymbol{\beta}}))[\mathbf{n} \times \boldsymbol{\beta}]}{cr^2(1 - \mathbf{n} \cdot \boldsymbol{\beta})^3} + \frac{[\mathbf{n} \times \dot{\boldsymbol{\beta}}]}{cr(1 - \mathbf{n} \cdot \boldsymbol{\beta})^2} \right\}_{\text{ret}}$$

and finally, for the magnetic field $\mathbf{B} = [\nabla \times \mathbf{A}]$,

$$\mathbf{B} = -\frac{\mu_0 e c}{4\pi} \left\{ \frac{(1 - \beta^2)[\mathbf{n} \times \boldsymbol{\beta}]}{r^2(1 - \mathbf{n} \cdot \boldsymbol{\beta})^3} + \frac{(\mathbf{n} \cdot \dot{\boldsymbol{\beta}})[\mathbf{n} \times \boldsymbol{\beta}] + (1 - \mathbf{n} \cdot \boldsymbol{\beta})[\mathbf{n} \times \dot{\boldsymbol{\beta}}]}{cr(1 - \mathbf{n} \cdot \boldsymbol{\beta})^3} \right\}_{\text{ret}}.$$

It can easily be verified that $\mathbf{B}$ is related to the electric field (2.13) by

$$\mathbf{B} = \frac{[\mathbf{n}_{\text{ret}} \times \mathbf{E}]}{c} \tag{2.16}$$

and is therefore perpendicular to $\mathbf{E}$ and $\mathbf{n}$. In the above equation the fields $\mathbf{B}$ and $\mathbf{E}$ are already functions of the observation time t and only the vector $\mathbf{n}$ has to be evaluated at the emission time t'. The electric and magnetic fields of a moving point charge e are now

$$\boxed{\begin{aligned} \mathbf{E}(t) &= \frac{e}{4\pi \epsilon_0} \left\{ \frac{(1 - \beta^2)(\mathbf{n} - \boldsymbol{\beta})}{r^2(1 - \mathbf{n} \cdot \boldsymbol{\beta})^3} + \frac{[\mathbf{n} \times [(\mathbf{n} - \boldsymbol{\beta}) \times \dot{\boldsymbol{\beta}}]]}{cr(1 - \mathbf{n} \cdot \boldsymbol{\beta})^3} \right\}_{\text{ret}} \\ \mathbf{B}(t) &= \frac{[\mathbf{n}_{\text{ret}} \times \mathbf{E}]}{c}. \end{aligned}} \tag{2.17}$$

These expressions for the electric and magnetic fields $\mathbf{E}(t)$ and $\mathbf{B}(t)$ of a moving charge are called retarded fields or also the Liénard–Wiechert equation [28, 29].

2.5 A discussion of the field equations

The Liénard–Wiechert expressions give the electric and magnetic fields created by a moving point charge. The field equation consists of two terms having different dependences on the distance r.

The first term decreases with the square of the distance r between the source and the point P of field observation and is often referred to as the *near field*. It does not depend on the acceleration but only on the velocity and position of the charge. Since it could be reduced to an electrostatic field by a Lorentz transformation, we conclude that it does not lead to an emission of radiation power. For the static case, $\dot{\boldsymbol{\beta}} = \boldsymbol{\beta} = 0$, we recover Coulomb's law

$$\mathbf{E} = \frac{e\mathbf{n}}{4\pi\epsilon_0 r^2}, \qquad \mathbf{B} = 0.$$

The second term is proportional to the acceleration $\dot{\boldsymbol{\beta}}$ and decreases with the first power of the distance r. It is referred to as the *far field* or, for reasons explained later, as the *radiation field*. This field is perpendicular to the vector $\mathbf{n}$ pointing from the position of the charge at time t' towards the observer P. These properties distinguish the far fields clearly from the electrostatic case. For both terms the magnetic field $\mathbf{B}$ is perpendicular to the electric field $\mathbf{E}$.

We will later calculate the radiated power and use the Poynting vector $\mathbf{S}$ which gives the directional energy flow per unit time t through a unit area measured in $\mathrm{V\,A\,m^{-2}}$,

$$\mathbf{S} = \frac{[\mathbf{E} \times \mathbf{B}]}{\mu_0}. \tag{2.18}$$

It is perpendicular to the two field vectors and can be expressed with the electric field alone by using the relation $\mathbf{B} = [\mathbf{n}_{\mathrm{ret}} \times \mathbf{E}]/c$:

$$\mathbf{S} = \frac{[\mathbf{E} \times \mathbf{B}]}{\mu_0} = \frac{1}{\mu_0 c}[\mathbf{E} \times [\mathbf{n}_{\mathrm{ret}} \times \mathbf{E}]] = \frac{1}{\mu_0 c}(E^2 \mathbf{n}_{\mathrm{ret}} - (\mathbf{n}_{\mathrm{ret}} \cdot \mathbf{E})\mathbf{E}).$$

The magnitude of the power flow is therefore proportional to E^2. The near-field contribution to it given by the first term in (2.17) will decrease with the distance as $1/r^4$, the far-field contribution decreases as $1/r^2$, and the cross term between the two fields leads to a contribution $\propto 1/r^3$. The total power can be obtained by calculating the power through a sphere of radius r around the source. Only the far field leads to a term that is independent of the radius of this sphere; the other terms can be neglected at large distances. The power flow given by the near field at small distances is due to local changes of the energy densities of the fields. Often only the radiated power is of interest, in which case only the far field has to be considered. This simplifies the calculation considerably. However, it will be shown later that the far field alone does not fulfill Maxwell's equations.

Before we treat synchrotron radiation, we will apply the Liénard–Wiechert fields (2.17) to some well-known electrodynamic problems. First we will treat the case of a uniformly moving charge to illustrate the relation between the two time scales t and t'. Secondly, we investigate the fields emitted by a charge executing a non-relativistic dipole oscillation to show the properties of the near and far fields. This will be useful later, to understand the radiation from a plane undulator.

2.6 Examples

2.6.1 The field of a charge moving with constant velocity

As a first example we consider a charge moving with constant velocity on a straight line. There is no acceleration, $\dot{\boldsymbol{\beta}} = 0$, and only the first term of the Liénard–Wiechert expression (2.17) is present:

$$\mathbf{E}(t) = \frac{e}{4\pi\epsilon_0} \left\{ \frac{r(1-\beta^2)(\mathbf{n}-\boldsymbol{\beta})}{r^3(1-\mathbf{n}\cdot\boldsymbol{\beta})^3} \right\}_{\mathrm{ret}}, \qquad \mathbf{B} = \frac{[\mathbf{n}_{\mathrm{ret}} \times \mathbf{E}]}{c}. \qquad (2.19)$$

This example is one of the few cases in which the relation between the two time scales has a simple closed expression. Its geometry is illustrated in Fig. 2.6 and uses a very general coordinate origin 0 to relate to the situations used in Figs. 2.1 and 2.4. The charge e at the time t' at point P' is determined by the radius vector $\mathbf{R}(t')$ and creates an electromagnetic field that is detected at the later time t by an observer P at a distance $r = c(t - t')$ from the source P'.

We treat the problem first geometrically. During the time $t - t'$ taken by the field to travel from the source P' to the observer P the charge itself moves by the distance $\beta c(t - t')$ from P' to the point A as described by the radius vector $\mathbf{R}(t)$. We introduce the vector $\mathbf{r}_b$ going from this position A to the observer P, which is given by the relation

$$\mathbf{r}_b = \mathbf{r} - r\boldsymbol{\beta} = r(\mathbf{n} - \boldsymbol{\beta}).$$

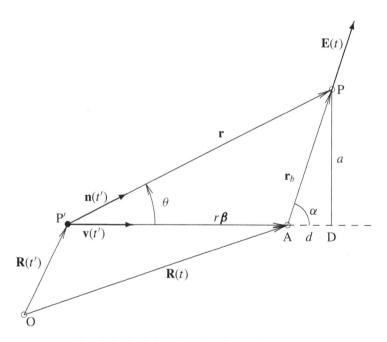

Fig. 2.6. The field of a uniformly moving charge.

With this we can express the electric field vector **E** in (2.19) as

$$\mathbf{E}(t) = \frac{e}{4\pi\epsilon_0\gamma^2} \frac{\mathbf{r}_b}{\{r^3(1 - \mathbf{n}\cdot\boldsymbol{\beta})^3_{\mathrm{ret}}\}}.$$

From the above field equation we observe the interesting fact that the electric field has the same direction as the vector $\mathbf{r}_b$; in other words, the direction of the electric field is as if it originated from the point A where the charge is located at the time t of observation.

To calculate the magnitude of the field we introduce the distance a between a straight particle trajectory and the observer. We also use the distance d between the point A and the projection D of P on the trajectory, which is related to r_b by

$$r_b^2 = r^2(\mathbf{n} - \boldsymbol{\beta})^2 = r^2(1 - 2(\mathbf{n}\cdot\boldsymbol{\beta}) + \beta^2) = a^2 + d^2.$$

We can express the denominator of (2.19) with r_b through the relation

$$r^2(1 - \mathbf{n}\cdot\boldsymbol{\beta})^2 = r^2(1 - 2(\mathbf{n}\cdot\boldsymbol{\beta}) + (\mathbf{n}\cdot\boldsymbol{\beta})^2) = r_b^2 - r^2\beta^2 + r^2(\mathbf{n}\cdot\boldsymbol{\beta})^2.$$

With $r^2(\mathbf{n}\cdot\boldsymbol{\beta})^2 = \beta^2(r^2 - a^2)$ we obtain from the triangle P′ D P

$$r^2(1 - \mathbf{n}\cdot\boldsymbol{\beta})^2 = a^2(1 - \beta^2) + d^2 = \frac{1}{\gamma^2}(a^2 + \gamma^2 d^2) \tag{2.20}$$

and for the electric field, (2.19),

$$\mathbf{E}(t) = \frac{e\gamma}{4\pi\epsilon_0} \frac{\mathbf{r}_b}{(a^2 + \gamma^2 d^2)^{3/2}}. \tag{2.21}$$

The magnetic field is obtained from the relation $\mathbf{B} = [\mathbf{n}_{\mathrm{ret}} \times \mathbf{E}]/c$ which contains the vector product

$$[\mathbf{n} \times \mathbf{r}_b] = [\mathbf{n} \times (\mathbf{r} - r\boldsymbol{\beta})] = [\boldsymbol{\beta} \times \mathbf{r}] = r\beta\sin\theta\,\mathbf{n}_\perp = a\beta\mathbf{n}_\perp,$$

where we define by $\mathbf{n}_\perp$ the unit vector in the direction $\boldsymbol{\beta} \times \mathbf{r}$, i.e. perpendicular to the plane O, A, P. Using $\sin\theta = a/r$ and $\epsilon_0\mu_0 = 1/c^2$, we obtain for the magnetic field

$$\mathbf{B} = \frac{e\gamma}{4\pi\epsilon_0 c} \frac{a\beta\mathbf{n}_\perp}{(a^2 + \gamma^2 d^2)^{3/2}}.$$

Using the angle α between the vector $\mathbf{r}_b$ and the trajectory we can express a and d by writing

$$a = r_b\sin\alpha, \qquad d = r_b\cos\alpha$$

and obtain for the electric field

$$\mathbf{E}(t) = \frac{e\gamma}{4\pi\epsilon_0 r_b^3} \frac{\mathbf{r}_b}{(\sin^2\alpha + \gamma^2\cos^2\alpha)^{3/2}} = \frac{e}{4\pi\epsilon_0\gamma^2 r_b^3} \frac{\mathbf{r}_b}{(1 - \beta^2\sin^2\alpha)^{3/2}}.$$

To express the result explicitly as a function of the time t we abandon the very general location of the coordinate origin used in Fig. 2.6 and change to better-adapted cylindrical

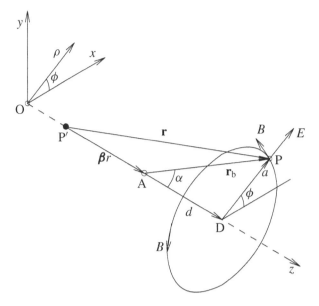

Fig. 2.7. The cylindrical geometry of the uniformly moving charge.

coordinates (ρ, ϕ, z) with the particle moving along the z-axis and the observer P being at (a, ϕ, z) as shown in Fig. 2.7. Owing to the rotational symmetry there is no dependence on the azimuthal angle and we set $\phi = 0$. We assume that a charge passes through the origin $z = 0$ at time $t = t' = 0$.

The vector $\mathbf{r}_p$, pointing from the origin to the observer P, and the normalized velocity $\boldsymbol{\beta}$ are

$$\mathbf{r}_p = [a, 0, z]$$
$$\boldsymbol{\beta} = [0, 0, \beta].$$

The time-dependent vectors $\mathbf{R}(t')$, pointing from the origin to the source, and $\mathbf{r}(t')$, pointing from the source to the observer, as functions of the emission time t' are, in cylindrical coordinates (ρ, z),

$$\mathbf{R}(t') = [0, 0, \beta c t'] \tag{2.22}$$
$$\mathbf{r}(t') = [a, 0, z - \beta c t'].$$

These are the parameters of the charge at time t' of emission which enter into the equation giving the fields at the location P. However, we would like to express them as functions of t. For this we need the relation between the two time scales. The distance $r = |\mathbf{r}|$ between source and observer is just the distance traveled by the radiation between the time t' of emission and the time t of observation:

$$r = c(t - t'). \tag{2.23}$$

With this we can express the position A of the charge $\mathbf{R}(t)$ at the time t, $\mathbf{R}(t) = \mathbf{R}(t') + r\boldsymbol{\beta}$, and the related vector $\mathbf{r}_b(t) = \mathbf{r}_p - \mathbf{R}(t)$:

$$\mathbf{R}(t) = [0, 0, \beta ct]$$
$$\mathbf{r}_b(t) = [a, 0, z - \beta ct].$$

The distance d between the point A and the projection D of the observer P on the axis is the z-component of $\mathbf{r}_b$:

$$d = z - \beta ct, \qquad \mathbf{r}_b = [a, 0, z - \beta ct] = r_b[\sin\alpha, 0, \cos\alpha] \tag{2.24}$$

as a function of t. On substituting this into (2.21) we obtain the electric field $\mathbf{E}$ and the magnetic field $\mathbf{B} = [\mathbf{n} \times \mathbf{E}]/c$ expressed in terms of the observation time t in cylindrical coordinates (E_ρ, E_ϕ, E_z):

$$\mathbf{E}(z, t) = \frac{e\gamma}{4\pi\epsilon_0} \frac{[a, 0, z - \beta ct]}{(a^2 + \gamma^2(z - \beta ct)^2)^{3/2}}$$
$$\mathbf{B}(z, t) = \frac{e\beta\gamma}{4\pi\epsilon_0 c} \frac{[0, a, 0]}{(a^2 + \gamma^2(z - \beta ct)^2)^{3/2}}. \tag{2.25}$$

From these fields we can calculate the energy flow given by the Poynting vector $\mathbf{S} = [\mathbf{E} \times \mathbf{B}]/\mu_0$ and the energy density w:

$$\mathbf{S}(z, t) = [S_\rho, S_\phi, S_z] = \frac{e^2\beta c\gamma^2}{16\pi^2\epsilon_0} \frac{[-a(z - \beta ct), 0, a^2]}{(a^2 + \gamma^2(z - \beta ct)^2)^3}$$

$$w(z, t) = \frac{\epsilon_0 E^2}{2} + \frac{B^2}{2\mu_0} = \frac{e^2\gamma^2}{32\pi^2\epsilon_0} \frac{a^2(1 + \beta^2) + (z - \beta ct)^2}{(a^2 + \gamma^2(z - \beta ct)^2)^3}.$$

Taking the time derivative of the second and the divergence of the first equation gives the continuity equation

$$\frac{\partial w}{\partial t} + \text{div}\,\mathbf{S} = 0.$$

This indicates that the power flow is just due to the motion of the field energy from one location to another and that no energy is radiated away to large distances.

To illustrate the field of the moving charge we discuss first its spatial field distribution and evaluate the electric field at the fixed time $t = 0$ when the charge is at the origin, in which case the emission time must be negative. The vector $\mathbf{r}_p$ has now the cylindrical coordinates $\mathbf{r}_p = (a, 0, z)$. The field vectors are, in cylindrical coordinates,

$$\mathbf{E}(z, 0) = \frac{e\gamma}{4\pi\epsilon_0} \frac{[a, 0, z]}{(a^2 + \gamma^2 z^2)^{3/2}}, \qquad \mathbf{B}(z, 0) = \frac{e\beta\gamma}{4\pi\epsilon_0 c} \frac{[0, a, 0]}{(a^2 + \gamma^2 z^2)^{3/2}},$$

which can also be expressed in terms of the angle α between the z-axis and the vector $\mathbf{r}_p$:

$$\mathbf{E}(z, 0) = \frac{e}{4\pi\epsilon_0\gamma^2} \frac{[\sin\alpha, 0, \cos\alpha]}{r_b^2(1 - \beta^2\sin^2\alpha)^{3/2}}$$

$$\mathbf{B}(z, 0) = \frac{e\beta}{4\pi\epsilon_0 c\gamma^2} \frac{[0, \sin\alpha, 0]}{r_b^2(1 - \beta^2\sin^2\alpha)^{3/2}}.$$

The absolute field values are

$$E = \frac{e}{4\pi\epsilon_0\gamma^2 r_b^2} \frac{1}{(1 - \beta^2\sin^2\alpha)^{3/2}}, \qquad B = \frac{e\beta}{4\pi\epsilon_0 c\gamma^2 r_b^2} \frac{1}{(1 - \beta^2\sin^2\alpha)^{3/2}}.$$

For $\beta \to 0$ we recover the electrostatic Coulomb field of a point charge:

$$\mathbf{E}_C = \frac{e}{4\pi\epsilon_0} \frac{\mathbf{r}_b}{r_b^3}, \qquad \mathbf{B} = 0.$$

This field distribution $\mathbf{E}/E_C$ as a function of the angle α is shown in Fig. 2.8 for various values of the Lorentz factor γ. With increasing velocity it becomes more concentrated into the transverse direction $\alpha = \pi/2$. The maximum field is in this direction and the minimum

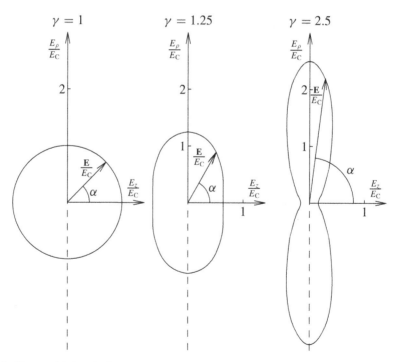

Fig. 2.8. The spatial electric-field distribution around the uniformly moving charge at the observation time t for various values of γ.

one is in the longitudinal direction $\alpha = 0$:

$$E_{max} = E_C \gamma, \qquad E_{min} = \frac{E_C}{\gamma^2},$$

where E_C is the corresponding Coulomb field. The flux of the electric field through a long cylinder of radius a around the z axis

$$2\pi a \int E_\rho \, dz = \frac{e}{\epsilon_0}$$

is independent of γ, as expected from the equation $\text{div } \mathbf{E} = \eta/\epsilon_0$.

The expression (2.25) gives the field as a function of the observation time t. In some cases we would like to know the time and location of the field emission. The distance r between the source at time t' of emission and the observer is obtained from the vector $\mathbf{r}$ (2.22) and the traveling time of the radiation (2.23):

$$r^2 = c^2(t - t')^2 = a^2 + (z - \beta ct')^2.$$

Since we are interested in the relation between the time scales at a fixed location, we put here the observer at $z = 0$ and obtain

$$c^2 t^2 - 2ctct' + c^2 t'^2 (1 - \beta^2) - a^2 = 0.$$

Solving for ct or ct' gives

$$ct = ct' \pm \sqrt{a^2 + \beta^2 c^2 t'^2}, \qquad ct' = \gamma^2 ct \mp \sqrt{\gamma^4 c^2 t^2 + \gamma^2(a^2 - c^2 t^2)}.$$

To determine the signs we set $t' = 0$ in the first equation, which must result in $t > 0$, and $t = 0$ in the second one, which must give $t' < 0$:

$$ct = ct' + \sqrt{a^2 + \beta^2 c^2 t'^2}, \qquad ct' = \gamma^2 ct - \sqrt{\gamma^4 c^2 t^2 + \gamma^2(a^2 - c^2 t^2)}.$$

These are relatively simple closed expressions, which are plotted in Fig. 2.9 for $\beta = 0.6$. Very early, when the charge is still at a large distance, $ct' \ll -a$, and very late, $ct' \gg a$, we have the asymptotic relations $ct \approx -|ct'|(1 - \beta)$ and $ct \approx ct'(1 + \beta)$, respectively. This is just a manifestation of the Doppler effect compressing a time interval during approach and expanding it during departure.

The distance $r(t')$ from the source to the observer is obtained from the triangle P′DP in Fig. 2.7:

$$r^2 = (\beta r + d)^2 + a^2 = \beta^2 r^2 + 2\beta rd + d^2 + a^2.$$

Using $\beta^2 \gamma^2 = \gamma^2 - 1$ and $d = z - \beta ct$, this gives

$$r = \beta \gamma^2 (z - \beta ct) + \sqrt{\gamma^4 (z - \beta ct)^2 + \gamma^2 a^2}. \qquad (2.26)$$

The sign of the square root has been determined to make $r > 0$.

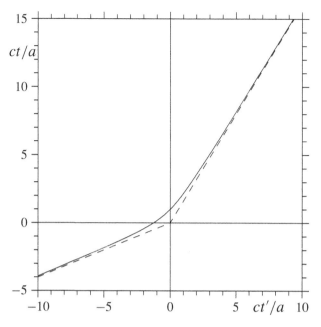

Fig. 2.9. The relation between the time t' of emission and the time t of observation for $\beta = 0.6$.

Using $z - \beta ct = r_b \cos \alpha$ from (2.24) we obtain for the distance between emission and observation

$$r = \gamma^2 r_b \left(\beta \cos \alpha + \sqrt{1 - \beta^2 \sin^2 \alpha} \right).$$

This becomes more transparent on choosing the observation position and time $z = 0$ and $t = 0$, which results in

$$r = a\gamma, \qquad ct' = -a\gamma, \qquad z' = -a\sqrt{\gamma^2 - 1} = -a\beta\gamma = -r\beta$$

for the distance $r(t')$, time t', and position $z'(t')$ on the axis of emission. The distance between the creation and observation of the field is therefore γ times larger than the distance a of the observer from the axis, which can be a large number for an ultra-relativistic particle. Fields observed at the same time t but different locations have been created at different times t' and distances r.

To study the time dependence of the field created by a uniformly moving charge, we set $z = 0$ in (2.25) to obtain

$$\mathbf{E}(0, t) = [E_\rho, E_\phi, E_z] = \frac{e\gamma}{4\pi\epsilon_0} \frac{[a, 0, -\beta ct]}{(a^2 + (\gamma\beta ct)^2)^{3/2}}$$

$$\mathbf{B}(0, t) = [B_\rho, B_\phi, B_z] = \frac{e\gamma\beta}{4\pi\epsilon_0 c} \frac{(0, a, 0)}{(a^2 + (\gamma\beta ct)^2)^{3/2}}.$$

The distance r between source and observation becomes, according to (2.26),

$$r = -\beta^2 \gamma^2 ct + \sqrt{\beta^2 \gamma^4 c^2 t^2 + \gamma^2 a^2}.$$

In Fig. 2.10 the vectors $\mathbf{r}$, $r\boldsymbol{\beta}$, and $\mathbf{E}$ for a uniformly moving charge are shown as functions of time to illustrate the two time scales t and t'. The charge is at point P' (full circle) at time t' of emission and at point A (empty circle) at time t when the emitted radiation reaches P. For $t = 0$ the fields have maximum values of

$$E(t = 0) = \frac{e\gamma}{4\pi \epsilon_0 a^2}, \qquad B(t = 0) = \frac{e\beta\gamma}{4\pi \epsilon_0 c a^2}.$$

The maximum electric field has only a ρ component and is a factor γ larger than the corresponding Coulomb field of a charge at rest.

We can also obtain the electric and magnetic fields of a uniformly moving charge by a Lorentz transformation of the static Coulomb field. This also indicates that a uniformly moving charge does not radiate any energy.

2.6.2 The field of a non-relativistic oscillating charge

We use the complete Liénard–Wiechert equations (2.17),

$$\mathbf{E}(t) = \frac{e}{4\pi \epsilon_0} \left\{ \frac{(1 - \beta^2)(\mathbf{n} - \boldsymbol{\beta})}{r^2(1 - \mathbf{n} \cdot \boldsymbol{\beta})^3} + \frac{[\mathbf{n} \times [(\mathbf{n} - \boldsymbol{\beta}) \times \dot{\boldsymbol{\beta}}]]}{cr(1 - \mathbf{n} \cdot \boldsymbol{\beta})^3} \right\}_{\text{ret}}$$

$$\mathbf{B}(t) = \frac{[\mathbf{n}_{\text{ret}} \times \mathbf{E}]}{c},$$

and investigate the radiation emitted by a charge executing a harmonic oscillation with frequency ω along the z-axis about the origin as shown in Fig. 2.11, which is often called dipole radiation. The position, velocity, and acceleration of the charge in Cartesian coordinates are

$$\mathbf{R}(t') = a \cos(\omega t')[0, \ 0, \ 1]$$

$$\boldsymbol{\beta}(t') = -\frac{a\omega}{c} \sin(\omega t')[0, \ 0, \ 1] \qquad (2.27)$$

$$\dot{\boldsymbol{\beta}}(t') = -\frac{a\omega^2}{c} \cos(\omega t')[0, \ 0, \ 1].$$

The radiation is detected by an observer P at a distance r_p from the origin given in Cartesian coordinates,

$$\mathbf{r}_p = r_p[\sin\theta \cos\phi, \ \sin\theta \sin\phi, \ \cos\theta], \qquad \mathbf{n}_p = \mathbf{r}_p / r_p,$$

with its unit vector $\mathbf{n}_p$. With this we obtain for the vector $\mathbf{r}$ pointing from the charge to the observer

$$\mathbf{r}(t') = \mathbf{r}_p - \mathbf{R}(t') = r_p \left[\sin\theta \cos\phi, \ \sin\theta \sin\phi, \ \cos\theta - \frac{a}{r_p} \cos(\omega t') \right].$$

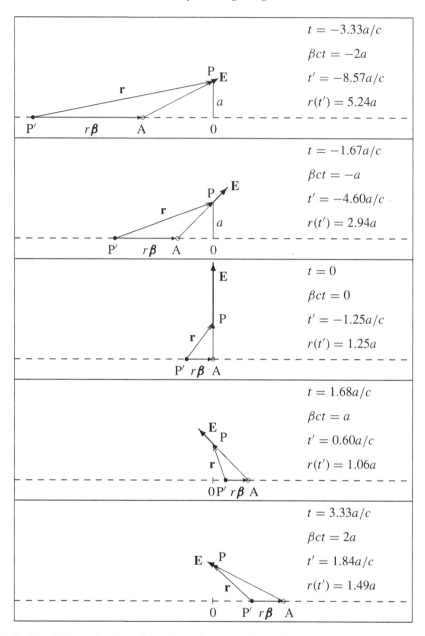

Fig. 2.10. The field as a function of time for a charge moving uniformly with $\beta = 0.6$ and $\gamma = 1.25$ and passing through the origin at $t = 0$.

We now make two assumptions. First, we assume that the harmonic oscillation should be non-relativistic, $\beta \ll 1$. According to (2.27) this means also that the amplitude of the oscillation is small compared with the wavelength of the radiation $a = \hat{\beta} c / \omega = \hat{\beta} \lambda / (2\pi)$. Secondly, the radiation should be observed from a distance that is large compared with the

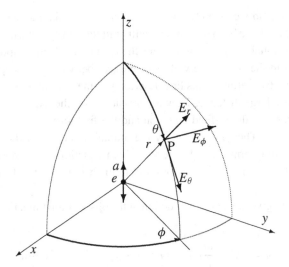

Fig. 2.11. Coordinates for the fields of an oscillating dipole.

wavelength λ and therefore also compared with the amplitude of oscillation, $\lambda/r_p \ll 1$, $a/r_p \ll 1$. We make the corresponding approximations and take only terms up to first order of these small quantities, leading to

$$r(t') \approx r_p\left(1 - \frac{\mathbf{n}_p \cdot \mathbf{R}(t')}{r_p}\right) = r_p\left(1 - \frac{a\cos\theta\cos(\omega t')}{r_p}\right)$$

$$\mathbf{n}(t') = \frac{\mathbf{r}(t')}{r(t')} \approx \mathbf{n}_p + \frac{\mathbf{n}_p(\mathbf{n}_p \cdot \mathbf{R}(t')) - \mathbf{R}(t')}{r_p}$$

$$= \mathbf{n}_p + \frac{a\cos(\omega t')}{r_p}[\cos\theta\sin\theta\cos\phi,\ \cos\theta\sin\theta\sin\phi,\ \cos^2\theta - 1]$$

$$1 - \mathbf{n} \cdot \boldsymbol{\beta} \approx 1 - \mathbf{n}_p \cdot \boldsymbol{\beta} = 1 + \frac{a\omega\cos\theta\sin(\omega t')}{c}.$$

With this we obtain from (2.17) the electric field as a function of the emission time t' in vector representation:

$$\mathbf{E}(t') = \frac{e}{4\pi\epsilon_0}\left(\frac{\mathbf{n}_p}{r_p^2} + \frac{3\mathbf{n}_p(\mathbf{n}_p \cdot \mathbf{R}(t')) - \mathbf{R}(t')}{r_p^3} + \frac{3\mathbf{n}_p(\mathbf{n}_p \cdot \boldsymbol{\beta}(t')) - \boldsymbol{\beta}(t')}{r_p^2}\right.$$

$$\left. + \frac{\mathbf{n}_p(\mathbf{n}_p \cdot \dot{\boldsymbol{\beta}}(t')) - \dot{\boldsymbol{\beta}}(t')}{cr_p}\right)$$

$$\mathbf{B}(t') = \frac{e}{4\pi\epsilon_0 c}\left(-\frac{[\mathbf{n}_p \times \boldsymbol{\beta}(t')]}{r_p^2} - \frac{[\mathbf{n}_p \times \dot{\boldsymbol{\beta}}(t')]}{cr_p}\right). \tag{2.28}$$

The first line in the upper equation of the electric-field expression refers to the near field. Its first term is just the static Coulomb term caused by the presence of the charge e, which is of little interest. Had we taken two opposite charges, oscillating against each other, this static term would not have been present. We will omit this term later. The second term in this first line is proportional to $1/r_p^3$. It represents the field of an electric dipole moment $eR(t')$. It is present even in the static case with vanishing velocity $\beta = 0$ except when $\mathbf{R}(t') = 0$. Therefore, the first two terms in this line describe just the field of a displaced charge in the approximation of a large distance. They are missing from the expression for the magnetic field. The third term in the E-field equation and the first term in the B-field equation are proportional to $1/r_p^2$. They are caused by the mutual induction of the two fields and are often called induction terms. The last term in both equations is proportional to $1/r_p$ and $\dot{\beta}$ and is called the far-field or radiation-field term. It is present only for an accelerated charge.

We would like to give the field as a function of the observation time. From $t = t' + r(t')/c$ we find

$$\omega t = \omega t' + \frac{\omega}{c} r(t') \approx \omega t' + \frac{\omega}{c} r_p - ka \cos\theta \cos(\omega t').$$

Since the oscillating terms in the field equations are multiplied by a factor of order β we can, within our approximation, neglect the last term. We use the wave number $k = 2\pi/\lambda = \omega/c$ and obtain for the time-scale relation

$$\omega t = \omega t' + kr_p, \qquad \omega t' = \omega t - kr_p.$$

The field (2.28) expressed in terms of t represents a wave propagating with velocity c, frequency ω, and wave number k in the $\mathbf{r}_p$-direction. We give it first in Cartesian coordinates:

$$
\begin{aligned}
\mathbf{E} = [E_x, E_y, E_z] = \frac{e}{4\pi\epsilon_0} \Bigg[& \frac{1}{r_p^2} [\sin\theta\cos\phi, \ \sin\theta\sin\phi, \ \cos\theta] \\
& \times \frac{a}{r_p^3} [3\cos\theta\sin\theta\cos\phi, \ 3\cos\theta\sin\theta\sin\phi, \ 3\cos^2\theta - 1]\cos(\omega t - kr_p) \\
& - \frac{a\omega}{r_p^2 c} [3\cos\theta\sin\theta\cos\phi, \ 3\cos\theta\sin\theta\sin\phi, \ 3\cos^2\theta - 1]\sin(\omega t - kr_p) \\
& - \frac{a\omega^2}{r_p c^2} [\cos\theta\sin\theta\cos\phi, \ \cos\theta\sin\theta\sin\phi, \ \cos^2\theta - 1]\cos(\omega t - kr_p) \Bigg]
\end{aligned}
$$

$$
\begin{aligned}
\mathbf{B} = (B_x, B_y, B_z) \\
= \frac{e}{4\pi\epsilon_0 c} \Bigg(& \frac{a\omega}{r_p^2 c} [\sin\theta\sin\phi, \ -\sin\theta\cos\phi, \ 0]\sin(\omega t - kr_p) \\
& + \frac{a\omega^2}{r_p c^2} [\sin\theta\sin\phi, \ -\sin\theta\cos\phi, \ 0]\cos(\omega t - kr_p) \Bigg).
\end{aligned}
$$

Owing to its symmetry the dipole radiation is better given in spherical coordinates (θ, ϕ, r) using the relations

$$E_\theta = E_x \cos \theta \cos \phi + E_y \cos \theta \sin \phi - E_z \sin \theta$$
$$E_\phi = -E_x \sin \phi + E_y \cos \phi$$
$$E_r = E_x \sin \theta \cos \phi + E_y \sin \theta \sin \phi + E_z \cos \theta.$$

We obtain from (2.28) the spherical-field components of an oscillating dipole without the Coulomb term $en/(4\pi\epsilon_0 r_p^2)$:

$$E_\theta = \frac{eak^3 \sin \theta}{4\pi\epsilon_0} \left(\frac{\cos(\omega t - kr_p)}{(kr_p)^3} - \frac{\sin(\omega t - kr_p)}{(kr_p)^2} - \frac{\cos(\omega t - kr_p)}{kr_p} \right)$$

$$E_r = \frac{2eak^3 \cos \theta}{4\pi\epsilon_0} \left(\frac{\cos(\omega t - kr_p)}{(kr_p)^3} - \frac{\sin(\omega t - kr_p)}{(kr_p)^2} \right) \tag{2.29}$$

$$B_\phi = \frac{eak^3 \sin \theta}{4\pi\epsilon_0 c} \left(\frac{-\sin(\omega t - kr_p)}{(kr_p)^2} - \frac{\cos(\omega t - kr_p)}{kr_p} \right)$$

$$E_\phi = B_\theta = B_r = 0.$$

At any location the electric far field has only a component in the θ-direction. Its near field consists of components in the r_p- and θ-directions. The magnetic field has only a ϕ-component. The radiation is therefore *linearly polarized*.

We would like to present this dipole-radiation field pattern by plotting the field lines, i.e. the curves which at any point have the electric-field direction as a tangent, and follow the method described in [11]. They are given in Cartesian and spherical coordinates by

$$\frac{dy}{dx} = \frac{E_y}{E_x}, \qquad \frac{1}{r_p}\frac{dr_p}{d\theta} = \frac{E_r}{E_\theta}.$$

Since the electric field is source free, div $\mathbf{E} = 0$, in the region of interest it can be derived from a vector potential $\mathbf{C}$ by putting

$$\mathbf{E} = [\nabla \times \mathbf{C}] = \operatorname{curl} \mathbf{C}.$$

The electric field has only components E_r and E_θ. The vector potential $\mathbf{C}$, being perpendicular to $\mathbf{E}$, has therefore only a ϕ-component with the relations

$$E_\theta = -\frac{1}{r_p}\frac{\partial(r_p C_\phi)}{\partial r_p}, \qquad E_r = \frac{1}{r_p \sin \theta}\frac{\partial(\sin \theta\, C_\phi)}{\partial \theta}. \tag{2.30}$$

This gives the equation of the field lines,

$$\frac{1}{r_p}\frac{dr_p}{d\theta} = \frac{E_r}{E_\theta} = -\frac{\dfrac{\partial(\sin \theta\, C_\phi)}{\partial \theta}}{\sin \theta \dfrac{\partial(r_p C_\phi)}{\partial r_p}},$$

which can be written as a total differential:

$$\frac{\partial(r_p \sin\theta \; C_\phi)}{\partial r_p} dr_p + \frac{\partial(r_p \sin\theta \; C_\phi)}{\partial\theta} d\theta = d(r_p \sin\theta \; C_\phi) = 0.$$

The field lines are therefore determined by the function

$$r \sin\theta \; C_\phi = \text{constant}.$$

The expression for the vector potential $C(\theta, r_p)$ can be determined from the field components (2.29) and the relations (2.30):

$$\frac{eak^3 \sin\theta}{4\pi\epsilon_0}\left(\frac{\cos(\omega t - kr_p)}{(kr_p)^3} - \frac{\sin(\omega t - kr_p)}{(kr_p)^2} - \frac{\cos(\omega t - kr_p)}{kr_p}\right) = -\frac{\partial(r_p C_\phi)}{r_p \partial r_p}$$

$$\frac{2eak^3 \cos\theta}{4\pi\epsilon_0}\left(\frac{\cos(\omega t - kr_p)}{(kr_p)^3} - \frac{\sin(\omega t - kr_p)}{(kr_p)^2}\right) = \frac{1}{r_p \sin\theta}\frac{\partial(\sin\theta \; C_\phi)}{\partial\theta},$$

giving

$$C_\phi(\theta, r_p) = \frac{e}{4\pi\epsilon_0}ak^2 \sin\theta\left(\frac{\cos(\omega t - kr_p)}{(kr_p)^2} - \frac{\sin(\omega t - kr_p)}{kr_p}\right).$$

The electric-field lines are obtained from $C_\phi(\theta, r_p) = $ constant and plotted in Fig. 2.12 for one period of the emitted dipole radiation.

We check the divergence of the electric field (2.29),

$$\text{div } \mathbf{E} = \frac{1}{r_p \sin\theta}\frac{\partial(\sin\theta \; E_\theta)}{\partial\theta} + \frac{1}{r_p^2}\frac{\partial(r_p^2 E_r)}{\partial r_p} = 0,$$

which vanishes as expected for a source-free region. However, if we take only the far field

$$E_{f\theta} = -\frac{eak^3 \sin\theta}{4\pi\epsilon_0}\frac{\cos(\omega t - kr_p)}{kr_p}$$

$$E_{fr} = 0 \tag{2.31}$$

as an approximation for large distances, we obtain for the divergence

$$\text{div}\mathbf{E}_f = -\frac{eak^3 \cos\theta}{2\pi\epsilon_0 r_p}\frac{\cos(\omega t - kr_p)}{kr_p}, \tag{2.32}$$

which does not vanish. The far field alone does not satisfy the equation div $\mathbf{E} = 0$ and is therefore not correct even at large distances. This is clearly visible from the field pattern in Fig. 2.12. The far field has only a θ-component for the electric field. The field lines cannot form closed loops but must end somewhere in free space, implying that there is a finite divergence, which is impossible without sources. Although the far field alone is proportional to $1/r_p$, the expression (2.32) for its divergence is proportional to $1/r_p^2$. It can be compensated by the radial component of the near field, which provides a divergence term proportional to $1/r_p^2$.

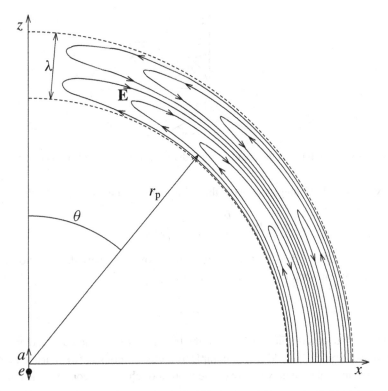

Fig. 2.12. Electric-field lines of the radiation emitted by an oscillating dipole.

The energy flow per unit time through a unit area is given by the Poynting vector **S**

$$\mathbf{S} = \frac{[\mathbf{E} \times \mathbf{B}]}{\mu_0} = \frac{[\mathbf{E} \times [\mathbf{n} \times \mathbf{E}]]}{\mu_0 c}.$$

For our far-field expression (2.31) we obtain

$$\mathbf{S} = [S_\theta, \, S_\phi, \, S_r] = \frac{e^2 a^2 \omega^4 \cos^2(\omega t - kr)}{(4\pi)^2 \epsilon_0 c^3 r_{\mathrm{p}}^2} [0, \, 0, \, \sin^2 \theta]$$

as the only term with a non-vanishing time average. Multiplying **S** by r^2 and averaging over time gives the radiated power per unit solid angle:

$$\frac{\mathrm{d}P}{\mathrm{d}\Omega} = r_{\mathrm{p}}^2 \langle S \rangle = \frac{e^2 a^2 \omega^4 \sin^2 \theta}{32\pi^2 \epsilon_0 c^3}. \tag{2.33}$$

This is the well-known distribution of dipole radiation. It is emitted mainly perpendicular to the motion of the charge ($\theta = \pi/2$) but with a large opening angle around it as shown in Fig. 2.13. No radiation is emitted in the direction of the motion, i.e. the z-axis. We calculate

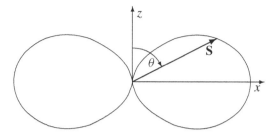

Fig. 2.13. A cut through the angular power distribution of the dipole radiation.

the total power emitted by integration over the solid-angle element $d\Omega = \sin\theta\, d\theta\, d\phi$,

$$P_u = \frac{\hat{P}}{2} = \frac{e^2 a^2 \omega^4}{12\pi \epsilon_0 c^3},$$

where we use $P_u = \langle P \rangle = P_0/2$ for the power averaged over one period. This allows us to write the angular distribution in the convenient form

$$\frac{dP}{d\Omega} = P_u \frac{3\sin^2\theta}{8\pi}.$$

We will come back to this dipole radiation for the treatment of the undulator radiation.

This example of a radiation field is made simple by the assumptions of a small excursion of the charge and a non-relativistic velocity of the motion of charge. It confines the radiating charge to a small volume in space and allows us to approximate the vector $\mathbf{r}$ pointing from the charge to the observer by the constant $\mathbf{r}_p$ in the triple vector product and the denominator $(1 - \mathbf{n} \cdot \boldsymbol{\beta})$ of (2.17), and leads to a simple relation between the emission time and observation time.

2.7 The near field and the far field

The Liénard–Wiechert equation (2.17) has two parts with different dependences on the distance r between the radiating charge and the observer. As we discussed before, the near field is proportional to $1/r^2$ and does not require an accelerated charge, whereas the far field decreases like $1/r$ and is proportional to the acceleration. Integrating the energy flux over the surface of a sphere that encloses the source gives an emitted power proportional to $1/r^2$ for the near field and a constant for the far field. Only the second leads to power being radiated at large distances, as we discussed before.

We saw in the example of the radiation emitted by an oscillating dipole that the near field should be included in order to be consistent with Maxwell's equations, in particular with the condition $\text{div }\mathbf{E} = 0$ in free space. Synchrotron radiation is emitted at rather high frequencies and experiments carried out with it are usually concerned with its power or the related photon flux; they only rarely measure the fields directly. Therefore, we need the power emitted by the synchrotron radiation and its distribution in angle, frequency, and polarization. Since the last two properties are evident in the expressions for the field, but not for the power, it is convenient to calculate first the radiation field (far field) in order to obtain these properties and determine the power later. The fact that div $\mathbf{E} \neq 0$ for the far field is of no concern as long as it is used only to obtain the power distribution.

2.8 The Fourier transform of the radiation field

2.8.1 The Fourier integral of the field

We derived the radiation field emitted by an accelerated charge,

$$\mathbf{E}(t) = \frac{e}{4\pi \epsilon_0 c} \left\{ \frac{[\mathbf{n} \times [(\mathbf{n} - \boldsymbol{\beta}) \times \dot{\boldsymbol{\beta}}]]}{r(1 - \mathbf{n} \cdot \boldsymbol{\beta})^3} \right\}_{\text{ret}}, \qquad \mathbf{B}(t) = \frac{1}{c}[\mathbf{n} \times \mathbf{E}]. \qquad (2.34)$$

As mentioned before, the difficulty in evaluating the above equations is due to the fact that the expression involving the particle motion has to be evaluated at the earlier time t' at which there was emitted the radiation which is being received by the observer now at the time t. The relation between the two time scales (2.1),

$$t = t' + \frac{r(t')}{c}, \qquad dt = (1 - \mathbf{n} \cdot \boldsymbol{\beta}) \, dt',$$

can be very complicated for a general motion of the particle. For this reason it is often advantageous to calculate directly the Fourier transform $\tilde{\mathbf{E}}(\omega)$ of the field as a function of frequency ω instead of the field $\mathbf{E}(t)$ as a function of time:

$$\tilde{\mathbf{E}}(\omega) = \frac{1}{\sqrt{2\pi}} \int_{-\infty}^{\infty} \mathbf{E}(t) e^{-i\omega t} \, dt. \qquad (2.35)$$

This integration involves the observation time t because we are interested in the spectrum of the radiation as seen by the observer. However, we can make a formal transformation of the integration variable t into t':

$$\tilde{\mathbf{E}}(\omega) = \frac{e}{4\pi \epsilon_0 c} \frac{1}{\sqrt{2\pi}} \int_{-\infty}^{\infty} \left(\frac{[\mathbf{n} \times [(\mathbf{n} - \boldsymbol{\beta}) \times \dot{\boldsymbol{\beta}}]]}{r(1 - \mathbf{n} \cdot \boldsymbol{\beta})^2} \right) e^{-i\omega(t' + r(t')/c)} \, dt'. \qquad (2.36)$$

We omitted the index 'ret' from the above equation since the variable of integration is now t' anyway. The two equations (2.34) and (2.36), giving the field as a function of time t or frequency ω, are equivalent. However, the second one can be simplified by making some approximations that are valid for most practical cases. We take now the ultra-relativistic case $\gamma \gg 1$ for which the emitted radiation is concentrated in a cone of small half opening angle of order $1/\gamma \ll 1$, as we saw from qualitative arguments in Chapter 1. The observer sees only radiation originating from a small part of the trajectory of approximate length $\ell_r \approx 2\rho/\gamma$, where ρ is the radius of curvature of the trajectory. The vector $\mathbf{r}$ pointing from the particle to the observer will in this case change little during the emission as long as it is observed from a large distance $r \gg 2\rho/\gamma$. We can therefore, to a good approximation, regard the vectors $\mathbf{r}$ and $\mathbf{n}$ in the large parentheses of (2.36) as being constant, but not the distance r appearing in the exponent. We can integrate (2.36) in parts,

$$\int \frac{\mathrm{d}U}{\mathrm{d}t'} V \, \mathrm{d}t' = UV - \int U \frac{\mathrm{d}V}{\mathrm{d}t'} \, \mathrm{d}t'$$

with

$$U = \frac{[\mathbf{n} \times [\mathbf{n} \times \boldsymbol{\beta}]]}{(1 - \mathbf{n} \cdot \boldsymbol{\beta})}, \qquad \frac{\mathrm{d}U}{\mathrm{d}t'} = \frac{[\mathbf{n} \times [(\mathbf{n} - \beta) \times \dot{\boldsymbol{\beta}}]]}{(1 - \mathbf{n} \cdot \boldsymbol{\beta})^2}$$

$$V = \mathrm{e}^{-\mathrm{i}\omega(t'+r(t')/c)}, \qquad \frac{\mathrm{d}V}{\mathrm{d}t'} = -\mathrm{i}\omega(1 - \mathbf{n} \cdot \boldsymbol{\beta})\mathrm{e}^{-\mathrm{i}\omega(t'+r(t')/c)}.$$

With this we obtain

$$\tilde{\mathbf{E}}(\omega) = \frac{e}{4\pi \sqrt{2\pi} \epsilon_0 c r_\mathrm{p}}$$

$$\times \left(\frac{[\mathbf{n} \times [\mathbf{n} \times \boldsymbol{\beta}]]}{(1 - \mathbf{n} \cdot \boldsymbol{\beta})} \mathrm{e}^{-\mathrm{i}\omega(t'+r(t')/c)} \Big|_{-\infty}^{\infty} + \mathrm{i}\omega \int_{-\infty}^{\infty} [\mathbf{n} \times [\mathbf{n} \times \boldsymbol{\beta}]] \mathrm{e}^{-\mathrm{i}\omega(t'+r(t')/c)} \, \mathrm{d}t' \right).$$

$$(2.37)$$

The first term on the right-hand side of the above equation can be neglected since it contains only an expression at $t' = \pm\infty$, which has no influence on the field seen by the observer at time t. We made earlier an approximation for $|\omega_0 t'| \ll 1$. It seem to be contradictory to integrate now over t' from $-\infty$ to $+\infty$. However, it can be shown that the term under the integral oscillates faster and faster as $|t'|$ increases. This provides nearly perfect cancelation such that the contribution to the integral becomes negligible for large values of $|t'|$. A more rigorous treatment of this point will be given in the next two subsections for a motion that is periodic or has a periodic velocity. We obtain for the Fourier-transformed radiation field

$$\boxed{\tilde{\mathbf{E}}(\omega) = \frac{\mathrm{i}\omega e}{4\pi \sqrt{2\pi} \epsilon_0 c r_\mathrm{p}} \int_{-\infty}^{\infty} [\mathbf{n} \times [\mathbf{n} \times \boldsymbol{\beta}]] \mathrm{e}^{-\mathrm{i}\omega(t'+r(t')/c)} \, \mathrm{d}t'.} \qquad (2.38)$$

It is still related to the magnetic field by

$$\tilde{\mathbf{B}}(\omega) = \frac{1}{c}[\mathbf{n} \times \tilde{\mathbf{E}}(\omega)].$$

The Fourier-transformed field equation (2.38) is much easier to use than the corresponding expression (2.34) in the time domain. However, one should remember that it involves the approximations that the particle is ultra-relativistic and that the observed radiation is emitted from a localized area such that the vector $\mathbf{n}$ can be treated as being constant. It is e.g. not very accurate for the radiation from a long undulator observed from a relatively small distance. The expression (2.38) is also less transparent, since it does not contain the acceleration $\dot{\boldsymbol{\beta}}$ explicitly. The integration from $-\infty$ to $+\infty$ may contain intervals having no acceleration, which should not contribute to the radiated power. It can be shown that the expression (2.38) treats such cases correctly.

2.8.2 The periodic motion

We observe the radiation from a distance much larger than the extent of the source $2\rho/\gamma$ and neglect the variation of the vectors $\mathbf{r}$ and $\mathbf{n}$ except in the relation between the time scales, $t = t' + r(t')/c$. We assume that the motion of the particle is a periodic function of time with period T_{rev}:

$$\mathbf{R}(t') = \mathbf{R}(t' + T_{\text{rev}}).$$

An example of such a motion is a charge moving on a closed orbit. For this motion the velocity averaged over a period vanishes, $\langle \boldsymbol{\beta}(t') \rangle = \boldsymbol{\beta}^* = 0$.

The distance r between charge and observer and the relation between emission time t' and observation time t,

$$r = |\mathbf{r}| = |r_{\text{p}} - \mathbf{R}(t')|, \qquad t = t' + \frac{r(t') - r_{\text{p}}}{c},$$

are also periodic functions with period T_{rev}. The changes in emission time $\delta t'$ and observation time δt during one period are the same, namely $\delta t' = \delta t = T_{\text{rev}}$. The radiation is described by the Liénard–Wiechert equations

$$\mathbf{E}(t) = \frac{e}{4\pi\epsilon_0 c} \left(\frac{[\mathbf{n} \times [(\mathbf{n} - \boldsymbol{\beta}) \times \dot{\boldsymbol{\beta}}]]}{r(1 - \mathbf{n} \cdot \boldsymbol{\beta})^3} \right)_{\text{ret}}, \qquad \mathbf{B}(t) = \frac{1}{c}[\mathbf{n} \times \mathbf{E}],$$

which now contain all periodic parameters. As a consequence also the radiation field is a periodic function with period T_{rev} in observation time t and can be developed in a Fourier series, leading to a line spectrum

$$\mathbf{E}(t) = \sum_{m=-\infty}^{\infty} \mathbf{E}_m e^{im\omega_{\text{rev}}t} \quad \text{with} \quad \mathbf{E}_m = \frac{\omega_{\text{rev}}}{2\pi} \int_0^{T_{\text{rev}}} \mathbf{E}(t) e^{-im\omega_0 t} \, dt$$

with $\omega_{\text{rev}} = 2\pi/T_{\text{rev}}$.

To calculate the Fourier components $\mathbf{E}_m$ we use the same method as in the previous section, i.e. we change the integration variable from t to t', approximate for ultra-relativistic velocity and observation from a large distance, and integrate in parts. This leads to a similar expression to the Fourier transform (2.37) but with ω replaced by the line frequencies $m\omega_{\text{rev}}$, a different factor in front, and different limits:

$$\mathbf{E}_m = \frac{e}{4\pi\epsilon_0 c T_{\text{rev}} r_{\text{p}}} \left(\frac{[\mathbf{n} \times [\mathbf{n} \times \boldsymbol{\beta}]]}{1 - \mathbf{n} \cdot \boldsymbol{\beta}} e^{-im\omega_{\text{rev}}(t' + r(t')/c)} \Big|_0^{T_{\text{rev}}} \right.$$

$$\left. + im\omega_{\text{rev}} \int_0^{T_{\text{rev}}} [\mathbf{n} \times [\mathbf{n} \times \boldsymbol{\beta}]] e^{-im\omega_{\text{rev}}(t' + r(t')/c)} \, dt' \right). \quad (2.39)$$

The first term on the right-hand side vanishes since it is periodic with period T_{rev} and takes the same value at $t' = 0$ and $t' = T_{\text{rev}}$. We obtain for the mth coefficient of the Fourier series

$$\boxed{\mathbf{E}_m = \frac{ime\omega_{\text{rev}}^2}{8\pi^2\epsilon_0 c r_p} \int_0^{T_{\text{rev}}} [\mathbf{n} \times [\mathbf{n} \times \boldsymbol{\beta}]] e^{-im\omega_{\text{rev}}(t' + r(t')/c)} \, dt'.} \tag{2.40}$$

To continue we need detailed knowledge of the orbit of the particle at the point where the observed radiation is emitted. This will be applied in Chapter 4 to determine the radiation emitted by a charge moving on a closed circular orbit.

2.8.3 The motion with a periodic velocity

Next we take the case of a motion with a periodic velocity that has a non-vanishing average value, $\langle \boldsymbol{\beta} \rangle = \boldsymbol{\beta}^* \neq 0$. The resulting motion consists of a periodic part and a constant drift velocity $\boldsymbol{\beta}^* c$,

$$\boldsymbol{\beta}(t') = \boldsymbol{\beta}(t' + T_{\text{rev}}), \qquad \mathbf{R}(t') = \mathbf{R}_{\text{periodic}}(t') + \boldsymbol{\beta}^* ct',$$

where T_{rev} is the period ln t' of the velocity. An example of such a motion is a charge moving through an undulator having a large number N_u of periods. The distance $\mathbf{r}(t')$ between the charge and observer consists also of a periodic part and a constant drift:

$$\mathbf{r}(t') = \mathbf{r}_p - \mathbf{R}_{\text{periodic}}(t) - \boldsymbol{\beta}^* ct'$$

and changes in one period by

$$\delta\mathbf{r} = -\delta\mathbf{R} = -\boldsymbol{\beta}^* c T_{\text{rev}}.$$

In the following we use the direction of the drift velocity $\boldsymbol{\beta}^*$ as the z-axis and measure the direction towards the observer by the angle θ with respect to it. We assume that the observation point is very far away, $r_p \gg \boldsymbol{\beta}^* c T_{\text{rev}} N_u$, such that the change of this angle θ in one period is small. The absolute value r of the distance between the radiating charge and the observer changes in one period by

$$\delta r = -\beta^* c \cos\theta \, T_{\text{rev}}. \tag{2.41}$$

From this we obtain also the increase T_p in observation time during a period T_{rev} of the particle velocity

$$T_p = T_{\text{rev}} + \frac{\delta r}{c} = (1 - \beta^* \cos\theta) T_{\text{rev}} \approx \frac{1 + \gamma^{*2}\theta^2}{2\gamma^{*2}} T_{\text{rev}} \tag{2.42}$$

approximated for the ultra-relativistic case and using $\gamma^* = 1/\sqrt{1 - \beta^{*2}}$. In the forward direction, $\theta \ll 1$, the interval of the observation time is much shorter than of the emission. This is just a manifestation of the Doppler effect as seen before.

On considering the observation from a large distance and making the corresponding approximation, we find that the observed radiation given by the Liénard–Wiechert equation is periodic with the period T_p. It can be developed into a Fourier series, resulting in a line spectrum with frequencies

$$m\omega_1 = m\frac{2\pi}{T_p} = m\frac{2\pi}{(1 - \beta^* \cos\theta)T_{rev}} \approx m\frac{2\pi}{T_{rev}}\frac{2\gamma^{*2}}{1 + \gamma^{*2}\theta^2}. \tag{2.43}$$

We obtain the Fourier components of the field

$$\mathbf{E}_m = \frac{1}{T_p}\int_0^{T_p} \mathbf{E}(t)e^{-i\omega t}\,dt.$$

As before, (2.36), we make a transformation from the variable t to t':

$$\mathbf{E}_m = \frac{1}{T_p}\int_0^{T_{rev}} \mathbf{E}(t')e^{-i\omega(t' + r(t')/c)}(1 - \mathbf{n}\cdot\boldsymbol{\beta})\,dt'.$$

Using the expression (2.34) for the field and integrating by parts as in (2.37), we obtain

$$\mathbf{E}_m = \frac{e}{4\pi\epsilon_0 c T_p r_p}\left(\frac{[\mathbf{n}\times[\mathbf{n}\times\boldsymbol{\beta}]]}{1 - \mathbf{n}\cdot\boldsymbol{\beta}}e^{-im\omega_1(t' + r(t')/c)}\Big|_0^{T_{rev}}\right.$$

$$\left. + im\omega_1\int_0^{T_{rev}} [\mathbf{n}\times[\mathbf{n}\times\boldsymbol{\beta}]]e^{-im\omega_1(t' + r(t')/c)}\,dt'\right).$$

The first term on the right-hand side consists, besides a constant factor, of a triple vector product divided by a denominator, all being periodic with period T_{rev}, and of an exponential. The latter is

$$e^{-im\omega_1(t' - r(t')/c)}\Big|_0^{T_{rev}} = e^{im\omega_1(T_{rev} + \delta r/c)} - 1 = e^{-im2\pi} - 1 = 0,$$

where we used (2.43) to express ω_1 and (2.41) to obtain the drift δr per period. We obtain for the Fourier component of the radiation field emitted by a charge moving with a periodic velocity a similar expression as for the case of a periodic motion, (2.38):

$$\mathbf{E}_m = \frac{ime\omega_1^2}{8\pi^2\epsilon_0 c r_p}\int_0^{T_{rev}} [\mathbf{n}\times[\mathbf{n}\times\boldsymbol{\beta}]]e^{-im\omega_1(t' + r(t')/c)}\,dt'. \tag{2.44}$$

Similarly to the case of the Fourier-transformed field, we have a problem with our approximation based on the observation from a large distance. To have a periodic velocity means a motion that lasts a very long time. Owing to the finite drift velocity the particle will move over a very large distance, which should be observed from an even larger distance. There is some approximation in this approach. The motion is not really periodic in the mathematical sense but it is assumed that it contains a large number N_u of periods. This case refers to an undulator having a finite length L_u containing many periods and that is observed from a distance $r_p \gg L_u$. This will be discussed later.

3

The emitted radiation field and power

3.1 Introduction

We investigate now the radiation field and power emitted by an accelerated charge. This was done already in the previous chapter for a charge executing a harmonic oscillation. However, in this case the motion was non-relativistic and confined to a small volume in space, which makes some approximations possible. We consider now a more general motion having an arbitrary velocity $\mathbf{v}$ and acceleration $\dot{\mathbf{v}}$.

We mentioned that a complete calculation of the field seen by the observer is made difficult by the relation between the emission time t' and the observation time t, which can be very complicated. In the first part of this chapter we will avoid this problem by concentrating on the field seen by the observer at some arbitrary fixed time $t' + r/c$ after the emission and do not attempt to give the field $\mathbf{E}(t)$ as a function of time. Therefore, we evaluate the radiation part of the Liénard–Wiechert equation (2.17) at the time t' and give the field for a fixed time later at the observer but not its time dependence. Obviously this will not give any information about the spectrum seen by the observer. For this calculation we need only the differential relation (2.6) between the time scales,

$$\frac{\mathrm{d}t}{\mathrm{d}t'} = 1 - \mathbf{n} \cdot \boldsymbol{\beta},$$

which is much simpler than obtaining the complete function $t = t(t')$.

We use this method in particular to obtain the total radiated power and its angular distribution. The energy emitted by a moving charge must be the same as the energy received by an observer, which covers the full solid angle. However, the emitted power can be different from that received at any particular instant. Owing to the different time scales t' and t, the energy radiated by a charge in a time interval $\Delta t'$ might be received by the observer within a much shorter interval Δt. Therefore, we have to distinguish between the power $P(t')$ emitted and the power $P_{\mathrm{p}}(t)$ received. The latter often has a temporal structure consisting of short but very intense pulses and a long dead time in between. This instantaneous received power is of little interest but its average is relevant. Obviously the average emitted and received powers are the same.

In connection with the instantaneous angular power density, we also investigate this angular spectral distribution. This assumes that the Fourier transform of the field $\tilde{\mathbf{E}}(\omega)$ has been calculated already, as discussed at the end of Chapter 2.

3.2 The emitted and received powers

The radiation fields have been derived in the previous chapter:

$$\mathbf{E}(t) = \frac{e}{4\pi\epsilon_0 c} \left\{ \frac{[\mathbf{n} \times [(\mathbf{n} - \boldsymbol{\beta}) \times \dot{\boldsymbol{\beta}}]]}{r(1 - \mathbf{n} \cdot \boldsymbol{\beta})^3} \right\}_{\text{ret}}, \qquad \mathbf{B}(t) = \frac{1}{c}[\mathbf{n} \times \mathbf{E}]. \qquad (3.1)$$

The power flux seen by the observer is given by the Poynting vector $\mathbf{S}$:

$$\mathbf{S} = \frac{1}{\mu_0}[\mathbf{E} \times \mathbf{B}] = \frac{1}{\mu_0 c}[\mathbf{E} \times [\mathbf{n} \times \mathbf{E}]] = \frac{1}{\mu_0 c}\left(E^2 \mathbf{n} - (\mathbf{n} \cdot \mathbf{E})\mathbf{E}\right).$$

For the radiation field (3.1) we have $\mathbf{E} \cdot \mathbf{n} = 0$ and obtain

$$\mathbf{S} = \frac{E^2}{\mu_0 c}\mathbf{n}.$$

The vector $\mathbf{S}$ has only a radial component, $S_r = \mathbf{n} \cdot \mathbf{S}$. Since the fields $\mathbf{E}$ and $\mathbf{B}$ are the ones seen by the observer at the time t, the Poynting vector $\mathbf{S}$ represents the energy U *received by the observer* per unit area and time interval Δt:

$$\mathbf{S} = \frac{1}{r^2}\frac{\mathrm{d}^2 U}{\mathrm{d}\Omega\,\mathrm{d}t}\mathbf{n} = \frac{1}{r^2}\frac{\mathrm{d}P_\mathrm{p}}{\mathrm{d}\Omega}\mathbf{n}.$$

The power *radiated by the particle* per unit solid angle is the energy emitted per unit solid angle and unit time interval $\Delta t'$:

$$\boxed{\frac{\mathrm{d}P}{\mathrm{d}\Omega} = \frac{\mathrm{d}^2 U}{\mathrm{d}\Omega\,\mathrm{d}t'} = \frac{\mathrm{d}^2 U}{\mathrm{d}\Omega\,\mathrm{d}t}\frac{\mathrm{d}t}{\mathrm{d}t'} = \frac{r^2|\mathbf{E}|^2}{\mu_0 c}(1 - \mathbf{n} \cdot \boldsymbol{\beta}).} \qquad (3.2)$$

To clarify the situation we express the energy radiated per solid angle both in terms of the emitted power P and in terms of the received power P_p:

$$\frac{\mathrm{d}U}{\mathrm{d}\Omega} = \int_{-\infty}^{\infty} \frac{\mathrm{d}P}{\mathrm{d}\Omega}\,\mathrm{d}t' = \int_{-\infty}^{\infty} \frac{\mathrm{d}P_\mathrm{p}}{\mathrm{d}\Omega}\,\mathrm{d}t = \frac{r^2}{\mu_0 c}\int_{-\infty}^{\infty} |\mathbf{E}(t)|^2\,\mathrm{d}t. \qquad (3.3)$$

Going from one time scale to the other can be regarded as a change of the integration variable. The total energy radiated by the particle is obtained by integrating over the full solid angle:

$$U = \int \frac{\mathrm{d}U}{\mathrm{d}\Omega}\,\mathrm{d}\Omega = \frac{r^2}{\mu_0 c}\int_{-\infty}^{\infty} \mathrm{d}t \int_0^{2\pi} \mathrm{d}\phi \int_{-\pi/2}^{\pi/2} E^2(t)\sin\theta\,\mathrm{d}\theta. \qquad (3.4)$$

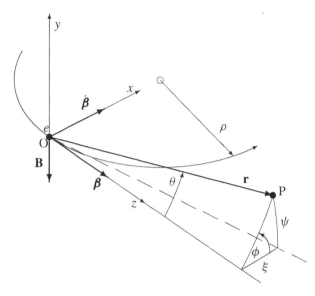

Fig. 3.1. The geometry of the radiation due to a transverse acceleration.

3.3 Transverse and longitudinal acceleration

All practical applications of radiation emitted by an ultra-relativistic charged particle are based on an acceleration perpendicular to the particle velocity. The use of a magnetostatic field for deflection excludes any longitudinal forces and electric fields are rarely used to deflect the beam. Here we will compare the radiation due to transverse acceleration and that due to longitudinal acceleration in order to illustrate some of the physics involved. The emitted power will be labeled in this section by an index 'T' or 'L' to distinguish between the two cases. In the rest of the book only transverse acceleration is considered and this subscript is omitted.

We calculate the angular distribution of the emitted radiation and use the coordinate system shown in Fig. 3.1. Compared with the one in Fig. 2.11 used before, it is rotated to adapt it better to the geometry used in experiments with synchrotron radiation. The particle moves momentarily in the z-direction and we use the angles θ and ϕ to describe the direction of emission. The unit vector $\mathbf{n}$, pointing from the particle to the observer, and the normalized velocity $\boldsymbol{\beta}$ are, in Cartesian coordinates,

$$\mathbf{n} = [\sin\theta\cos\phi, \ \sin\theta, \ \sin\phi, \ \cos\theta], \qquad \boldsymbol{\beta} = \beta[0, \ 0, \ 1].$$

3.3.1 The transverse acceleration

We take first the case of a *transverse* acceleration perpendicular to the velocity and pointing in the x-direction in Fig. 3.1. This is the case of ordinary synchrotron radiation emitted by a positive charge going through a magnetic field $\mathbf{B}$ pointing in the $-y$-direction, resulting in a trajectory curvature

$$\frac{1}{\rho} = \frac{eB}{m_0 c \beta \gamma}.$$

The normalized acceleration is perpendicular to the velocity and has, at the time it passes through the origin, only a component in the x-direction:

$$\boldsymbol{\beta} = \beta[0, 0, 1], \qquad \dot{\boldsymbol{\beta}} = \frac{\beta^2 c}{\rho}[1, 0, 0].$$

With (2.17) we obtain the radiation field in Cartesian coordinates:

$$\mathbf{E} = \frac{e}{4\pi\epsilon_0 c} \left\{ \frac{[\mathbf{n} \times [(\mathbf{n} - \boldsymbol{\beta}) \times \dot{\boldsymbol{\beta}}]]}{r(1 - \mathbf{n} \cdot \boldsymbol{\beta})^3} \right\}_{\text{ret}} = -\frac{e\beta^2}{4\pi\epsilon_0 r\rho}$$

$$\times \frac{[1 - \beta\cos\theta - \sin^2\theta\cos^2\phi, \; -\sin^2\theta\cos\phi\sin\phi, \; -\sin\theta\cos\phi(\cos\theta - \beta)]}{(1 - \beta\cos\theta)^3}.$$

$$(3.5)$$

In the forward direction $\theta = 0$ this field is

$$\mathbf{E}(\theta = 0) = -\frac{e\beta^2}{4\pi\epsilon_0 r\rho}\frac{[(1 - \beta), 0, 0]}{(1 - \beta)^3}.$$

Expressed in spherical coordinates the field is

$$\mathbf{E}_{[\theta,\phi,r]} = -\frac{e\beta^2}{4\pi\epsilon_0 r\rho}\frac{[(\cos\theta - \beta)\cos\phi, \; -\sin\phi(1 - \beta\cos\theta), \; 0]}{(1 - \beta\cos\theta)^3}. \qquad (3.6)$$

We combine equation (3.2) with the above expression and obtain the angular power distribution at the time t' when the particle passes through the origin:

$$\frac{dP_{\text{T}}}{d\Omega} = \frac{e^2\dot{\beta}^2}{(4\pi)^2\epsilon_0 c}\frac{(1 - \beta\cos\theta)^2 - (1 - \beta^2)\sin^2\theta\cos^2\phi}{(1 - \beta\cos\theta)^5}. \qquad (3.7)$$

Before we discuss this distribution we integrate over the solid angle $d\Omega = \sin\theta\,d\theta\,d\phi$ to obtain the total power:

$$P_{\text{T0}} = \frac{2e^2\dot{\beta}^2\gamma^4}{12\pi\epsilon_0 c} = \frac{2r_0 m_0 c^2\dot{\beta}^2\gamma^4}{3c} = \frac{2r_0 c\,m_0 c^2\beta^4\gamma^4}{3\rho^2} = \frac{2r_0\dot{p}^2\gamma^2}{3\,m_0 c}.$$

Here we introduced the classical particle radius

$$r_0 = \frac{e^2}{4\pi\epsilon_0 m_0 c^2} = \begin{cases} 2.818 \times 10^{-15}\,\text{m} & \text{for electrons} \\ 1.535 \times 10^{-18}\,\text{m} & \text{for protons.} \end{cases} \qquad (3.8)$$

It is a convenient parameter that collects some fundamental constants and will be used frequently throughout the book. It has its origin in a model of the electron as a sphere of radius r_0 with a surface charge e in which the electron mass is created by the electrostatic-field energy. We also use the relation between the time derivative $\dot{\mathbf{p}}$ of the momentum for a transverse acceleration having a constant γ:

$$\mathbf{p} = m_0 c \boldsymbol{\beta} \gamma, \qquad \dot{\mathbf{p}} = m_0 c \dot{\boldsymbol{\beta}} \gamma.$$

For a given curvature $1/\rho$ the radiated power is proportional to the fourth power of $\beta\gamma$, or, for a given derivative of the momentum, it is proportional to the square of γ. Usually the transverse acceleration is provided by the Lorentz force acting on a charged particle moving in the transverse magnetic field B. In this case the power can be expressed by writing

$$P_{\text{T0}} = \frac{2 r_0 c m_0 c^2 \beta^4 \gamma^4}{3 \rho^2} = \frac{2 r_0 e^2 c^3 B^2 \beta^2 \gamma^2}{3 m_0 c^2} = \frac{2 r_0 c^3 e^2 \beta^2 E_{\text{e}}^2 B^2}{3 (m_0 c^2)^3}. \tag{3.9}$$

It is proportional to the square of the energy $E_{\text{e}} = m_0 c^2 \gamma$ of the particle and to the square of the magnetic field B. The energy loss in one revolution on a closed circular orbit of bending radius ρ is called U_{s}:

$$U_{\text{s}} = \frac{2 \pi \rho}{\beta c} P_{\text{T0}} = \frac{4 \pi r_0 m_0 c^2 \beta^3 \gamma^4}{3 \rho}. \tag{3.10}$$

We use (3.9) and (3.10) to express the angular power and energy distribution for a transverse acceleration (3.7) in more comprehensive forms:

$$\frac{\mathrm{d} P_{\text{T}}}{\mathrm{d} \Omega} = \frac{3 P_{\text{T0}}}{8 \pi \gamma^4} \frac{(1 - \beta \cos \theta)^2 - (1 - \beta^2) \sin^2 \theta \cos^2 \phi}{(1 - \beta \cos \theta)^5}$$

$$\frac{\mathrm{d} U}{\mathrm{d} \Omega} = \frac{3 U_{\text{s}}}{8 \pi \gamma^4} \frac{(1 - \beta \cos \theta)^2 - (1 - \beta^2) \sin^2 \theta \cos^2 \phi}{(1 - \beta \cos \theta)^5}.$$

This distribution is shown in Fig. 3.2 as cuts through the (x, z)-plane (the plane of the particle trajectory) and through the (y, z)-plane for various values of the normalized velocity. With increasing β the distribution becomes more and more peaked in the forward direction. There are two directions in which no radiation is emitted. They lie in the (x, z)-plane $(\sin \phi = 0)$ at the angle given by

$$\cos \theta_0 = \beta \quad \text{or} \quad \theta_0 = \arccos \beta.$$

In the ultra-relativistic case this angle becomes

$$\theta_0 (\gamma \to \infty) = 1/\gamma.$$

As a quantitative measure for the concentration of the radiation into the forward direction with increasing velocity we calculate the fraction of the power radiated within a cone of half opening angle θ_0:

$$\frac{1}{P_{\text{T0}}} \int_0^{2\pi} \mathrm{d}\phi \int_0^{\theta_0} \sin \theta \, \mathrm{d}\theta \, \frac{\mathrm{d} P}{\mathrm{d} \Omega} = \frac{1}{2} \left(1 + \frac{9}{16} \beta \right).$$

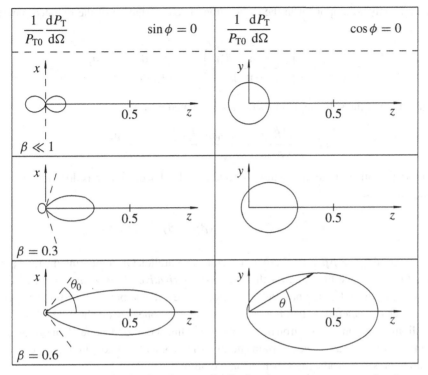

Fig. 3.2. The normalized radiated power distribution due to a transverse acceleration.

For very small velocities ($\beta \to 0$) this fraction is one half, as expected, and in the ultra-relativistic case ($\beta \to 1$) it becomes 25/32. In other words, at least half of the radiated power is concentrated within a cone of opening angle θ_0.

3.3.2 The longitudinal acceleration

Next we consider a longitudinal acceleration in the z-direction:

$$\boldsymbol{\beta} = \beta[0,\ 0,\ 1], \qquad \dot{\boldsymbol{\beta}} = \dot{\beta}[0,\ 0,\ 1].$$

This case is somewhat academic and of no interest for the production of synchrotron radiation but helps to illustrate the properties of the radiation emitted by a moving charge. Since $\boldsymbol{\beta}$ and $\dot{\boldsymbol{\beta}}$ are now parallel, we have $[\boldsymbol{\beta} \times \dot{\boldsymbol{\beta}}] = 0$. The electric field is given by the radiation term of (2.17) expressed in spherical coordinates,

$$\mathbf{E}_{[\theta,\phi,\phi]} = \frac{e}{4\pi\epsilon_0 c}\left\{\frac{[\mathbf{n} \times [\mathbf{n} \times \dot{\boldsymbol{\beta}}]]}{r(1 - \mathbf{n}\cdot\boldsymbol{\beta})^3}\right\}_{\text{ret}} = \frac{e\dot{\beta}}{4\pi\epsilon_0 cr}\frac{[\sin\theta,\ 0,\ 0]}{(1 - \beta\cos\theta)^3},$$

which has only a component in the θ-direction. The distribution of the radiated power is obtained from (3.2):

$$\frac{dP_L}{d\Omega} = \frac{e^2}{(4\pi)^2\epsilon_0 c} \left\{ \frac{[\mathbf{n} \times [\mathbf{n} \times \dot{\boldsymbol{\beta}}]]^2}{(1 - \mathbf{n} \cdot \boldsymbol{\beta})^5} \right\}_{ret} = \frac{e^2 \dot{\boldsymbol{\beta}}^2}{(4\pi)^2\epsilon_0 c} \frac{\sin^2 \theta}{(1 - \beta \cos \theta)^5}.$$

Integrating over the solid angle gives the total radiated power:

$$P_{L0} = \frac{e^2 \dot{\boldsymbol{\beta}}^2 \gamma^6}{6\pi \epsilon_0 c} = \frac{2 r_0 m_0 c^2 \dot{\boldsymbol{\beta}}^2 \gamma^6}{3c} = \frac{2 c r_0 \dot{\mathbf{p}}_L^2}{3 m_0 c^2}.$$

The time derivative of the momentum for a longitudinal acceleration is different from that for the transverse case,

$$\dot{\mathbf{p}}_L = \frac{d}{dt}(m_0 c \boldsymbol{\beta} \gamma) = m_0 c (\dot{\boldsymbol{\beta}} \gamma + \boldsymbol{\beta} \dot{\gamma}) = m_0 c \dot{\boldsymbol{\beta}} \gamma^3,$$

where we used $\boldsymbol{\beta} \dot{\beta} = \beta \dot{\boldsymbol{\beta}}$ since these two vectors are parallel. On comparing the total power for the two cases, we find that, for the *same time derivative of the momentum*, i.e. for the same value of the deflecting or accelerating force, the radiated power is γ^2 times larger for the transverse acceleration than it is for the longitudinal one. This is the main reason why, for colliding electron and positron beams, the use of linear colliders rather than storage rings becomes advantageous above a certain energy. On the other hand, longitudinal acceleration is of no interest for producing synchrotron radiation.

We now obtain for the angular distribution of the radiated power due to a longitudinal acceleration

$$\frac{dP_L}{d\Omega} = \frac{3 P_{L0}}{8\pi \gamma^6} \frac{\sin^2 \theta}{(1 - \beta \cos \theta)^5}.$$

Owing to the rotational symmetry with respect to the z-axis, this distribution is independent of the angle ϕ. In Fig. 3.3 the distribution of the radiated power for various values of the normalized velocity β is shown. To emphasize the symmetry we use $\rho = \sqrt{x^2 + y^2}$ as ordinate. The radiation is emitted in the vicinity of a cone around the z-axis with no radiation emitted along this axis itself. With increasing velocity the power distribution becomes more and more peaked in the forward direction, as in the case of transverse acceleration. The power radiated per unit solid angle has a maximum on a cone of opening angle θ_m given by

$$\cos \theta_m = \frac{\sqrt{1 + 15\beta^2} - 1}{3\beta}.$$

For vanishing velocity this angle becomes $\theta_m \approx \pi/2$, as expected.

In the ultra-relativistic approximation we obtain for the angular distribution and angle of its maximum

$$\frac{dP_L}{d\Omega} = P_{L0} \frac{12\gamma^2}{\pi} \frac{\gamma^2\theta^2}{(1 + \gamma^2\theta^2)^5}, \qquad \theta_m(\beta \to 1) = \frac{1}{2\gamma}.$$

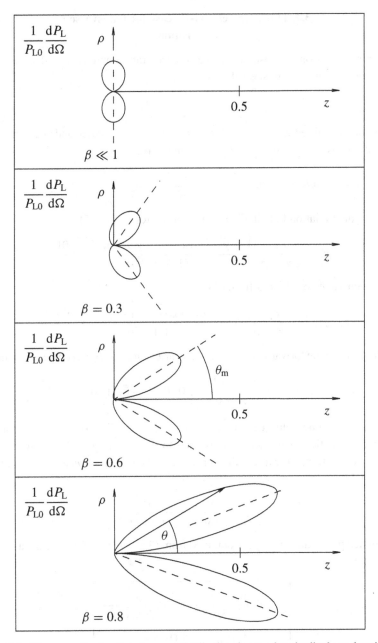

Fig. 3.3. The normalized radiated power distribution due to a longitudinal acceleration.

3.4 The ultra-relativistic case for transverse acceleration

We concentrate now on the case of transverse acceleration and assume that the radiating particle moves at ultra-relativistic velocity:

$$\beta \approx 1, \qquad \gamma \gg 1.$$

The radiation is emitted at small angles of order $1/\gamma$ or smaller and we can make the approximation $\cos\theta \approx 1 - \theta^2/2$ for $\theta \ll 1$, which gives

$$1 - \beta\cos\theta \approx 1 - \beta + \frac{\beta\theta^2}{2} = \frac{1 - \beta^2}{1 + \beta} + \frac{\beta\theta^2}{2} \approx \frac{1}{2\gamma^2}(1 + \gamma^2\theta^2).$$

We obtain the radiation field in Cartesian coordinates from (3.5),

$$\mathbf{E} = -\frac{e\gamma^4}{\pi\epsilon_0 r\rho} \frac{[1 - \gamma^2\theta^2\cos(2\phi), \ -\gamma^2\theta^2\sin(2\phi), \ 0]}{(1 + \gamma^2\theta^2)^3},$$

and that in spherical coordinates from (3.6),

$$\mathbf{E}_{(\theta,\phi,r)} = -\frac{e\gamma^4}{\pi\epsilon_0 r\rho} \frac{[(1 - \gamma^2\theta^2)\cos\phi, \ -(1 + \gamma^2\theta^2)\sin\phi, \ 0]}{(1 + \gamma^2\theta^2)^3}.$$

The field emitted in the forward direction has in Cartesian coordinates only an x-component,

$$\mathbf{E}(\theta = 0) = -\frac{e\gamma^4}{\pi\epsilon_0 r\rho}[1, 0, 0] = -\hat{E}_u[1, 0, 0], \tag{3.11}$$

with its absolute value being called $\hat{E}_u$ in view of its application for undulators. Since in most cases a static magnetic field B is used to deflect the particle and to provide the transverse acceleration, we express the curvature as $1/\rho = eB/(\gamma m_0 c)$ and obtain for the maximum field

$$\hat{E}_u = \frac{e\gamma^4}{\pi\epsilon_0 r\rho} = \frac{e^2 B\gamma^3}{\pi\epsilon_0 m_0 cr}. \tag{3.12}$$

The radiated power and energy are obtained from (3.9) and (3.10) for the ultra-relativistic case by setting $\beta = 1$,

$$\boxed{\begin{aligned} P_{T0} &= \frac{2cr_0 m_0 c^2\gamma^4}{3\rho^2} = \frac{2c^3 r_0 e^2 E_e^2 B^2}{3(m_0 c^2)^3} \\ U_s &= \frac{2\pi\rho}{c}P_{T0} = \frac{4\pi r_0 m_0 c^2\gamma^4}{3\rho}, \end{aligned}} \tag{3.13}$$

and the angular distribution is

$$\boxed{\frac{dP}{d\Omega} = P_{T0}\frac{3\gamma^2}{\pi}\frac{1 - 2\gamma^2\theta^2\cos(2\phi) + \gamma^4\theta^4}{(1 + \gamma^2\theta^2)^5}.} \tag{3.14}$$

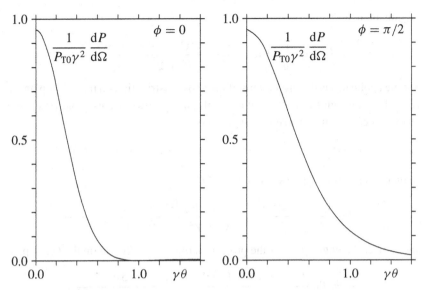

Fig. 3.4. Cuts through the power distribution for $\gamma \gg 1$.

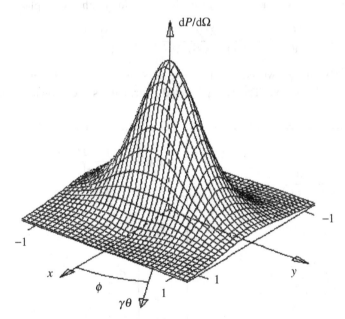

Fig. 3.5. The instantaneous angular distribution of the radiation emitted by a charge undergoing a transverse acceleration.

The form of the distribution depends on the product $\gamma\theta$ only and all angles θ scale as $1/\gamma$. It is shown in Fig. 3.4 as half cuts through the (x, z)-plane ($\phi = 0$) and the (y, z)-plane ($\phi = \pi/2$), and in Fig. 3.5 as a three-dimensional representation. The directions of vanishing radiation are in the (x, y)-plane at $\theta = 1/\gamma$, $\phi = 0$, and $\phi = \pi$. The variance of the opening angle is

$$\langle \theta^2 \rangle = \frac{3\gamma^2}{\pi} \int_0^{2\pi} d\phi \int_0^\infty \theta\, d\theta \, \frac{1 + 2\gamma^2\theta^2(1 - 2\cos^2\phi) + \gamma^4\theta^4}{(1 + \gamma^2\theta^2)^5} \theta^2 = \frac{1}{\gamma^2}.$$

For some applications the projections of the above distribution on the (x, z)-plane and on the (y, z)-plane are significant. We call the angles in these two planes ξ and ψ; see Fig. 3.1. The relations between the two sets of angles are

$$\psi = \theta \sin\phi, \qquad \xi = \theta \cos\phi, \qquad d\Omega = \theta\, d\phi\, d\theta = d\xi\, d\psi.$$

The angular distribution expressed in terms of ξ and ψ is

$$\frac{dP}{d\Omega} = P_{T0} \frac{3\gamma^2}{\pi} \frac{1 - 2\gamma^2\xi^2 + 2\gamma^2\psi^2 + \gamma^4\xi^4 + 2\gamma^2\xi^2\gamma^2\psi^2 + \gamma^4\psi^4}{(1 + \gamma^2\xi^2 + \gamma^2\psi^2)^5}.$$

Integrating over the angle ψ gives the projection of the distribution on the (x, z)-plane,

$$\frac{dP}{d\xi} = P_{T0} \frac{3\gamma}{32} \left(\frac{7\gamma^4\xi^4 - 16\gamma^2\xi^2 + 12}{(1 + \gamma^2\xi^2)^{9/2}} + \frac{5\gamma^2\xi^2}{(1 + \gamma^2\xi^2)^{7/2}} \right),$$

while the integration over the angle ξ gives the projection on the (y, z)-plane,

$$\frac{dP}{d\psi} = P_{T0} \frac{21\gamma}{32(1 + \gamma^2\psi^2)^{5/2}} \left(1 + \frac{5\gamma^2\psi^2}{7(1 + \gamma^2\psi^2)} \right). \tag{3.15}$$

We have expressed the two equations above as sums of two terms in view of a later discussion on polarization. The two projected distributions are shown in Fig. 3.6. We

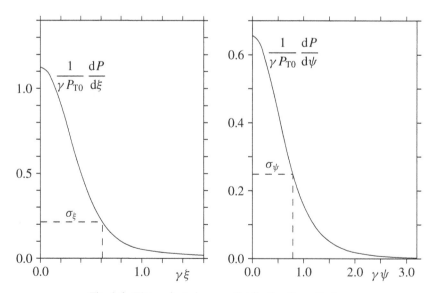

Fig. 3.6. The projected power distribution for $\gamma \gg 1$.

calculate the variances

$$\langle \xi^2 \rangle = \frac{3\gamma}{32} \int_{-\infty}^{\infty} \frac{12\gamma^4\xi^4 - 11\gamma^2\xi^2 + 12}{(1 + \gamma^2\xi^2)^{9/2}} \xi^2 \, d\xi = \frac{3}{8\gamma^2}$$

and

$$\langle \psi^2 \rangle = \frac{3\gamma}{32} \int_{-\infty}^{\infty} \frac{12\gamma^2 + 7}{(1 + \gamma^2\psi^2)^{7/2}} \psi^2 \, d\psi = \frac{5}{8\gamma^2}.$$

The sum of the variances of the angles ξ and ψ results in the variance of the angle θ:

$$\langle \xi^2 \rangle + \langle \psi^2 \rangle = \langle \theta^2 \rangle = 1/\gamma^2.$$

3.5 The angular spectral energy and power density

We calculated earlier the Fourier-transformed field $\tilde{\mathbf{E}}(\omega)$. We can use it to obtain the time-domain field $\mathbf{E}(t)$ by taking an inverse Fourier transform:

$$\mathbf{E}(t) = \frac{1}{\sqrt{2\pi}} \int_{-\infty}^{\infty} \tilde{\mathbf{E}}(\omega)e^{i\omega t} \, d\omega.$$

Substituting this into the expression (3.3) for the radiated energy gives

$$\frac{dU}{d\Omega} = \frac{r^2}{2\pi\mu_0 c} \int_{-\infty}^{\infty} dt \int_{-\infty}^{\infty} \int_{-\infty}^{\infty} \tilde{\mathbf{E}}(\omega)\tilde{\mathbf{E}}(\omega')e^{i(\omega+\omega')t} \, d\omega \, d\omega'.$$

Using the integral representation of the Dirac δ-function

$$\int_{-\infty}^{\infty} e^{iat} \, dt = 2\pi \, \delta(a),$$

we obtain for the integration over t

$$\frac{dU}{d\Omega} = \frac{r^2}{\mu_0 c} \int_{-\infty}^{\infty} \int_{-\infty}^{\infty} \tilde{\mathbf{E}}(\omega)\tilde{\mathbf{E}}(\omega')\delta(\omega + \omega') \, d\omega \, d\omega'$$

$$= \frac{r^2}{\mu_0 c} \int_{-\infty}^{\infty} \tilde{\mathbf{E}}(\omega)\tilde{\mathbf{E}}(-\omega) \, d\omega.$$

Since the field $\mathbf{E}(t)$ is a real function, its Fourier transform has the symmetry property

$$\tilde{\mathbf{E}}(-\omega) = \tilde{\mathbf{E}}^*(\omega), \qquad \tilde{\mathbf{E}}(\omega)\tilde{\mathbf{E}}(-\omega) = |\tilde{\mathbf{E}}(\omega)|^2 \tag{3.16}$$

with $\tilde{\mathbf{E}}^*(\omega)$ being the complex conjugate of $\tilde{\mathbf{E}}(\omega)$. This gives for the angular distribution of the radiation energy

$$\frac{dU}{d\Omega} = \frac{r^2}{\mu_0 c} \int_{-\infty}^{\infty} |\tilde{\mathbf{E}}(\omega)|^2 \, d\omega = \frac{2r^2}{\mu_0 c} \int_{0}^{\infty} |\tilde{\mathbf{E}}(\omega)|^2 \, d\omega.$$

Differentiating this with respect to ω gives the *angular spectral energy distribution* or *density* of the radiation:

$$\frac{\mathrm{d}^2 U}{\mathrm{d}\Omega\,\mathrm{d}\omega} = \frac{2r^2|\tilde{\mathbf{E}}(\omega)|^2}{\mu_0 c}. \tag{3.17}$$

We use here a factor of 2 on the right-hand side of the above equation, indicating that we use *positive frequencies* only. The spectral energy density is always an even function of ω. In a measurement only positive frequencies are observed. One therefore gives a spectral power or energy density in positive frequencies only. On the other hand, the electric-field components are not all even in ω and the symmetry relation with respect to ω, expressed by (3.16), can give some information about the properties of the field. For these reasons positive and negative frequencies are often used for the fields.

The spectral angular energy density (3.17) gives the energy received by the observer per unit frequency band and solid angle. This energy is given in the frequency domain, where the independent variable is ω and the time t does not appear. In this picture the radiation consists of constant narrow-frequency-band waves with a certain distribution in amplitude and phase. Sometimes one wants to give the spectral power distribution. This leads to a conceptual difficulty since the power is the energy radiated or received per unit time interval. However, if the energy radiated per unit time is *on average constant*, like for a closed circular orbit, the spectral power density is a useful quantity. In the case of a closed circular orbit in a storage ring the observers around the ring receive during every revolution of the orbiting charge a certain energy U in the form of a short flash. The *average* power received is then just

$$P = \frac{\omega_0}{2\pi} U_s = \frac{\beta c}{2\pi\rho} U_s.$$

We can therefore give a spectral angular power density:

$$\frac{\mathrm{d}^2 P}{\mathrm{d}\Omega\,\mathrm{d}\omega} = \frac{\omega_0}{2\pi}\frac{\mathrm{d}^2 U}{\mathrm{d}\Omega\,\mathrm{d}\omega} = \frac{2r^2\omega_0|\tilde{\mathbf{E}}(\omega)|^2}{2\pi\mu_0 c}. \tag{3.18}$$

If the ring has field-free straight sections, the above expression for the power relates only to the part of the orbit inside the magnetic field.

Sometimes we would also like to give the radiated power and its angular distribution for a single traversal in a magnet. In this case this refers to the power emitted *while* the particle traverses this magnet and is given by the above expression.

A related case is the undulator, which is a device that is often used to produce quasi-monochromatic radiation and will be discussed in Part III. It consists of a periodic (usually harmonic) magnetic field of total length L_u containing N_u periods of length λ_u. The charged particle traverses this undulator on a periodic trajectory with drift velocity $\beta^* c$ and frequency

$\Omega_u = 2\pi\beta^*c/\lambda_u$. We would like to give the average power emitted by the charge while it is passing through the undulator,

$$P_u = \frac{\beta^*c}{L_u}U_u = \frac{\beta^*c}{N_u\lambda_u}U_u = \frac{\Omega_u}{2\pi N_u}U_u,$$

with U_u being the energy lost in the undulator. This gives for the angular spectral power distribution

$$\frac{d^2P}{d\Omega\,d\omega} = \frac{\beta^*c}{N_u\lambda_u}\frac{d^2U}{d\Omega\,d\omega} = \frac{r^2\Omega_u}{\pi N_u\mu_0 c}|\tilde{E}(\omega)|^2.$$

Part II

Synchrotron radiation

4

Synchrotron radiation: basic physics

4.1 Introduction

The radiation emitted by a charged particle moving with constant, relativistic speed on a circular arc is called *synchrotron radiation*. It is sometimes also called *ordinary synchrotron radiation* or *bending-magnet radiation* to distinguish it from the more general case of a non-circular trajectory like undulator radiation. On some occasions this radiation will be abbreviated as SR.

Some approximations are made in treating synchrotron radiation. First we assume that the radiation is emitted in a long magnet that has a constant field providing a constant curvature $1/\rho$ over a distance $\ell_r > 2\rho/\gamma$. Secondly, the radiation is observed at a relatively large distance from the source $r_p \gg \rho/\gamma$. This approach is similar to the development of a general field into dipole, quadrupole, and higher-multipole components, for which the higher-order contributions become negligible at large distances. For an ultra-relativistic charge the opening angle of the emitted radiation is of order $1/\gamma$ and therefore very small. The radiation received originates from a small part of the trajectory and the observer is usually far away compared with this small source size. The third approximation assumes that the particle moves with ultra-relativistic velocity, $\gamma \gg 1$, which leads to some simplifications. All three approximations are satisfied for most practical sources of synchrotron radiation. As a further approximation a classical treatment of the radiation is given here and any quantum effects are kept for a later discussion. Here we neglect the effect on the particle of the sudden loss of energy during the emission of a photon. We will first consider the radiation emitted by a charge that makes a *single traversal* on a circular arc, resulting in a deflection much larger than the natural radiation opening angle of $1/\gamma$. Following the standard treatment, the emitted radiation is first calculated directly in the frequency domain. Since we are dealing with a single traversal of the particle, the resulting spectrum of the radiation is continuous. We will later also calculate the radiation field in the time domain and discuss its relation to the spectrum. This is of little practical significance but it helps us to understand the physics which determines the properties of synchrotron radiation.

Finally, we will also discuss the radiation emitted by a charge moving on a closed circle. Since this is a periodic motion, the emitted radiation spectrum consists of lines that are harmonics n of the frequency of revolution of the particle. In practice such lines are rarely observed. The emitted radiation is quantized and the charged particle will lose a finite amount of energy whenever a photon is emitted. As a result the motion is no longer periodic and the lines smear out.

In this chapter we concentrate on the derivation of the principal expressions for the field and the radiated power. We will explain the methods and approximations used and investigate the validity of the results. The properties of the radiation which are relevant for applications will be discussed in the next chapter.

The radiation field emitted by a charge moving on a circle was calculated early in the twentieth century [30, 31, 3]. Later, when accelerators approached energies for which this radiation is important, several theoretical investigations of synchrotron radiation were carried out [32–37]. At around this time also the first experimental observation was made [38].

4.2 The geometry and approximations

4.2.1 The particle motion

We consider a charge moving with a constant velocity on a full or partial circle with bending radius ρ and take the geometry illustrated in Fig. 4.1, where the trajectory $\mathbf{R}(t')$ lies in the (x, z)-plane and the particle moves through the origin at the time $t' = 0$. The radiation is received by an observer P at a fixed distance $\mathbf{r}_p$ from the origin and at a varying distance $\mathbf{r}(t') = \mathbf{r}_p - \mathbf{R}(t')$ from the radiating particle. The observer is assumed to be located in a vertical plane at a tangent to the circular trajectory at the origin at an angle ψ above the level of the orbit. This geometry of the particle motion has a cylindrical symmetry with the vertical axis going through the center of the circular orbit. It is therefore not necessary to consider an observer at a more general location. A second observer in a vertical plane at

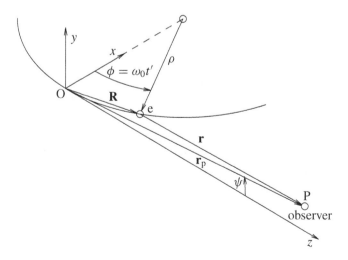

Fig. 4.1. The geometry used for the treatment of synchrotron radiation.

a tangent to the trajectory at a different point receives, apart from a time delay, the same radiation as that received by our first observer at P. We use the angular velocity of the charge,

$$\omega_0 = \beta c / \rho, \qquad (4.1)$$

where the constant bending radius ρ in a magnetic field B for a particle of momentum p is given by

$$\frac{1}{\rho} = \frac{eB}{p} = \frac{eB}{m_0 c \beta \gamma} \approx \frac{eBc}{m_0 c^2 \gamma}. \qquad (4.2)$$

The position $\mathbf{R}$, normalized velocity $\boldsymbol{\beta}$, and acceleration $\dot{\boldsymbol{\beta}}$ of the particle as functions of the time t' are

$$
\begin{aligned}
\mathbf{R}(t') &= \rho[(1 - \cos(\omega_0 t')), \; 0, \; \sin(\omega_0 t')] \\
\boldsymbol{\beta}(t') &= \beta[\sin(\omega_0 t'), \; 0, \; \cos(\omega_0 t')] \\
\dot{\boldsymbol{\beta}}(t') &= \dot{\beta}[\cos(\omega_0 t'), \; 0, \; -\sin(\omega_0 t')]
\end{aligned}
\qquad (4.3)
$$

with

$$\dot{\beta} = \beta \omega_0 = \beta^2 c / \rho.$$

We now choose $\mathbf{R}(0)$ as the origin and give the position of the observer P by the vector $\mathbf{r}_p$:

$$\mathbf{r}_p = r_p[0, \; \sin \psi, \; \cos \psi]. \qquad (4.4)$$

The vector $\mathbf{r}$ pointing from the particle to the observer and its absolute value r are

$$\mathbf{r} = \mathbf{r}_p - \mathbf{R} = [-\rho(1 - \cos(\omega_0 t')), \; r_p \sin \psi, \; r_p \cos \psi - \rho \sin(\omega_0 t')]$$

$$r = r_p \sqrt{1 - 2\frac{\rho}{r_p} \cos \psi \sin(\omega_0 t') + 2\left(\frac{\rho}{r_p}\right)^2 (1 - \cos(\omega_0 t'))}. \qquad (4.5)$$

4.2.2 The dipole approximation

At this point we make an approximation similar to that used for the development of a general field into multipoles. We develop the square root in (4.5) into powers of ρ / r_p:

$$r \approx r_p \left(1 - \frac{\rho}{r_p} \cos \psi \sin(\omega_0 t') + \frac{1}{2}\left(\frac{\rho}{r_p}\right)^2 (2 - 2\cos(\omega_0 t') - \cos^2 \psi \sin^2(\omega_0 t')) + \cdots \right). \qquad (4.6)$$

We will later neglect terms of higher than linear orders. In general this is justified only if $r_p \gg \rho$, i.e. if the radiation is observed at a distance r_p from the source that is much larger than the bending radius ρ of the trajectory. This is rarely fulfilled in practical cases. However, for an ultra-relativistic particle motion the condition for fast convergence of (4.6) is much less restrictive. The vertical and the instantaneous horizontal opening angles of the radiation are about $1/\gamma$ and therefore very small in the ultra-relativistic case, as explained in an earlier section. The observer P will receive only radiation that is emitted from a small section of the trajectory around the origin of angular length $\omega_0 \Delta t' \approx 2/\gamma$. We can estimate the distance r between the source and the observer from the magnitudes of the terms in (4.5) under the conditions

$$|\sin(\omega_0 t')| \lesssim \frac{1}{\gamma}; \qquad 1 - \cos(\omega_0 t') \lesssim \frac{1}{2\gamma^2}; \qquad |\sin \psi| \lesssim \frac{1}{\gamma},$$

where the symbol '$\lesssim$' indicates that the left-hand side is smaller than a few times the right-hand side. On applying this to (4.5) and neglecting terms of order higher than $1/\gamma^2$ we obtain

$$r \approx r_p \sqrt{1 - 2\frac{\rho}{r_p}\frac{1}{\gamma} + \left(\frac{\rho}{r_p}\right)^2 \frac{1}{\gamma^2}}.$$

Therefore, for the general case we have only to demand that

$$\frac{\rho}{r_p \gamma} \ll 1 \tag{4.7}$$

for the approximation

$$r \approx r_p \left(1 - \frac{\rho}{r_p} \cos \psi \sin(\omega_0 t')\right) = r_p - \rho \cos \psi \sin(\omega_0 t') \tag{4.8}$$

to be justified (a more detailed calculation shows that the neglected terms are of higher than the third power in $\rho/(r_p\gamma)$). The condition (4.7) is satisfied in most practical applications. The distance r_p is typically between 10 and 50 m whereas the bending radius lies between a few meters for wigglers and as much as a few hundred meters for the dipole magnets in a large machine. The Lorentz factor is around 1000 for small machines but can be over 10 000 for the very large machines which have also a large ρ. In all these cases the condition $r_p \gg \rho/\gamma$ is fulfilled. There might be special cases in which it is only marginally satisfied, such as for the observation of radiation emitted by protons in a large machine.

It should be noted that the condition $r_p \gg \rho/\gamma$ applies to the main part of the spectrum which is emitted within an opening angle of about $1/\gamma$. However, this angular distribution has tails containing mainly low frequencies. If very long wavelengths are observed the validity of the above approximations should be checked.

We assume now that the radiation is observed at a sufficiently large distance from the source that the condition (4.7) is satisfied and perform the calculation of the radiation field either in the time domain using the radiation part in (2.17),

$$\mathbf{E}(t) = \frac{e}{4\pi c \epsilon_0} \left\{ \frac{[\mathbf{n} \times [(\mathbf{n} - \boldsymbol{\beta}) \times \dot{\boldsymbol{\beta}}]]}{r(1 - \mathbf{n} \cdot \boldsymbol{\beta})^3} \right\}_{\text{ret}}, \tag{4.9}$$

or in the frequency domain with (2.38),

$$\tilde{\mathbf{E}}(\omega) = \frac{i\omega e}{4\pi \sqrt{2\pi} \epsilon_0 c r} \int_{-\infty}^{\infty} [\mathbf{n} \times [\mathbf{n} \times \boldsymbol{\beta}]] e^{-i\omega(t' + r(t')/c)} \, dt'. \tag{4.10}$$

For the derivation of this equation in Chapter 2 we already used implicitly the condition (4.7).

4.2.3 The relevant motion

We now approximate the various factors appearing in the field expressions to the lowest relevant order in $\rho/(r_p\gamma)$. Sometimes the first-order terms are canceled out or divided by γ and higher-order terms have to be included. Since this is difficult to foresee, we will often start a derivation including the higher-order terms and neglect them later.

We start with the two triple vector products $[\mathbf{n} \times [(\mathbf{n} - \boldsymbol{\beta}) \times \dot{\boldsymbol{\beta}}]]$ and $[\mathbf{n} \times [\mathbf{n} \times \boldsymbol{\beta}]]$. Here it is sufficient to express $\mathbf{n}$ to the lowest order, which we do by dividing the vector $\mathbf{r}$, given by (4.5), by its absolute value r approximated by (4.8):

$$\mathbf{n} = \frac{[-\rho(1 - \cos(\omega_0 t')),\ r_p \sin\psi,\ r_p \cos\psi - \rho\sin(\omega_0 t')]}{r_p - \rho\sin(\omega_0 t')} \approx [0,\ \sin\psi,\ \cos\psi]. \quad (4.11)$$

On using equations (4.3) for $\boldsymbol{\beta}$ and $\dot{\boldsymbol{\beta}}$ we obtain the triple vector products

$$[\mathbf{n} \times [(\mathbf{n} - \boldsymbol{\beta}) \times \dot{\boldsymbol{\beta}}]] \approx$$
$$\dot{\boldsymbol{\beta}}[-\cos(\omega_0 t') + \beta\cos\psi,\ -\cos\psi\sin\psi\sin(\omega_0 t'),\ \sin^2\psi\sin(\omega_0 t')]$$

and

$$[\mathbf{n} \times [\mathbf{n} \times \boldsymbol{\beta}]] \approx \beta[-\sin(\omega_0 t'),\ \sin\psi\cos\psi\cos(\omega_0 t'),\ -\sin^2\psi\cos(\omega_0 t')]. \quad (4.12)$$

Next we treat the relation between the observation time t and the emission time t' of the radiation which appears in the exponent of (4.10):

$$t = t' + \frac{r(t')}{c} \approx t' + \frac{r_p}{c} - \frac{\rho\cos\psi\sin(\omega_0 t')}{c}.$$

Usually one chooses for the origin of $\mathbf{R}(t')$ and t' a position and time that the particle takes in an arbitrarily chosen center of the source region. As a consequence the observation time t contains a delay r_p/c due to the distance between the center of the source and the observer, which is of little interest. We define a reduced observation time t_p that is shifted by this delay:

$$t_p = t - \frac{r_p}{c} = t' + \frac{r(t') - r_p}{c} \approx t' - \frac{\rho\cos\psi\sin(\omega_0 t')}{c}. \quad (4.13)$$

This relation between the two times is illustrated in Fig. 4.2 for the median plane $\psi = 0$. At the top, $\omega_0 t_p$ is plotted against $\omega_0 t'$ for the two cases $\beta = 0.5$ and $\beta = 0.8$. The slopes of these curves are small around the origin, which indicates that the radiation emitted over a time interval $\Delta t'$ is received by the observer within the smaller interval Δt_p. This time compression is best quantified in terms of the derivative dt'/dt_p shown at the bottom of Fig. 4.2. It increases with increasing β as the velocity of the particle approaches the speed of light. This gives a very short and intense pulse of radiation received by the observer.

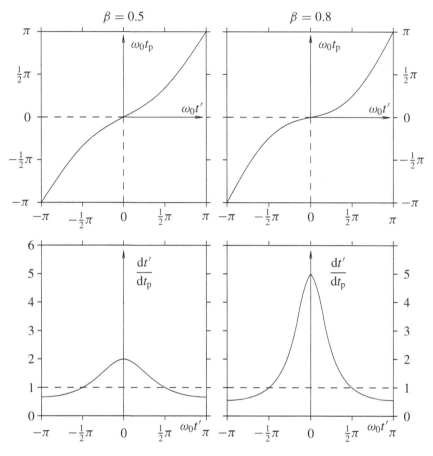

Fig. 4.2. The relation between the emission and observation time scales.

Note that so far we have made only an assumption about the minimum distance between the source and the observer, $r_p \gg \rho/\gamma$. Further approximations will be made below when we also assume that the particle moves at ultra-relativistic speed $\gamma \gg 1$.

4.2.4 The ultra-relativistic approximation

Here, we assume that the particle is ultra-relativistic:

$$\beta \approx 1, \qquad \gamma \gg 1.$$

This results in a small vertical opening angle:

$$\psi \lesssim \frac{1}{\gamma} \ll 1 \ \rightarrow \ \sin\psi \approx \psi, \qquad \cos\psi \approx 1 - \frac{\psi^2}{2}.$$

The instantaneous horizontal opening angle is also of order $1/\gamma$ and radiation is received only from a small portion of the trajectory of angular length $\Delta\phi \lesssim 1/\gamma \ll 1$. For the periodic motion of the charge on a closed circle, the observer receives radiation from each turn k emitted at the times $t' \approx \Delta t' + kT_0$, where $T_0 = 2\pi/\omega_0$ is the revolution time for the closed circle (or $T_{\rm rev} = 2\pi/\omega_{\rm rev}$ for a ring with straight sections). We concentrate on the revolution in which the particle goes through the origin at $t' = 0$. However, due to the periodic motion, these approximations are also valid for any other revolution. We have

$$\omega_0 t' = \frac{\beta c}{\rho} t' \lesssim \frac{1}{\gamma}$$

and can develop the trigonometric functions

$$\sin(\omega_0 t') \approx \omega_0 t' - \frac{(\omega_0 t')^3}{6} \quad \text{and} \quad \cos(\omega_0 t') \approx 1 - \frac{(\omega_0 t')^2}{2}.$$

We include terms up to the third power because in some applications the linear term becomes reduced such that its magnitude is of the same order as those of the higher-power terms. This will become clear shortly in the application. For the relation between the time t' of emission and the time $t_{\rm p}$ of observation we obtain

$$t_{\rm p} = t' + \frac{r(t') - r_{\rm p}}{c} = t' - \frac{\rho \cos\psi \, \sin(\omega_0 t')}{c} \approx t'\left(1 - \beta + \beta\frac{\psi^2}{2}\right) + \frac{c^2\beta^3 t'^3}{6\rho^2}.$$

With $1 - \beta \approx 1/(2\gamma^2)$ and $\beta \approx 1$ this becomes

$$t_{\rm p} = t'\frac{1 + \gamma^2\psi^2}{2\gamma^2} + \frac{c^2 t'^3}{6\rho^2} = \frac{(1 + \gamma^2\psi^2)t'}{2\gamma^2}\left(1 + \frac{\gamma^2(\omega_0 t')^2}{3(1 + \gamma^2\psi^2)}\right). \tag{4.14}$$

The linear term has been divided by γ^2 and becomes comparable to the third-power term. The two time scales are related by a third-order equation. We take the derivative

$$\frac{dt_{\rm p}}{dt'} = \frac{1 + \gamma^2\psi^2}{2\gamma^2} + \frac{\omega_0^2 t'^2}{2} = \frac{1}{2\gamma^2}(1 + \gamma^2\psi^2 + \gamma^2(\omega_0 t')^2).$$

Since $\psi \lesssim 1/\gamma$ and $\omega_0 t' \lesssim 1/\gamma$ the radiation emitted during a time interval $\Delta t'$ is received by the observer over a much shorter time interval $\Delta t_{\rm p} \approx \Delta t'/\gamma^2$. This time compression is strongest in the median plane $\psi = 0$ and at the time $t' = 0$ when the particle goes through the origin, namely $\Delta t_{\rm p} = \Delta t'/(2\gamma^2)$. We solve (4.14) for t' using the standard solution of third-order equations,

$$\omega_0 t' \approx \frac{\sqrt{1 + \gamma^2\psi^2}}{\gamma}\left(\left(\frac{2\omega_c t_{\rm p}}{(1 + \gamma^2\psi^2)^{3/2}} + \sqrt{1 + \frac{(2\omega_c t_{\rm p})^2}{(1 + \gamma^2\psi^2)^3}}\right)^{1/3}\right.$$

$$\left. + \left(\frac{2\omega_c t_{\rm p}}{(1 + \gamma^2\psi^2)^{3/2}} - \sqrt{1 + \frac{(2\omega_c t_{\rm p})^2}{(1 + \gamma^2\psi^2)^3}}\right)^{1/3}\right),$$

where we introduced the critical frequency ω_c:

$$\omega_c = \frac{3c\gamma^3}{2\rho} = \frac{3}{2}\omega_0\gamma^3.$$ (4.15)

It is, apart from a numerical factor, equal to the typical frequency (1.3), which we found in Chapter 1 to be representative for the spectrum of the radiation on the basis of qualitative arguments. We will meet ω_c again later and explain its physical meaning. We treat it for the time being just as a convenient parameter.

Using the relation

$$\ln\left(w + \sqrt{1 + w^2}\right) = \operatorname{arcsinh} w \quad \text{or} \quad \pm w + \sqrt{1 + w^2} = e^{\pm \operatorname{arcsinh} w}$$

with

$$w = \frac{2\omega_c t_p}{(1 + \gamma^2\psi^2)^{3/2}}$$

we obtain

$$\frac{\gamma\omega_0 t'}{\sqrt{1 + \gamma^2\psi^2}} \approx (e^{(\operatorname{arcsinh} w)/3} - e^{-(\operatorname{arcsinh} w)/3}) = 2\sinh\left(\frac{1}{3}\operatorname{arcsinh} w\right),$$

or

$$t' = \frac{2\sqrt{1 + \gamma^2\psi^2}}{\gamma\omega_0}\sinh\left(\frac{1}{3}\operatorname{arcsinh}\left(\frac{2\omega_c t_p}{(1 + \gamma^2\psi^2)^{3/2}}\right)\right).$$

For the ultra-relativistic case we now have the relation between the two time scales expressed in a symmetric form:

$$\frac{\omega_c t_p}{(1 + \gamma^2\psi^2)^{3/2}} = \frac{3\gamma\omega_0 t'}{4\sqrt{1 + \gamma^2\psi^2}}\left(1 + \frac{1}{3}\left(\frac{\gamma\omega_0 t'}{\sqrt{1 + \gamma^2\psi^2}}\right)^2\right)$$

$$\frac{\gamma\omega_0 t'}{\sqrt{1 + \gamma^2\psi^2}} = 2\sinh\left(\frac{1}{3}\operatorname{arcsinh}\left(\frac{2\omega_c t_p}{(1 + \gamma^2\psi^2)^{3/2}}\right)\right).$$ (4.16)

The ultra-relativistic approximations for the first triple vector products (4.11) and (4.12) and the denominator of (4.9) are

$$[\mathbf{n} \times [(\mathbf{n} - \boldsymbol{\beta}) \times \dot{\boldsymbol{\beta}}]] = \frac{\dot{\beta}}{2\gamma^2}[-(1 + \gamma^2\psi^2 - \gamma^2(\omega_0 t')^2), -2\gamma\omega_0 t'\gamma\psi, 0]$$

$$[\mathbf{n} \times [\mathbf{n} \times \boldsymbol{\beta}]] = [-\omega_0 t', \psi, 0]$$ (4.17)

$$(1 - \mathbf{n} \cdot \boldsymbol{\beta}) = \frac{1}{2\gamma^2}(1 + \gamma^2\psi^2 + \gamma^2(\omega_0 t')^2).$$

4.3 The continuous spectrum radiated on a circular arc

4.3.1 The Fourier-transformed field

We first treat the case of a *single* traversal of a charge moving with *ultra-relativistic* velocity on a circular arc with bending radius ρ. The vertical and the instantaneous horizontal opening angles are both of order $1/\gamma$ and small, a fact we used before to make the approximations. For our treatment to be valid this arc has to have a minimum length $\ell_r \gtrsim 2\rho/\gamma$ since the observer receives radiation from a part of the trajectory having about this length.

Because the charge makes a single traversal, a single pulse of radiation is emitted, and the observed spectrum will be *continuous*. We calculate the Fourier transform of the radiated field. We saw before that it is more convenient to use the reduced observation time t_p,

$$t_p = t - \frac{r_p}{c} = t' + \frac{r(t') - r_p}{c},$$

instead of t and adapt the Fourier transform (2.38) to $\mathbf{E}(t_p)$:

$$\mathbf{E}(\omega) = \frac{1}{\sqrt{2\pi}} \int_{-\infty}^{\infty} \mathbf{E}(t_p) e^{-i\omega t_p} \, dt_p$$

$$= \frac{i\omega e}{4\pi \sqrt{2\pi} \epsilon_0 c r_p} \int_{-\infty}^{\infty} [\mathbf{n} \times [\mathbf{n} \times \boldsymbol{\beta}]] e^{-i\omega(t' + (r(t') - r_p)/c)} \, dt'.$$

Since we are considering an ultra-relativistic particle, the radiation received has to be emitted close to the origin and we can use the approximations discussed before. We also approximate the distance r appearing in the denominator of the above equation,

$$r = r_p - \rho \sin(\omega_0 t') \cos \psi \approx r_p \left(1 - \frac{\rho}{r_p \gamma}\right) \approx r_p,$$

simply by the constant $r \approx r_p$. With this we obtain for the Fourier-transformed field

$$\tilde{\mathbf{E}}(\omega) = \frac{i\omega e}{4\pi \sqrt{2\pi} \epsilon_0 c r_p} \int_{-\infty}^{\infty} [-\omega_0 t', \ \psi, \ 0] \exp\left(-i\omega \left(\frac{t'(1 + \gamma^2 \psi^2)}{2\gamma^2} + \frac{c^2 t'^3}{6\rho^2}\right)\right) dt'.$$

This expression gives the radiation field as a function of the *frequency ω as measured by the observer*. The earlier time t' of emission can be regarded here as a variable of integration. We express the exponential of an imaginary quantity in terms of the corresponding trigonometric functions. Since the x-component of the triple vector product is odd and the y-component even, only the sine term contributes to $\tilde{E}_x(\omega)$, and only the cosine term to $\tilde{E}_y(\omega)$. The two field components are

$$\tilde{E}_x(\omega) = \frac{-e\omega}{4\pi \sqrt{2\pi} \epsilon_0 c r_p} \int_{-\infty}^{\infty} \omega_0 t' \sin\left(\omega t' \frac{1 + \gamma^2 \psi^2}{2\gamma^2} + \frac{\omega c^2 t'^3}{6\rho^2}\right) dt'$$

$$\tilde{E}_y(\omega) = \frac{ie\omega}{4\pi \sqrt{2\pi} \epsilon_0 c r_p} \int_{-\infty}^{\infty} \psi \cos\left(\omega t' \frac{1 + \gamma^2 \psi^2}{2\gamma^2} + \frac{\omega c^2 t'^3}{6\rho^2}\right) dt'.$$

They have the symmetry relations

$$\tilde{E}_x(-\omega) = \tilde{E}_x(\omega), \qquad \tilde{E}_y(-\omega) = -\tilde{E}_y(\omega). \tag{4.18}$$

We substitute a new variable u for t',

$$t' = \left(\frac{2\rho^2}{|\omega|c^2}\right)^{1/3} u = \left(\frac{2}{|\omega|\omega_0^2}\right)^{1/3} u,$$

where we use the absolute sign to conserve the symmetry relation (4.18), and use the critical frequency ω_c (4.15),

$$\omega_c = \frac{3c\gamma^3}{2\rho} = \frac{3\omega_0\gamma^3}{2},$$

to bring the integrals into a standard form:

$$\tilde{E}_x = \frac{-e\gamma}{(2\pi)^{3/2}\epsilon_0 cr_p} \left(\frac{3|\omega|}{4\omega_c}\right)^{1/3} \int_{-\infty}^{\infty} u \sin\left(\left(\frac{3\omega}{4\omega_c}\right)^{2/3}(1+\gamma^2\psi^2)u + \frac{u^3}{3}\right) du$$

$$\tilde{E}_y = \frac{ie\gamma^2\psi(\omega/|\omega|)}{(2\pi)^{3/2}\epsilon_0 cr_p} \left(\frac{3\omega}{4\omega_c}\right)^{2/3} \int_{-\infty}^{\infty} \cos\left(\left(\frac{3\omega}{4\omega_c}\right)^{2/3}(1+\gamma^2\psi^2)u + \frac{u^3}{3}\right) du. \tag{4.19}$$

They resemble the integral representation (A.1) of the Airy function and its derivative discussed in Appendix A:

$$\mathrm{Ai}(v) = \frac{1}{2\pi}\int_{-\infty}^{\infty} \cos\left(vt + \frac{t^3}{3}\right) dt$$

$$\mathrm{Ai}'(v) = \frac{d\mathrm{Ai}(v)}{dv} = -\frac{1}{2\pi}\int_{-\infty}^{\infty} t \sin\left(vt + \frac{t^3}{3}\right) dt.$$

On comparing these definitions with the integral expressions for the two field components we obtain

$$\boxed{\begin{aligned}
\tilde{E}_x(\omega) &= \frac{e\gamma}{\sqrt{2\pi}\,\epsilon_0 cr_p} \left(\frac{3|\omega|}{4\omega_c}\right)^{1/3} \mathrm{Ai}'\left(\left(\frac{3\omega}{4\omega_c}\right)^{2/3}(1+\gamma^2\psi^2)\right) \\
\tilde{E}_y(\omega) &= \frac{ie\gamma(\omega/|\omega|)}{\sqrt{2\pi}\,\epsilon_0 cr_p} \left(\frac{3|\omega|}{4\omega_c}\right)^{2/3} \gamma\psi\,\mathrm{Ai}\left(\left(\frac{3\omega}{4\omega_c}\right)^{2/3}(1+\gamma^2\psi^2)\right)
\end{aligned}} \tag{4.20}$$

There is a relation (A.3) between the Airy functions and the modified Bessel functions of orders 2/3 and 1/3:

$$\mathrm{Ai}(x) = \frac{1}{\pi}\sqrt{\frac{x}{3}}K_{1/3}\left(\frac{2x^{3/2}}{3}\right)$$

$$\mathrm{Ai}'(x) = -\frac{1}{\pi}\frac{x}{\sqrt{3}}K_{2/3}\left(\frac{2x^{3/2}}{3}\right).$$

We can therefore express the electric-field components also with the modified Bessel functions:

$$\tilde{E}_x(\omega) = \frac{-\sqrt{3}e\gamma}{(2\pi)^{3/2}\epsilon_0 c r_\mathrm{p}}\left(\frac{|\omega|}{2\omega_\mathrm{c}}\right)(1+\gamma^2\psi^2)K_{2/3}\left(\frac{\omega}{2\omega_\mathrm{c}}(1+\gamma^2\psi^2)^{3/2}\right)$$

$$\tilde{E}_y(\omega) = \frac{i\sqrt{3}e\gamma}{(2\pi)^{3/2}\epsilon_0 c r_\mathrm{p}}\left(\frac{\omega}{2\omega_\mathrm{c}}\right)\gamma\psi\sqrt{1+\gamma^2\psi^2}K_{1/3}\left(\frac{\omega}{2\omega_\mathrm{c}}(1+\gamma^2\psi^2)^{3/2}\right).$$

Without any further discussion, the Airy and Bessel functions represent only names for the integrals (4.19). However, when we discuss the characteristics of synchrotron radiation, we profit from many relations of these functions that are presented in the standard literature. In Appendix A we summarize the properties of the Airy functions which are most relevant for our application. They are smooth and well behaved as the plot in Fig. A.1 shows. We can give the expressions describing synchrotron radiation either in terms of Airy functions or in terms of modified Bessel functions. We choose here to use the first, but will express all important equations also in terms of Bessel functions.

The Fourier-transformed horizontal field $E_x(\omega)$ is large in the median plane $\psi = 0$ and has its maximum close to the critical frequency ω_c as we expected from qualitative arguments. The vertical field $\tilde{E}_y(\omega)$ vanishes in the median plane. We will later compare this Fourier-transformed electric field with the corresponding presentation in the time domain.

4.3.2 The spectral power density of the radiation

In Chapter 3 we derived the average spectral power density (3.18) of the radiation emitted by a charge. We found for the power emitted per unit solid angle and frequency

$$\frac{d^2 P}{d\Omega\,d\omega} = \frac{\omega_0}{2\pi}\frac{d^2 U}{d\Omega\,d\omega} \approx \frac{2r^2|\tilde{E}(\omega)|^2}{2\pi\mu_0\rho}.$$

The radiated power or energy is a scalar with only one component. However, it is calculated from a field having two components, E_x and E_y, which correspond to the two modes of the linear polarization. The first one has the electric field parallel to the plane of the particle motion and is often called the σ-mode. For the second one the electric field is perpendicular to the plane of the orbit and it is called the π-mode. The radiated power consists of two parts corresponding to the two modes of polarization. We keep them separated in order to obtain the power in each mode and to make the physics of synchrotron radiation more transparent:

$$\frac{d^2 P}{d\Omega\, d\omega} = \frac{d^2 P_\sigma}{d\Omega\, d\omega} + \frac{d^2 P_\pi}{d\Omega\, d\omega} = \frac{2r^2}{2\pi\mu_0\rho}\left(\left|\tilde{E}_x(\omega)\right|^2 + \left|\tilde{E}_y(\omega)\right|^2\right). \tag{4.21}$$

On substituting the expressions (4.20) for the two field components we obtain for the spectral angular power distributions of the two polarization modes

$$\frac{d^2 P_\sigma}{d\Omega\, d\omega} = \frac{2r_0 m_0 c^2 \gamma^2}{\pi\rho}\left(\frac{3\omega}{4\omega_c}\right)^{2/3} \mathrm{Ai}'^2\left(\left(\frac{3\omega}{4\omega_c}\right)^{2/3}(1+\gamma^2\psi^2)\right)$$

$$\frac{d^2 P_\pi}{d\Omega\, d\omega} = \frac{2r_0 m_0 c^2 \gamma^2}{\pi\rho}\left(\frac{3\omega}{4\omega_c}\right)^{4/3} \gamma^2\psi^2\, \mathrm{Ai}^2\left(\left(\frac{3\omega}{4\omega_c}\right)^{2/3}(1+\gamma^2\psi^2)\right),$$

where we use again the classical electron radius (3.8).

We calculated the total radiated power for a general transverse acceleration in Chapter 3 using a more direct approach. We obtained in the ultra-relativistic approximation (3.13)

$$P_s = \frac{2r_0 c m_0 c^2 \gamma^4}{3\rho^2},$$

and call it now P_s to indicate that it refers to synchrotron radiation, i.e. a transverse acceleration in long magnets. We use it and the critical frequency (4.15) to present the above expressions for the angular spectral power distribution in a more transparent way:

$$\boxed{\begin{aligned}\frac{d^2 P_\sigma}{d\Omega\, d\omega} &= \frac{P_s \gamma}{\omega_c}\frac{9}{2\pi}\left(\frac{3\omega}{4\omega_c}\right)^{2/3} \mathrm{Ai}'^2\left(\left(\frac{3\omega}{4\omega_c}\right)^{2/3}(1+\gamma^2\psi^2)\right)\\[2mm] \frac{d^2 P_\pi}{d\Omega\, d\omega} &= \frac{P_s \gamma}{\omega_c}\frac{9}{2\pi}\left(\frac{3\omega}{4\omega_c}\right)^{2/3} \gamma^2\psi^2\, \mathrm{Ai}^2\left(\left(\frac{3\omega}{4\omega_c}\right)^{2/3}(1+\gamma^2\psi^2)\right).\end{aligned}} \tag{4.22}$$

4.4 The radiation emitted on a circular arc in the time domain

4.4.1 The radiation field in the time domain

The field radiated by a charge moving on a circular arc is now calculated in the time domain. We start with the Liénard–Wiechert expression

$$\mathbf{E}(t_p) = \frac{e}{4\pi c\epsilon_0 r_p}\left\{\frac{[\mathbf{n}\times[(\mathbf{n}-\boldsymbol{\beta})\times\dot{\boldsymbol{\beta}}]]}{(1-\mathbf{n}\cdot\boldsymbol{\beta})^3}\right\}_{\mathrm{ret}}, \tag{4.23}$$

where we approximated $r \approx r_p$ in the denominator. The expression in the curly brackets has to be evaluated at the time t' of emission, which is related to the observation time by

$$t_p = t' + \frac{r(t')-r_p}{c}.$$

Earlier we calculated the triple vector product and the denominator of (4.23) in the ultra-relativistic approximation (4.17), giving for the field

$$\mathbf{E}(t_p) = \frac{e\omega_0\gamma^4}{\pi\epsilon_0 c r_p} \frac{[-(1+\gamma^2\psi^2 - (\gamma\omega_0 t')^2),\ -2\gamma\omega_0 t'\,\gamma\psi,\ 0]}{(1+\gamma^2\psi^2 + (\gamma\omega_0 t')^2)^3}.$$

We express this with (4.16) in terms of the observation time t_p:

$$\gamma\omega_0 t' = 2\sqrt{1+\gamma^2\psi^2}\,\sinh\left(\frac{1}{3}\operatorname{arcsinh}\left(\frac{2\omega_c t_p}{(1+\gamma^2\psi^2)^{3/2}}\right)\right).$$

To prevent the expressions from becoming too lengthy, we introduce the dimensionless time variable

$$\tau = \frac{2\omega_c t_p}{(1+\gamma^2\psi^2)^{3/2}},$$

which leads to the fields in the time domain,

$$E_x(t_p) = -\frac{e\omega_0\gamma^4}{\pi\epsilon_0 c r_p} \frac{\left(1 - 4\sinh^2\left(\frac{1}{3}\operatorname{arcsinh}\tau\right)\right)}{(1+\gamma^2\psi^2)^2\left(1 + 4\sinh^2\left(\frac{1}{3}\operatorname{arcsinh}\tau\right)\right)^3}$$

$$E_y(t_p) = -\frac{e\omega_0\gamma^4}{\pi\epsilon_0 c r_p} \frac{4\gamma\psi\sinh\left(\frac{1}{3}\operatorname{arcsinh}\tau\right)}{(1+\gamma^2\psi^2)^{5/2}\left(1 + 4\sinh^2\left(\frac{1}{3}\operatorname{arcsinh}\tau\right)\right)^3},$$

where $E_z(t_p)$ vanishes in this approximation. The field components have the following symmetry properties with respect to time t_p and angle ψ:

$$E_x(-t_p, \psi) = E_x(t_p, \psi), \qquad E_y(-t_p, \psi) = -E_y(t_p, \psi)$$
$$E_x(t_p, -\psi) = E_x(t_p, \psi), \qquad E_y(t_p, -\psi) = -E_y(t_p, \psi).$$

The horizontal field value has a maximum at $t_p = 0$, $\psi = 0$ of

$$E_{\max} = \frac{e\omega_0\gamma^4}{\pi\epsilon_0 c r_p} = \frac{4 r_0 m_0 c^2 \gamma^4}{e\rho r_p}.$$

The two field components, normalized with respect to $E_{\max}$, are plotted in Fig. 4.3 as functions of the observation time multiplied by the critical frequency $\omega_c t_p$ for various vertical angles ψ. Since the radiation field has no DC component the time integral and the related Fourier-transformed field at zero frequency both vanish:

$$\int_{-\infty}^{\infty} \mathbf{E}(t_p)\,dt_p = \sqrt{2\pi}\,\tilde{\mathbf{E}}(0) = 0.$$

For the vertical component this is automatically fulfilled by its asymmetry. The horizontal component has a positive center and a long negative tail. Some interesting discussions about this point can be found in [39, 40]. It is worthwhile to note that the horizontal field in the median plane goes through zero at $t_p = 1/\omega_c$, which is a very short time. At larger vertical angles ψ the field pulse becomes longer, in agreement with the behavior of the spectrum.

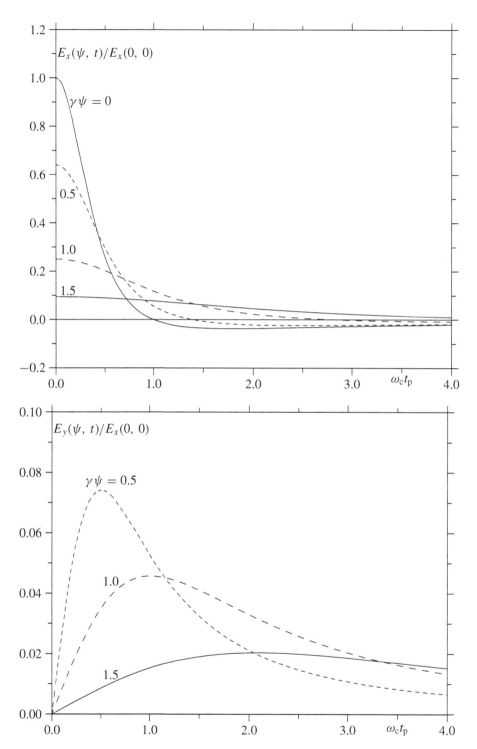

Fig. 4.3. The normalized horizontal (top) and vertical (bottom) electric fields as functions of the observation time t_p for various vertical angles ψ.

4.4.2 The radiated energy and power in the time domain

We derived in Chapter 3 the expression (3.2) for the radiated energy received by the observer per unit solid angle, which must be equal to the energy emitted per unit solid angle by the charge:

$$\frac{dU}{d\Omega} = \frac{r_p^2}{\mu_0 c} \int_{-\infty}^{\infty} |\mathbf{E}^2(t_p)| \, dt_p = \frac{r_p^2}{\mu_0 c} \int_{-\infty}^{\infty} |\mathbf{E}^2(t')|(1 - \mathbf{n} \cdot \boldsymbol{\beta}) \, dt'.$$

The second form of the integral is easier to evaluate:

$$\frac{dU}{d\Omega} = \frac{e^2 \omega_0^2 \gamma^6}{2\pi^2 \epsilon_0 c} \int_{-\infty}^{\infty} \frac{[(1 + \gamma^2 \psi^2 - (\gamma \omega_0 t')^2)^2 + (2\gamma \omega_0 t')^2 \gamma^2 \psi^2]}{(1 + \gamma^2 \psi^2 + (\gamma \omega_0 t')^2)^5} \, dt'.$$

The first term in the square brackets is due to the x-component of the field and represents the σ-mode while the second term gives the π-mode of the polarization. We first have a look at the constant factor

$$\frac{e^2 \omega_0^2 \gamma^6}{2\pi^2 \epsilon_0 c} = \frac{2 c r_0 m_0 c^2 \gamma^6}{\pi \rho^2} = \frac{3\gamma^2 P_s}{\pi},$$

where we use the classical electron radius r_0 (3.8) and the total radiated power P_s (3.13). The integration gives

$$\frac{dU}{d\Omega} = P_s \frac{21}{32} \frac{\rho}{c} \frac{\gamma}{(1 + \gamma^2 \psi^2)^{5/2}} \left(1 + \frac{5}{7} \frac{\gamma^2 \psi^2}{1 + \gamma^2 \psi^2}\right).$$

This is the energy per unit solid angle radiated by the charge in one traversal. If this charge is actually circulating in a storage ring with the Larmor frequency ω_0, this energy will be radiated each turn, giving an average radiated power per solid angle of

$$\frac{dP}{d\Omega} = \frac{c}{2\pi\rho} \frac{dU}{d\Omega} = \frac{P_s}{2\pi} \frac{21}{32} \frac{\gamma}{(1 + \gamma^2 \psi^2)^{5/2}} \left(1 + \frac{5}{7} \frac{\gamma^2 \psi^2}{1 + \gamma^2 \psi^2}\right).$$

Since we are here treating the synchrotron radiation in the time domain we are also interested in the instantaneous power P_p which the observer receives per unit solid angle as a function of the time t_p:

$$\frac{dP_p(t_p)}{d\Omega} = \frac{d^2 U}{d\Omega \, dt_p} = \frac{r_p^2 E^2(t_p)}{\mu_0 c}.$$

With the expression for the field we obtain

$$\frac{dP_{p\sigma}}{d\Omega} = P_s \frac{6\gamma^4}{\pi} \frac{(1 - 4\sinh^2(\frac{1}{3}\operatorname{arcsinh}\tau))^2}{(1 + \gamma^2 \psi^2)^4 (1 + 4\sinh^2(\frac{1}{3}\operatorname{arcsinh}\tau))^6}$$

$$\frac{dP_{p\pi}}{d\Omega} = P_s \frac{6\gamma^4}{\pi} \frac{4\gamma\psi \sinh^2(\frac{1}{3}\operatorname{arcsinh}\tau)}{(1 + \gamma^2 \psi^2)^5 (1 + 4\sinh^2(\frac{1}{3}\operatorname{arcsinh}\tau))^6}.$$

The flash of the radiation received by the observer lasts a time of the order of $1/\omega_c$. The instantaneous power is therefore about $\omega_c/\omega_0 \approx \gamma^3$ times larger than the average power.

4.4.3 The radiation field in the time and frequency domains

The normalized horizontal field component in the median plane $\psi = 0$ in the time and frequency domains is shown in Fig. 4.4. This field decreases rapidly with time and reaches half its maximum value at about $\omega_c t_p = 0.32$. It goes through zero at $t_p = 1/\omega_c$ and ends with a long tail. The Fourier-transformed field $\tilde{E}_x(\omega)$ extends to rather low frequencies due to the long tail in the time domain. It has a maximum slightly below the critical frequency ω_c, as we already expected from qualitative arguments discussed in Chapter 1. At higher frequencies the spectrum decreases slowly but extends quite far, corresponding to the relatively sharp

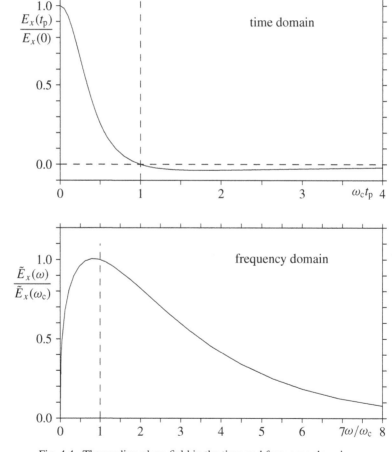

Fig. 4.4. The median-plane field in the time and frequency domains.

peak at the origin of the time-domain representation. At $\omega = 0$ the Fourier-transformed field vanishes, in other words, $\tilde{E}_x(\omega)$ has no DC part and the time-averaged field is zero.

4.5 The line spectrum radiated on closed circles

4.5.1 The relevant motion

We consider now a charge moving with constant velocity on a *closed circle* of radius ρ and approximate again for observation at a large distance. At first we do not assume that the particle moves with relativistic velocity. This makes the derivation only slightly more complicated and not only leads to a more general result but also clarifies some of the approximations used in earlier sections. This case of a charge moving on a closed circle represents a *periodic motion* with the period

$$T_0 = \frac{2\pi}{\omega_0} = \frac{2\pi\rho}{\beta c}.$$

The field of the emitted radiation is therefore also periodic and its spectrum is expected to consist of *lines* with frequency

$$\omega_n = n\omega_0 = n\frac{\beta c}{\rho}$$

using n for the harmonic number to distinguish this case from that of strong undulators, which is to be discussed later.

We assume that the radiation is observed from a large distance r_p and use the approximation discussed in Section 4.2. Since we do not start with an ultra-relativistic approximation, an observation distance $r_p \gg \rho$ is in principle required in order for the dipole approximation to be valid, which is not very realistic. On the other hand, the results will be applied to a relativistic case and this condition can be relaxed.

Within these approximations we have from (4.8) the distance r, from (4.12) the triple vector product, and from (4.13) the relation between the emission and observation times:

$$r = r_p\left(1 - \frac{\rho}{r_p}\cos\psi\,\sin(\omega_0 t')\psi\right) = r_p - \rho\cos\psi\,\sin(\omega_0 t')$$

$$[\mathbf{n} \times [\mathbf{n} \times \boldsymbol{\beta}]] = \beta[-\sin(\omega_0 t'),\ \sin\psi\,\cos\psi\,\cos(\omega_0 t'),\ -\sin^2\psi\,\cos(\omega_0 t'),\ 0]$$

$$t_p = t - \frac{r_p}{c} = t' + \frac{r(t') - r_p}{c} \approx t' - \frac{\rho\cos\psi\,\sin(\omega_0 t')}{c}. \tag{4.24}$$

Since the electric field is perpendicular to the direction of propagation, two components are sufficient to describe it. We use the coordinates ϕ and ψ as shown in Fig. 4.5. Since we always observe the radiation in the (y, z)-plane, i.e. at $\phi = 0$, we have the relations

$$E_\phi = E_x, \qquad E_\psi = -E_\Theta = E_y\cos\psi - E_z\sin\psi, \qquad E_r = E_y\sin\psi + E_z\cos\psi.$$

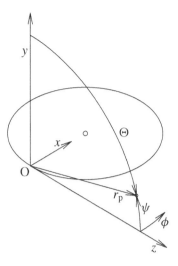

Fig. 4.5. Spherical and Cartesian coordinates.

We express the triple vector product in these coordinates,

$$[\mathbf{n} \times [\mathbf{n} \times \boldsymbol{\beta}]]_{\phi,\psi,r} = \beta[-\sin(\omega_0 t'), \cos(\omega_0 t')\sin\psi, 0],$$

which has no component in the radial direction.

4.5.2 The line spectrum of the electric field

The particle motion is periodic with period T_0 and we develop the radiation field into a *Fourier series*. Following the treatment carried out in Chapter 2, we start with the radiation field in the time domain (2.34) using now the reduced observation time t_p:

$$\mathbf{E}(t_p) = \frac{e}{4\pi\epsilon_0 c} \left\{ \frac{[\mathbf{n} \times [(\mathbf{n}-\boldsymbol{\beta})\times\dot{\boldsymbol{\beta}}]]}{r(1-\mathbf{n}\cdot\boldsymbol{\beta})^3} \right\}_{\text{ret}}, \qquad \mathbf{B}(t) = \frac{1}{c}[\mathbf{n}\times\mathbf{E}].$$

This periodic field is expressed in terms of the Fourier series (2.39):

$$\mathbf{E}(t_p) = \sum_{n=-\infty}^{\infty} \mathbf{E}_n e^{in\omega_0 t_p} \text{ with } \mathbf{E}_n = \frac{1}{T_0}\int_0^{T_0} \mathbf{E}(t)e^{-in\omega_0 t_p}\,dt_p.$$

The Fourier components of the field are given by integration over t_p, which we express in terms of t' using the relation given above (4.24):

$$\mathbf{E}_n = \frac{in\omega_0 e}{4\pi\epsilon_0 c r_p T_0}\int_0^{T_0} [\mathbf{n}\times[\mathbf{n}\times\boldsymbol{\beta}]]e^{-in\omega_0(t'+(r(t')-r_p)/c)}\,dt'.$$

This result was already obtained as (2.40) for a general periodic motion and observation time t.

The triple vector product, given by (4.12) in Cartesian coordinates, is

$$[\mathbf{n} \times [\mathbf{n} \times \boldsymbol{\beta}]]_{\phi, \theta, r} = \beta[-\sin(\omega_0 t'), \sin \psi \cos(\omega_0 t'), 0]$$

in the $[\phi, \theta, r]$ system. We express the distance r in the exponent by use of (4.24), use $\omega_0 = \beta c / \rho$, and rewrite the exponential in (4.5) in terms of trigonometric functions:

$$e^{-in\omega(t'+(r(t')-r_p)/c)} = e^{-in(\omega_0 t' - \beta \cos \psi \sin(\omega_0 t'))}$$
$$= \cos(n\omega_0 t' - n\beta \cos \psi \sin(\omega_0 t')) - i \sin(n\omega_0 t' - n\beta \cos \psi \sin(\omega_0 t')). \quad (4.25)$$

This is multiplied by the above triple vector product and integrated over t'. Since the ϕ-component of the triple vector product is an odd function in t' and the ψ-component is even, only the sine term of (4.25) contributes to $E_{n\phi}$ and only the cosine term to $E_{n\psi}$. We obtain

$$E_{n\phi} = \frac{ne\omega_0\beta}{4\pi\epsilon_0 cr_p} \frac{1}{T_0} \int_0^{T_0} \sin(\omega_0 t') \sin(n\omega_0 t' - n\beta \cos \psi \sin(\omega_0 t')) \, dt'$$

$$E_{n\psi} = \frac{ine\omega_0\beta}{4\pi\epsilon_0 cr_p} \frac{\sin \psi}{T_0} \int_0^{T_0} \cos(\omega_0 t') \cos(n\omega_0 t' - n\beta \cos \psi \sin(\omega_0 t')) \, dt'$$

$$E_{nr} = 0.$$

We use the integral representation (B.1) of the Bessel function and the relation (B.2) given in Appendix B to express the integrals appearing above in terms of $J_n(z)$ and its derivative $J'_n(z)$:

$$I_1 = \int_0^{T_0} \sin(\omega_0 t') \sin(n\omega_0 t' - n\beta \cos \psi \sin(\omega_0 t')) \, dt'$$
$$= \frac{1}{2} \int_0^{T_0} [\cos(n\beta \cos \psi \sin(\omega_0 t') - (n-1)\omega_0 t')$$
$$\qquad - \cos(n\beta \cos \psi \sin(\omega_0 t') - (n+1)\omega_0 t')] \, dt'$$
$$= \frac{\pi}{\omega_0}[J_{n-1}(n\beta \cos \psi) - J_{n+1}(n\beta \cos \psi)] = \frac{2\pi}{\omega_0} J'_n(n\beta \cos \psi)$$

$$(4.26)$$

$$I_2 = \int_0^{T_0} \cos(\omega_0 t') \cos(n\omega_0 t' - n\beta \cos \psi \sin(\omega_0 t')) \, dt'$$
$$= \frac{1}{2} \int_0^{T_0} [\cos(n\beta \cos \psi \sin(\omega_0 t') - (n-1)\omega_0 t')$$
$$\qquad + \cos(n\beta \cos \psi \sin(\omega_0 t') - (n+1)\omega_0 t')] \, dt'$$
$$= \frac{\pi}{\omega_0}[J_{n-1}(n\beta \cos \psi) + J_{n+1}(n\beta \cos \psi)] = \frac{2\pi}{\omega_0\beta \cos \psi} J_n(n\beta \cos \psi).$$

We can now write the Fourier components of the electric field:

$$E_{n\phi} = \frac{e\omega_0\beta}{4\pi\,\epsilon_0 cr_p} n J_n'(n\beta\cos\psi)$$

$$E_{n\psi} = i\frac{e\omega_0\tan\psi}{4\pi\,\epsilon_0 cr_p} n J_n(n\beta\cos\psi).$$

These Fourier components exist both for positive and for negative harmonics $-\infty < n < \infty$. The electric field is now represented by the Fourier series

$$E_\phi(t_p) = \frac{e\omega_0\beta}{4\pi\,\epsilon_0 cr_p}\sum_{n=-\infty}^{\infty} n J_n'(n\beta\cos\psi)e^{in\omega_0 t_p}$$

$$E_\psi(t_p) = i\frac{e\omega_0}{4\pi\,\epsilon_0 cr_p}\tan\psi\sum_{n=-\infty}^{\infty} n J_n(n\beta\cos\psi)e^{in\omega_0 t_p}.$$

On combining terms with positive and negative n and using the symmetry relations of the Bessel functions (B.5)–(B.7),

$$J_n(na) = J_{-n}(-na), \qquad J_n'(na) = -J_{-n}'(-na),$$

we obtain a sum over positive frequencies only, which contains trigonometric functions instead of exponentials:

$$E_\phi(t_p) = \frac{e\omega_0\beta}{2\pi\,\epsilon_0 cr_p}\sum_{n=1}^{\infty} n J_n'(n\beta\cos\psi)\cos(n\omega_0 t_p)$$

$$E_\psi(t_p) = -\frac{e\omega_0}{2\pi\,\epsilon_0 cr_0}\tan\psi\sum_{n=1}^{\infty} n J_n(n\beta\cos\psi)\sin(n\omega_0 t_p).$$

The spectrum radiated by a charge going through a circular arc has a maximum close to the critical frequency ω_c. We expect therefore that the line spectrum obtained from a charge moving on a closed circle peaks around a harmonic number n_c given by

$$n_c = \frac{\omega_c}{\omega_0} = \frac{3}{2}\gamma^3.$$

For a relativistic particle most of the spectrum lines are at very high harmonics of the Larmor frequency, $n \gg 1$. Unless one is very familiar with the Bessel functions of very high order and argument it is difficult to visualize the spectrum given by the above equation. We will later approximate these expressions for $n \gg 1$ and compare them with the results obtained for the continuous spectrum.

4.5.3 The power of the line spectrum

The energy radiated per unit area and time is given by the Poynting vector

$$\mathbf{S} = \frac{1}{\mu_0}[\mathbf{E} \times \mathbf{B}] = \frac{1}{\mu_0 c}E^2\mathbf{n} = \frac{1}{\mu_0 c}\left(E_\phi^2 + E_\psi^2\right)\mathbf{n}.$$

It points in the direction $\mathbf{n}$ of propagation and consists of two parts given by the two directions of polarization. Since we have expressed the field as a Fourier series, the Poynting vector also has a component at each harmonic n. Since they have the same value for positive and negative frequency, we sum the pairs with the same absolute value of n and use $n > 0$ as we usually did for the power:

$$\mathbf{S}_n = \frac{1}{\mu_0 c}\left(E_{-n}^2 + E_n^2\right)\mathbf{n} = \frac{2}{\mu_0 c}E_n^2\mathbf{n} = \frac{2}{\mu_0 c}\left[E_{n\phi}^2 + E_{n\psi}^2\right]\mathbf{n}.$$

The *average* power radiated by the charge at a single harmonic n into a unit solid angle is

$$\frac{dP_n}{d\Omega} = \frac{r_p^2(\mathbf{n} \cdot \mathbf{S}_n)}{2} = \frac{r_p^2}{\mu_0 c}E_n^2$$

$$= \frac{r_0 m_0 c^2 \omega_0^2}{2\pi c}n^2\left(\beta^2 J_n'^{\,2}(n\beta \cos\psi) + \tan^2\psi\, J_n^2(n\beta \cos\psi)\right),$$

where we expressed some fundamental constants by using the classical electron radius $r_0 = e^2/(4\pi\epsilon_0 m_0 c^2)$. We know the total emitted power from the general expression (3.9),

$$P_s = \frac{2cr_0 m_0 c^2 \beta^4 \gamma^4}{3\rho^2},$$

and use it to express the factor in front of the power distribution,

$$\frac{r_0 m_0 c^2 \omega_0^2 \beta^2}{2\pi c} = P_s\frac{3}{4\pi \gamma^4},$$

where we also expressed the angular velocity by using $\omega_0 = \beta c/\rho$. Finally we integrate over the solid angle and can cross check the total power. We distinguish between the two parts of the radiated power due to the field components E_ϕ and E_ψ which correspond to the σ- and π-modes of polarization. For the first one the direction of the electric field is parallel to the plane of the orbit of the particle and for the second one it is perpendicular to it. The angular distributions of the two modes are

$$\boxed{\begin{aligned}
\frac{dP_{n\sigma}}{d\Omega} &= P_s\frac{3}{4\pi \gamma^4}n^2 J_n'^{\,2}(n\beta \cos\psi) \\[2mm]
\frac{dP_{n\pi}}{d\Omega} &= P_s\frac{3}{4\pi \gamma^4}n^2 J_n^2(n\beta \cos\psi)\tan^2\psi/\beta^2 \\[2mm]
\frac{dP_n}{d\Omega} &= P_s\frac{3}{4\pi \beta^2 \gamma^4}n^2\left(J_n'^{\,2}(n\beta \cos\psi) + J_n^2(n\beta \cos\psi)\tan^2\psi/\beta^2\right).
\end{aligned}}$$

(4.27)

The power radiated per unit solid angle at all frequencies is obtained by summing over the contribution of each harmonic given by (4.27). For this we need sums (B.17) and (B.19), which are derived in Appendix B:

$$\sum_{n=1}^{\infty} n^2 J_n^2(nz) = \frac{z^2(4+z^2)}{16(1-z^2)^{7/2}}, \qquad \sum_{n=1}^{\infty} n^2 J_n'^2(nz) = \frac{4+3z^2}{16(1-z^2)^{5/2}}.$$

With $z = \beta \cos \psi$ we obtain

$$\boxed{\begin{aligned}
\frac{dP_\sigma}{d\Omega} &= \sum_{n=1}^{\infty} \frac{dP_{n\sigma}}{d\Omega} = P_s \frac{3}{4\pi\gamma^4} \frac{4 + 3\beta^2 \cos^2 \psi}{16(1 - \beta^2 \cos^2 \psi)^{5/2}} \\
\frac{dP_\pi}{d\Omega} &= \sum_{n=1}^{\infty} \frac{dP_{n\pi}}{d\Omega} = P_s \frac{3}{4\pi\gamma^4} \frac{\sin^2 \psi \, (4 + \beta^2 \cos^2 \psi)}{16(1 - \beta^2 \cos^2 \psi)^{7/2}}.
\end{aligned}}$$

Finally, we obtain the total power radiated for the two polarization components by integrating the above equations over the solid angle $d\Omega = \sin \Theta \, d\Theta \, d\phi = \cos \psi \, d\psi \, d\phi$, which is just $d\Omega = 2\pi \cos \psi \, d\psi$ since there is no dependence on the azimuthal angle ϕ:

$$P_\sigma = P_s \frac{3}{2\gamma^4} \int_{-\pi/2}^{\pi/2} \frac{4 + 3\beta^2 \cos^2 \psi}{16(1 - \beta^2 \cos^2 \psi)^{5/2}} \cos \psi \, d\psi = P_s \frac{6 + \beta^2}{8}$$

$$P_\pi = P_s \frac{3}{2\gamma^4} \int_{-\pi/2}^{\pi/2} \frac{\sin^2 \psi (4 + \beta^2 \cos^2 \psi)}{16(1 - \beta^2 \cos^2 \psi)^{7/2}} \cos \psi \, d\psi = P_s \frac{2 - \beta^2}{8}.$$

The above integrals were calculated with the substitution $\sin \psi = x$. Adding the two polarization components gives the total power P_s. The distribution into the two polarization modes is, in the ultra-relativistic case $\gamma \gg 1$,

$$\frac{P_\sigma}{P_s} = \frac{7}{8}, \qquad \frac{P_\pi}{P_s} = \frac{1}{8},$$

and, for a non-relativistic charge for which $\beta \ll 1$,

$$\frac{P_\sigma}{P_s} = \frac{3}{4}, \qquad \frac{P_\sigma}{P_s} = \frac{1}{4}.$$

The total energy U_s radiated in one revolution is obtained from the power P_s by multiplying it by the time $T_0 = 2\pi\rho/(\beta c)$ taken for the particle to pass around a closed circle:

$$P_s = \frac{2c r_0 m_0 c^2 \beta^4 \gamma^4}{3\rho^2}, \qquad U_s = P_s \frac{2\pi\rho}{\beta c} = \frac{4\pi r_0 m_0 c^2 \beta^3 \gamma^4}{3\rho}.$$

So far we have assumed that we have an electron moving on a closed circle with bending radius ρ. If there are field-free straight sections between bending magnets the motion is still periodic but the revolution time is longer, $T_{\text{rev}} > T_0$, and the corresponding frequency smaller, $\omega_{\text{rev}} < \omega_0 = \beta c/\rho$. As a consequence the spectrum consists of lines with frequency $\omega_n = n\omega_{\text{rev}}$. The energy U_s radiated per turn is not changed but the power P_s now refers to its value taken while the electron traverses the magnet and the average power is $\langle P \rangle = P_s \omega_{\text{rev}} / \omega_0$.

4.5.4 The relation between the continuous and the line spectra

We take now the ultra-relativistic case with $\beta \approx 1$ and $\gamma \gg 1$. As a consequence we also have a small vertical opening angle $\psi \ll 1$ and we can approximate the argument of the Bessel functions:

$$\beta \approx 1 - \frac{1}{2\gamma^2}, \qquad \cos \psi \approx 1 - \frac{\psi^2}{2}, \qquad n\beta \cos \psi \approx n \left(1 - \frac{1 + \gamma^2 \psi^2}{2\gamma^2} \right).$$

The line spectrum emitted by a charge moving on a closed circle was expressed in terms of Bessel functions of order equal to the Larmor-frequency harmonics n. For most of the radiation this order is rather large, as indicated by the critical harmonic $n_c = 3\gamma^3/2$ and the value of the Bessel function is not easy to evaluate. However, approximations for Bessel functions of large order and arguments are given in the standard literature and are derived in Appendix B as (B.12) and (B.13) for $\gamma \gg 1$ and $\psi \ll 1$:

$$n J_n(n\beta \cos \psi) \approx 2 \left(\frac{n}{2} \right)^{2/3} \text{Ai} \left(\left(\frac{n}{2\gamma^3} \right)^{2/3} (1 + \gamma^2 \psi^2) \right)$$

$$n J_n'(n\beta \cos \psi) \approx -2 \left(\frac{n}{2} \right)^{1/3} \text{Ai}' \left(\left(\frac{n}{2\gamma^3} \right)^{2/3} (1 + \gamma^2 \psi^2) \right).$$

Using $n_c = 3\gamma^3/2$ and substituting these approximations into the expression (4.27) for the angular distribution of the nth harmonic, we obtain

$$\frac{\mathrm{d} P_{n\sigma}}{\mathrm{d}\Omega} = P_s \frac{3}{\pi \gamma^2} \left(\frac{3n}{4n_c} \right)^{2/3} \text{Ai}'^2 \left(\left(\frac{3n}{4n_c} \right)^{2/3} (1 + \gamma^2 \psi^2) \right)$$

$$\frac{\mathrm{d} P_{n\pi}}{\mathrm{d}\Omega} = P_s \frac{3}{\pi \gamma^2} \gamma^2 \psi^2 \left(\frac{3n}{4n_c} \right)^{4/3} \text{Ai}^2 \left(\left(\frac{3n}{4n_c} \right)^{2/3} (1 + \gamma^2 \psi^2) \right).$$

We can find an approximate expression for the angular spectral density since each spectral line gives the power in one frequency bin of width ω_0 and frequency $\omega \approx n\omega_0$,

$$\frac{\mathrm{d}^2 P}{\mathrm{d}\Omega \, \mathrm{d}\omega} \approx \frac{1}{\omega_0} \frac{\mathrm{d} P_n}{\mathrm{d}\Omega},$$

giving, with $\omega_c = 3\omega_0 \gamma^3/2$,

$$\frac{\mathrm{d}^2 P_\sigma}{\mathrm{d}\Omega \, \mathrm{d}\omega} = \frac{P_s}{\omega_c} \frac{9\gamma}{2\pi} \left(\frac{3n}{4n_c} \right)^{2/3} \text{Ai}'^2 \left(\left(\frac{3n}{4n_c} \right)^{2/3} (1 + \gamma^2 \psi^2) \right)$$

$$\frac{\mathrm{d} P_{n\pi}}{\mathrm{d}\Omega} = \frac{P_s}{\omega_c} \frac{9\gamma}{2\pi} \gamma^2 \psi^2 \left(\frac{3n}{4n_c} \right)^{4/3} \text{Ai}^2 \left(\left(\frac{3n}{4n_c} \right)^{2/3} (1 + \gamma^2 \psi^2) \right).$$

This is identical to the corresponding expression (4.22) for the continuous spectrum.

A charge moving on a closed circle with constant angular velocity represents a periodic motion and the emitted radiation is expected to be represented by a line spectrum. However, this is strictly correct only in the classical treatment, in which the emission of radiation is continuous and its effect on the particle motion is neglected. In reality the radiation is emitted as photons and the particle suffers a sudden loss of energy in the process. We will show later that the frequency of revolution in a storage ring depends on the particle energy. The calculated line spectrum consists of harmonics of a fixed frequency of revolution. A change of the latter after the emission of each photon will smear the lines out, particularly at the high frequencies. The synchrotron-radiation spectrum from a realistic storage ring is therefore continuous except at very low harmonics.

5

Synchrotron radiation: properties

5.1 Introduction

On the basis of the equations derived in the previous chapter, we present now the properties of synchrotron radiation that are relevant for applications. We use here the ultra-relativistic case, observe the traversal of a single particle from a large distance, and describe it in the frequency domain as a continuous spectrum. This is also a very good approximation for a charge moving on a closed circle, since the line spectrum is in practice almost always smeared out by quantum excitation.

The spectral properties of the synchrotron radiation can be described using either modified Bessel functions $K_{1/3}(z)$, $K_{2/3}(z)$, and $\int K_{5/3}(z)\,dz$ or the Airy function $\mathrm{Ai}(z)$, its derivative $\mathrm{Ai}'(z)$, and the integral $\int \mathrm{Ai}(z)\,dz$. It does not matter which set of functions is used. The Airy functions are easier to find in tables [41] and will be chosen here for all derivations, but the important results are also given in terms of modified Bessel functions. The mathematical properties and applications of the Airy functions are discussed in Appendix A.

5.2 The total radiated power and energy

In Chapter 3 we derived the power (3.13) radiated by a particle of rest mass m_0 and charge e undergoing a transverse acceleration $\beta^2 c^2/\rho$ by moving with velocity βc on a circular arc of bending radius ρ,

$$P_s = \frac{2 r_0 c m_0 c^2 \beta^4 \gamma^4}{3 \rho^2},\tag{5.1}$$

with the classical particle radius $r_0 = e^2/(4\pi \epsilon_0 m_0 c^2)$. In most cases the acceleration is due to the Lorentz force provided by a magnetic field B perpendicular to the plane of the orbit. This results in a curvature of

$$\frac{1}{\rho} = \frac{eB}{p} = \frac{eB}{m_0 c \beta \gamma}$$

and in an expression for the radiated power:

$$P_s = \frac{2 r_0 c^3 e^2 \beta^2 \gamma^2 B^2}{3 m_0 c^2} = \frac{2 r_0 c^3 e^2 \beta^2 E_e^2 B^2}{3 (m_0 c^2)^3}.\tag{5.2}$$

From now on we assume that the particle moves with ultra-relativistic velocity, $\beta \approx 1$, $\gamma \gg 1$, and make the corresponding approximations. The total power P_s emitted by one electron on a trajectory of bending radius ρ is

$$P_s = \frac{2r_0 c m_0 c^2 \gamma^4}{3\rho^2} = \frac{2r_0 c^3 e^2 \gamma^2 B^2}{3\,m_0 c^2} = \frac{2r_0 c^3 e^2 E_e^2 B^2}{3(m_0 c^2)^3}. \tag{5.3}$$

The derivation of this equation was very general, involving, apart from the properties of the particle, only the instantaneous curvature $1/\rho$ or the magnetic field B. It is also valid for an inhomogeneous field as long as $1/\rho(s)$ refers to the local curvature.

The local loss of energy by a particle per unit time element dt', length ds, and bending angle $d\phi$ is

$$\frac{dU}{dt'} = P_s, \qquad \frac{dU}{ds} = \frac{P_s}{c} = \frac{2r_0 m_0 c^2 \gamma^4}{3\rho^2}, \qquad \frac{dU}{d\phi} = P_s \frac{\rho}{c} = \frac{2r_0 m_0 c^2 \gamma^4}{3\rho}. \tag{5.4}$$

Of importance is the energy U_s lost by a particle in one revolution, which is obtained from the integral

$$U_s = \oint \frac{P(s)}{c}\, ds = \frac{2r_0 m_0 c^2 \gamma^4}{3} \oint \frac{1}{\rho^2}\, ds = \frac{2r_0 m_0 c^2 \gamma^4}{3} I_{s2}. \tag{5.5}$$

Since this case has to be evaluated often, a special *synchrotron-radiation integral* has been introduced [42]:

$$I_{s2} = \oint \frac{1}{\rho^2}\, ds. \tag{5.6}$$

A ring with magnets all having the same field B and radius of curvature ρ and field-free regions in addition is called *isomagnetic*. Its synchrotron-radiation integral and the loss of energy are

$$I_{s2} = \frac{2\pi\rho}{\rho^2} = \frac{2\pi}{\rho}, \qquad U_s = \frac{4\pi r_0 m_0 c^2 \gamma^4}{3\rho}.$$

Owing to the field-free straight section such a ring has a circumference C_{rev} that is larger than $2\pi\rho$ and has a correspondingly large revolution time T_{rev} and frequency ω_{rev}:

$$\omega_{rev} = \frac{2\pi c}{C_{rev}} = \frac{2\pi}{T_{rev}} \leq \frac{c}{\rho} = \omega_0.$$

This does not affect the instantaneous power (5.3) or the energy lost per turn (5.5). However, in some rare cases one is interested in the emitted power averaged over one revolution:

$$\langle P_s \rangle = \frac{U_s}{T_{rev}} = P_s \frac{\omega_{rev}}{\omega_0}.$$

So far we have been concerned with the energy or power radiated by a single particle of charge e. We will now consider a storage ring with a large number N_e of particles representing an average beam current $I = eN_e/T_{rev}$. In most practical applications the radiation of one particle is independent of that emitted by other particles. The radiation has in this case no *time coherence*. To obtain the total radiated power we simply add the contributions of each particle. If U_s is the energy lost by one particle in one revolution the total average power emitted in an isomagnetic ring by a current I is

$$P_I = \frac{N_e U_s}{T_{\text{rev}}} = \frac{I U_s}{e} = \frac{4\pi r_0 m_0 c^2 \gamma^4 I}{3e\rho}.$$

This equation can also be applied to a part of the storage ring, e.g. a single magnet. In the following sections we will investigate the angular and spectral distributions of the synchrotron radiation. We will write most expressions such that the total radiated power P_s appears as a factor. Depending on whether one is interested in the distribution of the radiated energy or the power emitted by one or many electrons, this factor P_s can be exchanged for the energy U_s or P_I.

5.3 The angular spectral distribution

5.3.1 The general distribution

We discuss now the so-called angular spectral power or energy distribution which is the power or energy radiated per unit solid-angle element and frequency band. For the ultra-relativistic case the angles are small and the solid-angle element becomes

$$d\Omega = d\psi \, d\phi.$$

The radiation is uniform with respect to the azimuthal angle ϕ and we have to consider only the ψ-dependence of the distribution. In the previous chapter we derived an expression (4.22) for this angular spectral power distribution, which we present now in a compact form,

$$\boxed{\frac{d^2 P}{d\Omega \, d\omega} = \frac{P_s \gamma}{\omega_c}[F_{s\sigma}(\omega, \psi) + F_{s\pi}(\omega, \psi)] = \frac{P_s \gamma}{\omega_c} F_s(\omega, \psi),} \tag{5.7}$$

by giving the dependences on angle and frequency with two dimensionless functions that can be expressed with the Airy functions,

$$
\begin{aligned}
F_{s\sigma}(\omega, \psi) &= \frac{9}{2\pi}\left(\frac{3\omega}{4\omega_c}\right)^{2/3} \text{Ai}'^2\left(\left(\frac{3\omega}{4\omega_c}\right)^{2/3}(1 + \gamma^2\psi^2)\right) \\[2mm]
F_{s\pi}(\omega, \psi) &= \frac{9}{2\pi}\left(\frac{3\omega}{4\omega_c}\right)^{4/3} \gamma^2\psi^2 \, \text{Ai}^2\left(\left(\frac{3\omega}{4\omega_c}\right)^{2/3}(1 + \gamma^2\psi^2)\right),
\end{aligned}
\tag{5.8}
$$

or, using the relations (A.3), by use of modified Bessel functions,

$$F_{s\sigma}(\omega, \psi) = \left(\frac{3}{2\pi}\right)^3 \left(\frac{\omega}{2\omega_c}\right)^2 (1 + \gamma^2\psi^2)^2 K_{2/3}^2 \left(\frac{\omega}{2\omega_c}(1 + \gamma^2\psi^2)^{3/2}\right)$$

$$F_{s\pi}(\omega, \psi) = \left(\frac{3}{2\pi}\right)^3 \left(\frac{\omega}{2\omega_c}\right)^2 \gamma^2\psi^2(1 + \gamma^2\psi^2) K_{1/3}^2 \left(\frac{\omega}{2\omega_c}(1 + \gamma^2\psi^2)^{3/2}\right).$$

$$(5.9)$$

These two functions give the distributions of the two modes of polarization; the σ-mode having the electric field in the plane of the particle trajectory and the π-mode with the electric

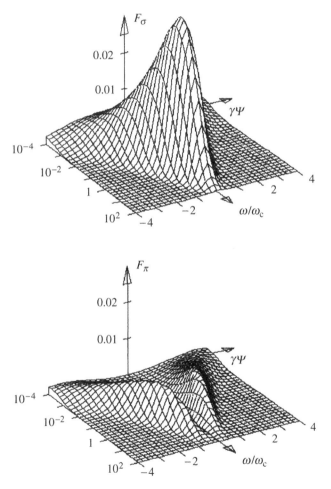

Fig. 5.1. Normalized angular spectral power density for the horizontal (top) and vertical (bottom) polarization modes of synchrotron radiation.

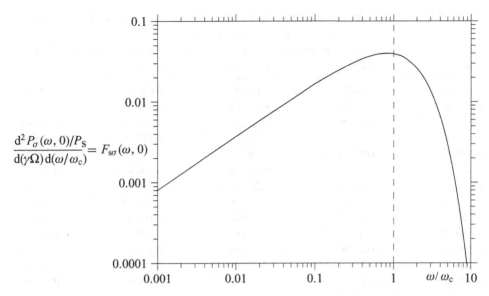

Fig. 5.2. The angular spectral power density in the median plane.

field perpendicular to it. As we will see later, these functions fulfill the normalization

$$\int_0^{2\pi} d\phi \int_{-\infty}^{\infty} d(\gamma\psi) \int_0^{\infty} F_s(\omega, \psi) \, d(\omega/\omega_c) = 1. \tag{5.10}$$

To illustrate the form of the distribution the normalized spectral angular distributions $F_{s\sigma}$ and $F_{s\pi}$ of the two modes of polarization are plotted in Fig. 5.1 against the angle ψ and the logarithm of the frequency ω. Both modes are symmetric with respect to ψ and their angular distributions are narrow at high frequencies and wide at low frequencies. The σ-mode is large in the median plane $\psi = 0$, with a peak close to the critical frequency ω_c. The π-mode vanishes in the median plane.

In Fig. 5.2 the angular spectral power density in the median plane $\psi = 0$ is plotted against frequency on a double-logarithmic scale and the normalized function $F_s(0)$ is listed in Table 5.1. At the critical frequency ω_c it is $F_{s\sigma}(\omega_c, 0) = 0.0396$ and it reaches a maximum value of 0.0401 at $\omega \approx 0.83 \, \omega_c$. The angular distribution of three selected frequencies is shown in Fig 5.3.

5.3.2 The distribution at low frequencies

Next we investigate the angular spectral power density at low frequencies and assume that $\omega \ll \omega_c$. In this case the argument of the Airy functions is very small unless $\gamma^2\psi^2$ becomes very large. Only then will the value of these functions change and we make a small error

Table 5.1. *The normalized angular power density $F_s(\omega/\omega_c)$ in the median plane and the normalized power spectrum $S_s(\omega/\omega_c)$*

ω/ω_c	$F_s(\omega/\omega_c, 0)$	$S_{s\sigma}(\omega/\omega_c)$	$S_{s\pi}(\omega/\omega_c)$	$S_s(\omega/\omega_c)$
0.001 00	0.000 79	0.099 40	0.032 80	0.132 00
0.002 00	0.001 26	0.125 00	0.040 90	0.166 00
0.004 00	0.001 99	0.156 00	0.050 70	0.207 00
0.006 00	0.002 61	0.178 00	0.057 30	0.236 00
0.008 00	0.003 16	0.195 00	0.062 30	0.258 00
0.010 00	0.003 67	0.210 00	0.066 40	0.276 00
0.020 00	0.005 81	0.260 00	0.079 80	0.339 00
0.040 00	0.009 15	0.318 00	0.093 30	0.411 00
0.060 00	0.011 90	0.355 00	0.100 00	0.455 00
0.080 00	0.014 30	0.381 00	0.104 00	0.485 00
0.100 00	0.016 40	0.401 00	0.106 00	0.507 00
0.200 00	0.024 60	0.454 00	0.106 00	0.560 00
0.400 00	0.034 20	0.468 00	0.091 50	0.559 00
0.600 00	0.038 70	0.441 00	0.075 00	0.516 00
0.800 00	0.040 10	0.400 00	0.060 60	0.460 00
1.000 00	0.039 60	0.355 00	0.048 70	0.404 00
2.000 00	0.026 60	0.171 00	0.016 10	0.187 00
4.000 00	0.006 79	0.030 90	0.001 83	0.032 80
6.000 00	0.001 35	0.005 01	0.000 22	0.005 23
8.000 00	0.000 24	0.000 77	0.000 03	0.000 80

by replacing $1 + \gamma^2\psi^2$ by $\gamma^2\psi^2$ and obtain for the argument

$$\left(\frac{3\omega}{4\omega_c}\right)^{2/3}(1+\gamma^2\psi^2) \approx \left(\frac{3\omega}{4\omega_c}\right)^{2/3}\gamma^2\psi^2 = \left(\frac{\omega}{2\omega_0}\right)^{2/3}\psi^2 = \left(\frac{\pi\rho}{\lambda}\right)^{1/3}\psi^2,$$

where we expressed the critical frequency $\omega_c = 3\omega_0\gamma^3/2$ in terms of the angular velocity ω_0. With this we obtain from (4.20) the radiation field at low frequencies,

$$\tilde{E}_\perp(\psi, \lambda) = \frac{e}{\sqrt{2\pi}\,\epsilon_0 cr}$$

$$\times \left(\frac{\pi\rho}{\lambda}\right)^{1/3}\left[\text{Ai}'\left(\left(\frac{\pi\rho}{\lambda}\right)^{2/3}\psi^2\right),\ i\psi\left(\frac{\pi\rho}{\lambda}\right)^{1/3}\text{Ai}\left(\left(\frac{\pi\rho}{\lambda}\right)^{2/3}\psi^2\right)\right], \quad (5.11)$$

and the angular spectral power distribution

$$\frac{d^2P}{d\Omega\,d\omega} = \frac{P_s\gamma}{\omega_c}[F_{s\sigma}(\omega, \psi) + F_{s\pi}(\omega, \psi)] = \frac{P_s\gamma}{\omega_c}F_s(\omega, \psi)$$

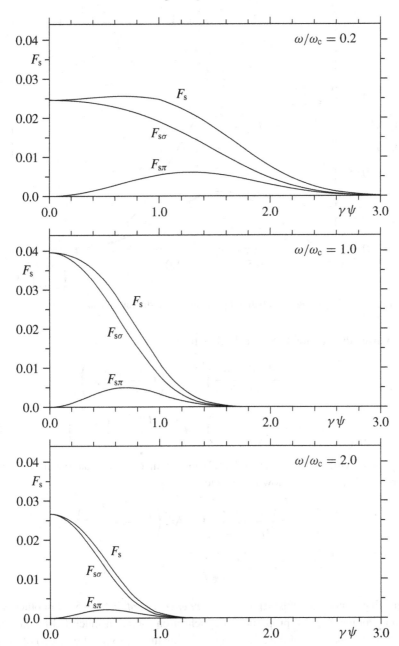

Fig. 5.3. The angular spectral power densities of the total radiation and the two modes of polarization versus the vertical angle for various frequencies.

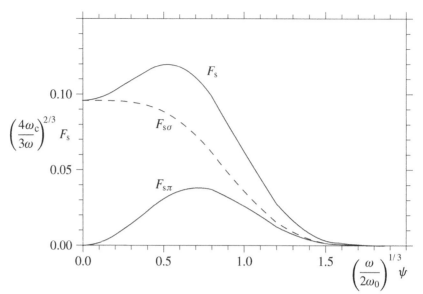

Fig. 5.4. The angular distribution of synchrotron radiation at small frequencies.

with normalized distribution functions for low frequencies:

$$F_{s\sigma}(\omega, \psi) \approx \frac{9}{2\pi} \frac{1}{\gamma^2} \left(\frac{\omega}{2\omega_0}\right)^{2/3} \mathrm{Ai}'^2\left(\left(\frac{\omega}{2\omega_0}\right)^{2/3} \psi^2\right) \tag{5.12}$$

$$F_{s\pi}(\omega, \psi) \approx \frac{9}{2\pi} \frac{1}{\gamma^2} \left(\frac{\omega}{2\omega_0}\right)^{4/3} \psi^2 \mathrm{Ai}^2\left(\left(\frac{\omega}{2\omega_0}\right)^{2/3} \psi^2\right). \tag{5.13}$$

We express ω_c with (4.15) and P_s with (5.3) and obtain explicit equations for the angular spectral power distributions at low frequencies, $\omega \ll \omega_c$:

$$\frac{\mathrm{d}^2 P_\sigma}{\mathrm{d}\Omega\,\mathrm{d}\omega} = \frac{2r_0 m_0 c^2}{\pi\rho} \left(\frac{\omega}{2\omega_0}\right)^{2/3} \mathrm{Ai}'^2\left(\left(\frac{\omega}{2\omega_0}\right)^{2/3} \psi^2\right)$$

$$\frac{\mathrm{d}^2 P_\pi}{\mathrm{d}\Omega\,\mathrm{d}\omega} = \frac{2r_0 m_0 c^2}{\pi\rho} \left(\frac{\omega}{2\omega_0}\right)^{4/3} \psi^2 \mathrm{Ai}^2\left(\left(\frac{\omega}{2\omega_0}\right)^{2/3}\right). \tag{5.14}$$

It is interesting to note that this expression is *independent of γ* (Fig. 5.4); in other words, for a given storage ring the properties of synchrotron radiation much below the critical frequency are independent of the beam energy. Using the approximation (A.5) of the Airy function for large arguments given in Appendix A, we find that, for large angles ψ, the two modes of polarization have the same intensity. This will be discussed in more detail later.

This low-frequency part of the spectrum is often used for beam diagnostics. To obtain information about the cross section of the beam, one forms an image using the visible part of the synchrotron radiation. In most storage rings the critical frequency is much above the visible spectrum and the above approximations are justified.

5.3.3 The distribution at high frequencies

At the other end of the spectrum we can make an approximation for $\omega \gg \omega_c$ using the development (A.5) of the Airy function at a large argument $x \gg 1$ explained in Appendix A:

$$\text{Ai}(x \to \infty) \approx \frac{1}{2\sqrt{\pi}\sqrt{x}}e^{-2x^{3/2}/3}, \qquad \text{Ai}'(x \to \infty) \approx \frac{1}{2}\sqrt{\frac{\sqrt{x}}{\pi}}e^{-2x^{3/2}/3}.$$

This gives for the normalized distributions

$$F_{s\sigma}(\omega \gg \omega_c) \approx \frac{27\omega}{32\pi^2\omega_c}\sqrt{1+\gamma^2\psi^2}e^{-(\omega/\omega_c)(1+\gamma^2\psi^2)^{3/2}}$$

$$F_{s\pi}(\omega \gg \omega_c) \approx \frac{27\omega}{32\pi^2\omega_c}\frac{\gamma^2\psi^2}{\sqrt{1+\gamma^2\psi^2}}e^{-(\omega/\omega_c)(1+\gamma^2\psi^2)^{3/2}}.$$

For this case of very high frequencies the opening angle of the radiation is small, $\gamma^2\psi^2 \ll 1$, and we can approximate further:

$$F_{s\sigma}(\omega \gg \omega_c) \approx \frac{27}{32\pi^2}\frac{\omega}{\omega_c}e^{-\omega/\omega_c}e^{-3(\omega/\omega_c)\gamma^2\psi^2/2}, \qquad F_{s\pi}(\omega \gg \omega_c) \approx 0.$$

For very high frequencies the contribution of the π-mode becomes negligible and the σ-mode approaches a Gaussian angular distribution with RMS opening angle

$$\gamma\psi_{\text{RMS}}(\omega \gg \omega_c) = \sqrt{\frac{\omega_c}{3\omega}}.$$

5.4 The spectral distribution

5.4.1 The general spectrum

For many applications of synchrotron radiation the vertical angular distribution is not resolved and one is interested in the spectral density only. It is obtained by integrating the angular spectral power density over the solid angle:

$$\boxed{\frac{dP}{d\omega} = \int \frac{d^2P}{d\Omega\,d\omega}\,d\Omega = \frac{P_s}{\omega_c}\left[S_{s\sigma}\left(\frac{\omega}{\omega_c}\right) + S_{s\pi}\left(\frac{\omega}{\omega_c}\right)\right] = \frac{P_s}{\omega_c}S_s\left(\frac{\omega}{\omega_c}\right).} \qquad (5.15)$$

We describe the normalized spectral power density by using a dimensionless function S_s that depends only on the ratio between the frequency ω and the critical frequency ω_c. It has two parts, $S_{s\sigma}$ and $S_{s\pi}$, which correspond to the two modes of polarization. Since the synchrotron radiation is independent of ϕ the integration over this angle results simply in a factor of 2π. To obtain these functions $S_{s\sigma}$ and $S_{s\pi}$ we integrate the normalized angular spectral power density (5.8) over the solid angle $d\phi\,d\psi$:

$$S_{s\sigma}\left(\frac{\omega}{\omega_c}\right) = 9\left(\frac{3\omega}{4\omega_c}\right)^{2/3}\int_{-\infty}^{\infty} \mathrm{Ai}'^2\left(\left(\frac{3\omega}{4\omega_c}\right)^{2/3}(1+\gamma^2\psi^2)\right)d(\gamma\psi)$$

$$S_{s\pi}\left(\frac{\omega}{\omega_c}\right) = 9\left(\frac{3\omega}{4\omega_c}\right)^{4/3}\gamma^2\psi^2\int_{-\infty}^{\infty} \mathrm{Ai}^2\left(\left(\frac{3\omega}{4\omega_c}\right)^{2/3}(1+\gamma^2\psi^2)\right)d(\gamma\psi).$$

These integrals are obtained from some general expressions, (A.13) and (A.22), derived in Appendix A by setting $a = b = (3\omega/4\omega_c)^{2/3}$, giving for the functions S_s, using Airy functions,

$$S_{s\sigma}\left(\frac{\omega}{\omega_c}\right) = \frac{27}{16}\frac{\omega}{\omega_c}\left(-3\frac{\mathrm{Ai}'(z)}{z} - \frac{1}{3} + \int_0^z \mathrm{Ai}(z')\,dz'\right)$$

$$S_{s\pi}\left(\frac{\omega}{\omega_c}\right) = \frac{27}{16}\frac{\omega}{\omega_c}\left(-\frac{\mathrm{Ai}'(z)}{z} - \frac{1}{3} + \int_0^z \mathrm{Ai}(z')\,dz'\right) \qquad (5.16)$$

$$S_s\left(\frac{\omega}{\omega_c}\right) = \frac{54}{16}\frac{\omega}{\omega_c}\left(-2\frac{\mathrm{Ai}'(z)}{z} - \frac{1}{3} + \int_0^z \mathrm{Ai}(z')\,dz'\right)$$

with

$$z = \left(\frac{3\omega}{2\omega_c}\right)^{2/3},$$

or, using modified Bessel functions,

$$S_{s\sigma}\left(\frac{\omega}{\omega_c}\right) = \frac{9\sqrt{3}}{16\pi}\frac{\omega}{\omega_c}\left(\int_{\omega/\omega_c}^{\infty} K_{5/3}(z')\,dz' + K_{2/3}\left(\frac{\omega}{\omega_c}\right)\right)$$

$$S_{s\pi}\left(\frac{\omega}{\omega_c}\right) = \frac{9\sqrt{3}}{16\pi}\frac{\omega}{\omega_c}\left(\int_{\omega/\omega_c}^{\infty} K_{5/3}(z')\,dz' - K_{2/3}\left(\frac{\omega}{\omega_c}\right)\right) \qquad (5.17)$$

$$S_s\left(\frac{\omega}{\omega_c}\right) = \frac{9\sqrt{3}}{8\pi}\frac{\omega}{\omega_c}\int_{\omega/\omega_c}^{\infty} K_{5/3}(z')\,dz'.$$

These normalized power-spectrum functions are some of the most important expressions characterizing the properties of synchrotron radiation. They are shown in Fig. 5.5 on a double-logarithmic plot and in Fig. 5.6 on a linear scale as functions of ω/ω_c. They are also listed in Table 5.1. At the critical frequency their values are

$$S_{s\sigma}(1) = 0.3554, \qquad S_{s\pi}(1) = 0.0487, \qquad S_s(1) = 0.4040.$$

The maxima of the three functions occur somewhat below the critical frequency. The power spectrum is rather broad and smooth, as expected from qualitative arguments made earlier.

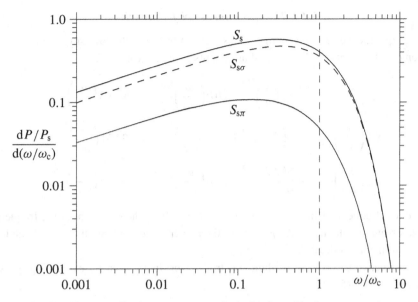

Fig. 5.5. The normalized power spectrum in double-logarithmic representation.

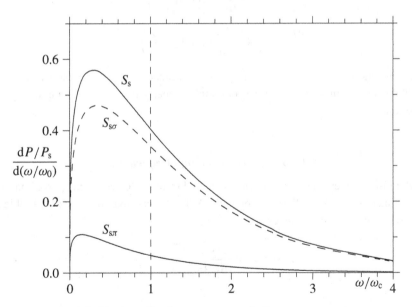

Fig. 5.6. The normalized power spectrum in linear representation.

5.4.2 The spectrum at low frequencies

At low frequency we can approximate the spectrum by using the lowest term of the Airy-function developments (A.4) for small arguments given in Appendix A:

$$S_{s\sigma}\left(\frac{\omega}{\omega_c}\right) \approx -3\frac{27}{16}\left(\frac{2}{3}\right)^{2/3} \mathrm{Ai}'(0)\left(\frac{\omega}{\omega_c}\right)^{1/3} = 0.99993\left(\frac{\omega}{\omega_c}\right)^{1/3}$$

$$S_{s\pi}\left(\frac{\omega}{\omega_c}\right) \approx -\frac{27}{16}\left(\frac{2}{3}\right)^{2/3} \mathrm{Ai}'(0)\left(\frac{\omega}{\omega_c}\right)^{1/3} = 0.33331\left(\frac{\omega}{\omega_c}\right)^{1/3} \qquad (5.18)$$

$$S_s\left(\frac{\omega}{\omega_c}\right) \approx -4\frac{27}{16}\left(\frac{2}{3}\right)^{2/3} \mathrm{Ai}'(0)\left(\frac{\omega}{\omega_c}\right)^{1/3} = 1.33323\left(\frac{\omega}{\omega_c}\right)^{1/3}.$$

At these low frequencies the spectral power increases with the third root of the frequency. It is interesting to note that three quarters of the power is radiated into the σ-mode and one quarter into the π-mode of the polarization.

It is instructive to give the explicit spectral power distribution by expressing ω_c with (4.15) and P_s with (5.2):

$$\frac{\mathrm{d}P_\sigma}{\mathrm{d}\omega} = -3\frac{r_0 m_0 c^2}{2\rho} \mathrm{Ai}'(0)\left(\frac{\omega}{\omega_0}\right)^{1/3}$$

$$\frac{\mathrm{d}P_\sigma}{\mathrm{d}\omega} = -\frac{r_0 m_0 c^2}{2\rho} \mathrm{Ai}'(0)\left(\frac{\omega}{\omega_0}\right)^{1/3}$$

$$\frac{\mathrm{d}P_\sigma}{\mathrm{d}\omega} = -4\frac{r_0 m_0 c^2}{2\rho} \mathrm{Ai}'(0)\left(\frac{\omega}{\omega_0}\right)^{1/3}.$$

For a given ring the spectrum at low frequencies is independent of the energy of the particle. This is not astonishing, since we found this property before for the angular spectral distribution.

5.4.3 The spectrum at high frequencies

For the high-frequency end of the spectrum we find from the lowest-term approximation of the Airy function given in (A.5) of Appendix A that the π-mode becomes negligible and the σ-mode decays with frequency like

$$S_s \approx S_{s\sigma} \approx \frac{27\sqrt{2}}{16\sqrt{3\pi}}\sqrt{\frac{\omega}{\omega_c}}\mathrm{e}^{-\omega/\omega_c}.$$

5.4.4 The spectrum integrated up to a given frequency

Sometimes the total power radiated below (or above) a given frequency is of interest. It is obtained by integrating the spectral power density over the frequency using the integrals

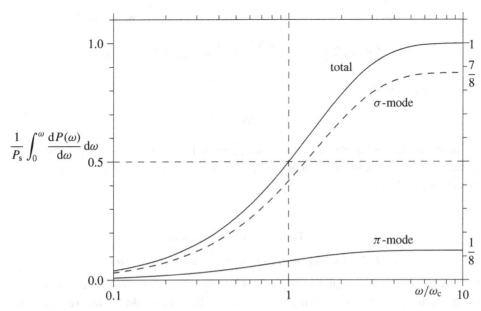

Fig. 5.7. The spectral power density integrated from 0 to ω.

(A.27) and (A.28) given in Appendix A:

$$\int_0^{\omega/\omega_c} S_{s\sigma}\, d\left(\frac{\omega}{\omega_c}\right) = \frac{7}{8} - \frac{3z^2}{8}\, \text{Ai}'(z) - \frac{21}{8} z\, \text{Ai}(z) - \frac{21 + 3z^3}{8} \int_z^\infty \text{Ai}(z')\, dz'$$

$$\int_0^{\omega/\omega_c} S_{s\pi}\, d\left(\frac{\omega}{\omega_c}\right) = \frac{1}{8} - \frac{3z^2}{8}\, \text{Ai}'(z) - \frac{3}{8} z\, \text{Ai}(z) - \frac{3 + 3z^3}{8} \int_z^\infty \text{Ai}(z')\, dz'$$

$$\int_0^{\omega/\omega_c} S_s\, d\left(\frac{\omega}{\omega_c}\right) = 1 - \frac{3z^2}{4}\, \text{Ai}'(z) - 3z\, \text{Ai}(z) - \frac{12 + 3z^3}{4} \int_z^\infty \text{Ai}(z')\, dz'$$

$$\text{with} \quad z = \left(\frac{3\omega}{2\omega_c}\right)^{2/3}.$$

This integrated power spectrum is shown in Fig. 5.7.

5.4.5 The integral over all frequencies

The integration over all frequencies gives the distribution of the total power emitted into the two modes of polarization and confirms that the total normalized power-spectrum function $S_s(\omega/\omega_c)$ has a normalized area

$$\int_0^\infty \left(S_{s\sigma}\left(\frac{\omega}{\omega_c}\right) + S_{s\pi}\left(\frac{\omega}{\omega_c}\right) \right) d\left(\frac{\omega}{\omega_c}\right) = \frac{7}{8} + \frac{1}{8} = 1. \tag{5.19}$$

If the integration is carried out up to the critical frequency only, we obtain

$$\int_0^{\omega_c} \left(S_{s\sigma}\left(\frac{\omega}{\omega_c}\right) + S_{s\pi}\left(\frac{\omega}{\omega_c}\right) \right) d\left(\frac{\omega}{\omega_c}\right) = 0.42 + 0.08 = 0.5.$$

The critical frequency

$$\omega_c = 3\omega_0 \gamma^3 / 2$$

divides the total power spectrum into *two equal parts*. So far we have treated it as a convenient parameter with which to characterize the spectrum. With the above relation it receives a precise physical meaning.

5.5 The angular distribution

5.5.1 The angular distribution as a function of frequency

In the discussion of the spectral angular power density we gave the angular distribution for some selected frequencies shown in Figs. 5.3 and 5.4. Here we would like to investigate the overall behavior of the vertical opening angle and calculate its variance as a function of ω:

$$\langle \gamma^2 \psi^2 \rangle_\sigma = \frac{\int \gamma^2 \psi^2 \dfrac{d^2 P_\sigma}{d\Omega\, d\omega} d\Omega}{\int \dfrac{d^2 P_\sigma}{d\Omega\, d\omega} d\Omega} = \frac{2\pi}{S_{s\sigma}} \int_{-\infty}^{\infty} \gamma^2 \psi^2 F_{s\sigma}(\omega, \psi) d(\gamma\psi)$$

$$\langle \gamma^2 \psi^2 \rangle_\pi = \frac{\int \gamma^2 \psi^2 \dfrac{d^2 P_\pi}{d\Omega\, d\omega} d\Omega}{\int \dfrac{d^2 P_\sigma}{d\Omega\, d\omega} d\Omega} = \frac{2\pi}{S_{s\pi}} \int_{-\infty}^{\infty} \gamma^2 \psi^2 F_{s\pi}(\omega, \psi) d(\gamma\psi).$$

We substitute the expressions (5.8) for the functions $F_{s\sigma}$ and $F_{s\pi}$ and obtain two integrals that can be solved from the general forms (A.14) and (A.23) given in Appendix A by setting $a = b = (3\omega/4\omega_c)^{2/3}$:

$$\langle \gamma^2 \psi^2 \rangle_\sigma = \frac{1}{S_{s\sigma}} \frac{27\,\omega}{64\omega_c} \left(5\frac{\mathrm{Ai}(z)}{z^2} + \frac{\mathrm{Ai}'(z)}{z} + \int_z^{\infty} \mathrm{Ai}(z')\,dz' \right)$$

$$\langle \gamma^2 \psi^2 \rangle_\pi = \frac{1}{S_{s\pi}} \frac{81\,\omega}{64\omega_c} \left(\frac{\mathrm{Ai}(z)}{z^2} + \frac{\mathrm{Ai}'(z)}{z} + \int_z^{\infty} \mathrm{Ai}(z')\,dz' \right).$$

$$(5.20)$$

The variance of the vertical opening angle for the total radiation is obtained by adding the properly weighted values for the two components of the polarization:

$$\langle \gamma^2 \psi^2 \rangle = \frac{S_{s\sigma} \langle \gamma^2 \psi^2 \rangle_\sigma + S_{s\pi} \langle \gamma^2 \psi^2 \rangle_\pi}{S_{s\sigma} + S_{s\pi}}.$$

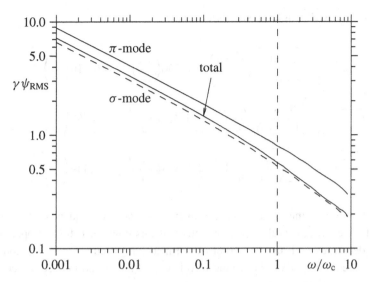

Fig. 5.8. The vertical RMS opening angle versus the frequency.

The RMS opening angle is the square root of the variance,

$$(\gamma\psi)_{\text{RMS}} = \sqrt{\langle\gamma^2\psi^2\rangle},$$

and is shown in Fig. 5.8 as a function of the frequency ω. Figures 5.3 and 5.4 show the σ-mode being large in the median plane $\psi = 0$ and decaying smoothly for larger angles, which can be characterized quite well by a Gaussian fit and a RMS opening angle. This is much less the case for the distribution of the π-mode, which has a minimum at the center. Still, the RMS opening angle can give a rough idea about its size. At the critical frequency the RMS opening angles are approximately

$$\sqrt{\langle\gamma^2\psi^2\rangle}_\sigma \approx 0.53, \qquad \sqrt{\langle\gamma^2\psi^2\rangle}_\pi \approx 0.80, \qquad \sqrt{\langle\gamma^2\psi^2\rangle} \approx 0.57.$$

They are a little smaller then the value of $1/\gamma$ usually considered as being the typical opening angle of synchrotron radiation.

To obtain the opening angle for very small frequencies, $\omega \ll \omega_c$, we can either approximate the expressions (5.20) for small values of z or form the proper integrals over the angular spectral power density (5.14). Using for the ratio between the Airy function and its derivative at the origin

$$\sqrt{-\frac{\text{Ai}(0)}{\text{Ai}'(0)}} = \sqrt{\frac{3^{1/3}\Gamma\left(\frac{1}{3}\right)}{3^{2/3}\Gamma\left(\frac{2}{3}\right)}} = 3^{1/3}\Gamma\left(\frac{1}{3}\right)\sqrt{\frac{\sin(\pi/3)}{3\pi}} = 1.1712,$$

we obtain for the RMS angles at small frequencies

$$\sqrt{\langle\psi^2\rangle_\sigma} = \sqrt{-\frac{5\mathrm{Ai}(0)}{12\mathrm{Ai}'(0)}}\left(\frac{\omega_0}{\omega}\right)^{1/3} = 0.756\left(\frac{\omega_0}{\omega}\right)^{1/3} = 0.410\left(\frac{\lambda}{\rho}\right)^{1/3}$$

$$\sqrt{\langle\psi^2\rangle_\pi} = \sqrt{-\frac{9\mathrm{Ai}(0)}{12\mathrm{Ai}'(0)}}\left(\frac{\omega_0}{\omega}\right)^{1/3} = 1.014\left(\frac{\omega_0}{\omega}\right)^{1/3} = 0.550\left(\frac{\lambda}{\rho}\right)^{1/3} \qquad (5.21)$$

$$\sqrt{\langle\psi^2\rangle} = \sqrt{-\frac{6\mathrm{Ai}(0)}{12\mathrm{Ai}'(0)}}\left(\frac{\omega_0}{\omega}\right)^{1/3} = 0.828\left(\frac{\omega_0}{\omega}\right)^{1/3} = 0.449\left(\frac{\lambda}{\rho}\right)^{1/3}.$$

We found before that the angular spectral distribution at low frequencies is independent of the particle energy; it is therefore obvious that this holds also for the opening angle. The latter decreases with the third root of the increasing frequency ω. Obviously these expressions are no longer valid once the wavelength of the radiation becomes comparable to the radius of curvature.

5.5.2 The frequency-integrated angular distribution

To obtain the angular distribution of the total radiation we integrate the spectral angular power distribution over all frequencies:

$$\frac{\mathrm{d}P}{\mathrm{d}\Omega} = \int_0^\infty \frac{\mathrm{d}^2 P}{\mathrm{d}\Omega\,\mathrm{d}\omega}\,\mathrm{d}\omega = \frac{P_\mathrm{s}\gamma}{\omega_\mathrm{c}}\int_0^\infty [F_{\mathrm{s}\sigma}(\omega,\psi) + F_{\mathrm{s}\pi}(\omega,\psi)]\,\mathrm{d}\omega.$$

Using the expressions (5.8) for the distribution functions $F_{\mathrm{s}\sigma}$ and $F_{\mathrm{s}\pi}$, we obtain integrals that can be solved by substituting $p = 3\omega/(4\omega_\mathrm{c})$ and $b = 1 + \gamma^2\psi^2$ into equation (A.26) in Appendix A,

$$\frac{\mathrm{d}P_\sigma}{\mathrm{d}\Omega} = \frac{P_\mathrm{s}\gamma}{2\pi}\frac{21}{32}\frac{1}{(1+\gamma^2\psi^2)^{5/2}}, \qquad \frac{\mathrm{d}P_\pi}{\mathrm{d}\Omega} = \frac{P_\mathrm{s}\gamma}{2\pi}\frac{15}{32}\frac{\gamma^2\psi^2}{(1+\gamma^2\psi^2)^{7/2}},$$

and, for the total radiation after integrating over ϕ,

$$\frac{\mathrm{d}P}{\mathrm{d}(\gamma\psi)} = P_\mathrm{s}\frac{21}{32}\frac{1}{(1+\gamma^2\psi^2)^{5/2}}\left(1 + \frac{5}{7}\frac{\gamma^2\psi^2}{(1+\gamma^2\psi^2)}\right). \qquad (5.22)$$

The latter result was derived in Chapter 3 in a more direct way, (3.15). These angular distributions of the frequency-integrated radiation are shown in Fig. 5.9. By integrating

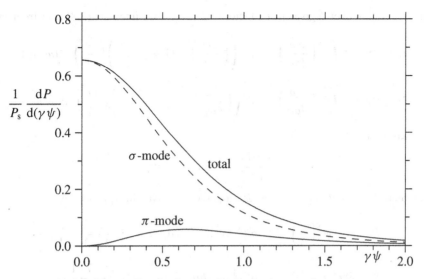

Fig. 5.9. The angular distribution after integrating over frequencies.

over the normalized angle $\gamma\psi$ we regain the partition (5.19) of the total radiation into the two modes of polarization $P_\sigma = 7P_s/8$ and $P_\pi = P_s/8$.

The variance of the frequency-integrated angular distribution is

$$\langle \gamma^2\psi^2 \rangle_\sigma = \frac{1}{P_\sigma} \int_{-\infty}^{\infty} \frac{dP_\sigma}{d\psi} \gamma^2\psi^2 \, d(\gamma\psi) = \frac{1}{2}$$

$$\langle \gamma^2\psi^2 \rangle_\pi = \frac{1}{P_\pi} \int_{-\infty}^{\infty} \frac{dP_\pi}{d\psi} \gamma^2\psi^2 \, d(\gamma\psi) = \frac{3}{2}$$

$$\langle \gamma^2\psi^2 \rangle = \frac{1}{P_s} \int_{-\infty}^{\infty} \frac{dP}{d\psi} \gamma^2\psi^2 \, d(\gamma\psi) = \frac{5}{8}.$$

The square root of the variance gives the RMS values of the normalized vertical opening angle $(\gamma\psi)_{\mathrm{RMS}}$, these being 0.707 for the σ-mode, 1.225 for the π-mode, and 0.791 for the total radiation.

Later we will calculate the vertical beam size due to quantum excitation during emission of synchrotron radiation. For this we need the variance of the product of the angle and the frequency, $\langle \psi^2\omega^2 \rangle$. We scale the frequency and angle with ω_c and $1/\gamma$, and use the angular spectral distribution function (5.8) with the normalization (5.10) to obtain

$$\left\langle \left(\frac{\omega}{\omega_c}\gamma\psi \right)^2 \right\rangle = \int_0^{2\pi} d\phi \int_{-\infty}^{\infty} d(\gamma\psi) \int_0^{\infty} F_s(\omega, \psi)(\omega/\omega_c)^2 \gamma^2\psi^2 \, d(\omega/\omega_c). \tag{5.23}$$

The integration over ϕ gives just a factor of 2π. Next we integrate (5.8) over frequency:

$$I_1 = 9\gamma^2\psi^2 \int_0^\infty \left(\frac{3\omega}{4\omega_c}\right)^{2/3} \mathrm{Ai}'^2\left(\left(\frac{3\omega}{4\omega_c}\right)^{2/3}(1+\gamma^2\psi^2)\right)\left(\frac{\omega}{\omega_c}\right)^2 \mathrm{d}(\omega/\omega_c)$$

$$I_2 = 9\gamma^4\psi^4 \int_0^\infty \left(\frac{3\omega}{4\omega_c}\right)^{4/3} \mathrm{Ai}^2\left(\left(\frac{3\omega}{4\omega_c}\right)^{2/3}(1+\gamma^2\psi^2)\right)\left(\frac{\omega}{\omega_c}\right)^2 \mathrm{d}(\omega/\omega_c).$$

The substitution

$$p = \frac{3\omega}{4\omega_c}, \qquad b = 1+\gamma^2\psi^2$$

brings the integrals into the forms (A.25) and (A.21) solved in Appendix A,

$$I_1 = 9\gamma^2\psi^2 \left(\frac{4}{3}\right)^3 \int_0^\infty p^{8/3}\,\mathrm{Ai}'^2\left(bp^{2/3}\right)\mathrm{d}p = \gamma^2\psi^2 \frac{5\cdot 7\cdot 23}{2^7\cdot 3\cdot b^{11/2}}$$

$$I_2 = 9\gamma^4\psi^4 \left(\frac{4}{3}\right)^3 \int_0^\infty p^{10/3}\,\mathrm{Ai}^2\left(bp^{2/3}\right)\mathrm{d}p = \gamma^4\psi^4 \frac{5\cdot 7\cdot 11}{2^7\cdot b^{13/2}}$$

with the sum

$$I_1 + I_2 = \frac{5\cdot 7\cdot 23}{2^7\cdot 3}\frac{\gamma^2\psi^2}{(1+\gamma^2\theta^2)^{11/2}}\left(1 + \frac{33}{23}\frac{\gamma^2\psi^2}{1+\gamma^2\theta^2}\right).$$

Integrating over the angle gives the variance:

$$\left\langle\left(\frac{\omega}{\omega_c}\gamma^2\psi^2\right)^2\right\rangle = \frac{5\cdot 7\cdot 23}{2^7\cdot 3}\int_{-\infty}^\infty \left(\frac{\gamma^2\psi^2}{(1+\gamma^2\theta^2)^{11/2}} + \frac{33}{23}\frac{\gamma^4\psi^4}{(1+\gamma^2\theta^2)^{13/2}}\right)\mathrm{d}(\gamma\psi) = \frac{8}{27}.$$

$$(5.24)$$

5.6 The polarization

5.6.1 The description of linear and circular polarization

As an introduction we start with a general description of linear and elliptical polarizations of synchrotron radiation. To make the underlying physics more transparent, we give the field in real notation. A more general treatment using complex functions can be found in many books on optics and electrodynamics, e.g. [9, 12].

We consider an electromagnetic wave with field components $[E_x,\ E_y]$, wave number k and frequency ω propagating in the z-direction,

$$\mathbf{E}(z,t) = [E_x\cos(kz - \omega t),\ E_y\cos(kz - \omega t - \varphi)]$$
$$= [\hat{E}_x\cos(kz - \omega t)\,\boldsymbol{\eta}_x + \hat{E}_y\cos(kz - \omega t - \varphi)\,\boldsymbol{\eta}_y],$$

where we introduced two unit vectors $\boldsymbol{\eta}_x$ and $\boldsymbol{\eta}_y$ in the x- and y-directions:

$$\boldsymbol{\eta}_x = (1, 0), \qquad \boldsymbol{\eta}_y = (0, 1), \qquad |\boldsymbol{\eta}_x|^2 = |\boldsymbol{\eta}_y|^2 = 1, \qquad (\boldsymbol{\eta}_x \cdot \boldsymbol{\eta}_y) = 0. \qquad (5.25)$$

The phase shift φ between the two field components determines the type of polarization. These components are $\pm\pi/2$ apart for synchrotron radiation emitted in dipole magnets and, as we will see later, also for the radiation from a helical undulator. On the other hand, we have $\varphi = 0$ for the radiation emitted in a plane undulator. We restrict ourself to these two cases.

We start with $\varphi = 0$, which gives a field with polarization

$$\mathbf{E}(z, t) = [\hat{E}_x \cos(kz - \omega t)\,\boldsymbol{\eta}_x + \hat{E}_y \cos(kz - \omega t)\,\boldsymbol{\eta}_y]$$

oscillating in a plane at an angle of $\arctan(\hat{E}_y/\hat{E}_x)$ relative to the median plane. At a fixed location $z = r_p$ we can express the field as a function of the time $t_p = t - r_p/c$:

$$\mathbf{E}(t_p) = [\hat{E}_x \cos(\omega t_p)\,\boldsymbol{\eta}_x + \hat{E}_y \cos(\omega t_p)\,\boldsymbol{\eta}_y].$$

The power radiated by this wave, averaged over one oscillation period, is of interest. From (3.2) we obtain for the average angular power density

$$\left\langle \frac{dP}{d\Omega} \right\rangle = \frac{r^2(1 - \mathbf{n} \cdot \boldsymbol{\beta})}{\mu_0 c} \langle |\mathbf{E}|^2 \rangle = \frac{r^2(1 - \mathbf{n} \cdot \boldsymbol{\beta})}{2\mu_0 c} [\hat{E}_x^2 + \hat{E}_y^2].$$

Next we choose $\varphi = \pm\pi/2$ and obtain elliptical polarization:

$$\mathbf{E}(z, t) = [\hat{E}_x \cos(kz - \omega t)\,\boldsymbol{\eta}_x \pm \hat{E}_y \sin(kz - \omega t)\,\boldsymbol{\eta}_y].$$

The field vector rotates around the z-axis as a function of z or t. Its head describes an ellipse with main axes in the $\boldsymbol{\eta}_x$- and $\boldsymbol{\eta}_y$-directions as a function of $kz - \omega t$.

To illustrate this we consider first the case $\hat{E}_x = \hat{E}_y = E_0$, giving circular polarization:

$$\mathbf{E}(z, t) = E_0[\cos(kz - \omega t)\,\boldsymbol{\eta}_x \pm \sin(kz - \omega t)\,\boldsymbol{\eta}_y].$$

We take first the minus sign and set $t = 0$, giving the field as a function of position z:

$$\mathbf{E}(z) = E_0[\cos(kz)\,\boldsymbol{\eta}_x - \sin(kz)\,\boldsymbol{\eta}_y].$$

With increasing longitudinal coordinate z the head of the field vector describes a left-handed circular helix (anti-corkscrew) with period $\lambda = 2\pi/k$. On setting $z = 0$, we obtain the field at a fixed location as a function of the time t:

$$\mathbf{E}(t) = E_0[\cos(\omega t)\,\boldsymbol{\eta}_x + \sin(\omega t)\,\boldsymbol{\eta}_y].$$

For an observer looking towards the source the head of the field vector rotates anti-clockwise as a function of time with frequency ω. We call the light left-handed circularly polarized. However, since, on looking in the direction of propagation of the wave, the electric-field vector rotates like a corkscrew having positive helicity, we label it with $\mathbf{E}_+$. For the opposite sign, the wave has right-handed circular polarization and a negative helicity. We have therefore the field for the two modes of polarization as a function of time:

$$\mathbf{E}_+ = E_0[\cos(\omega t)\,\boldsymbol{\eta}_x + \sin(\omega t)\,\boldsymbol{\eta}_y], \qquad \mathbf{E}_- = E_0[\cos(\omega t)\,\boldsymbol{\eta}_x - \sin(\omega t)\,\boldsymbol{\eta}_y]. \qquad (5.26)$$

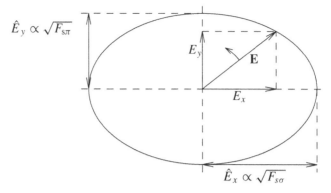

Fig. 5.10. Elliptical polarization at $\omega = \omega_c$, $\psi = 1/\gamma$.

We concentrate on observing the field as a function of time at a fixed location, choosing still $\varphi = \pm\pi/2$ but allowing different values for the component amplitudes $\hat{E}_x$ and $\hat{E}_y$:

$$\mathbf{E}(t) = [\hat{E}_x \cos(\omega t)\,\boldsymbol{\eta}_x \pm \hat{E}_y \sin(\omega t)\,\boldsymbol{\eta}_y].$$

The electric-field vector rotates with frequency ω and positive or negative helicity while its head describes an upright ellipse with main axes $\hat{E}_x$ and $\hat{E}_y$ as indicated in Fig. 5.10. To give a quantitative description of the elliptical polarization we split the field into the two circular polarization helicities with amplitudes $\hat{E}_+$ and $\hat{E}_-$:

$$\begin{aligned}
\mathbf{E}(t) &= \mathbf{E}_+(t) + \mathbf{E}_-(t) \\
&= \hat{E}_+ \frac{\cos(\omega t)\,\boldsymbol{\eta}_x + \sin(\omega t)\,\boldsymbol{\eta}_y}{\sqrt{2}} + \hat{E}_- \frac{\cos(\omega t)\,\boldsymbol{\eta}_x - \sin(\omega t)\,\boldsymbol{\eta}_y}{\sqrt{2}}
\end{aligned} \qquad (5.27)$$

with

$$\hat{E}_+ = \frac{\hat{E}_x + \hat{E}_y}{\sqrt{2}}, \qquad \hat{E}_- = \frac{\hat{E}_x - \hat{E}_y}{\sqrt{2}}. \qquad (5.28)$$

This decomposition of the field into two modes of circular polarization of opposite helicities is illustrated in Fig. 5.11.

We described the linear polarization in terms of two oscillating vectors, $\cos(\omega t)\,\boldsymbol{\eta}_x$ and $\sin(\omega t)\,\boldsymbol{\eta}_y$, each with RMS value $1/\sqrt{2}$, which are (in terms of the time average) orthogonal. We now describe the elliptical polarization with the two rotating vectors

$$\frac{\cos(\omega t)\,\boldsymbol{\eta}_x + \sin(\omega t)\,\boldsymbol{\eta}_y}{\sqrt{2}} \quad \text{and} \quad \frac{\cos(\omega t)\,\boldsymbol{\eta}_x - \sin(\omega t)\,\boldsymbol{\eta}_y}{\sqrt{2}},$$

which have also RMS values of $1/\sqrt{2}$ and are (in terms of the time average) orthogonal to each other.

The electric vector $\mathbf{E}(t)$ is the same in both cases but it is decomposed into different components. Therefore we can also present the angular power distribution (3.2) of the

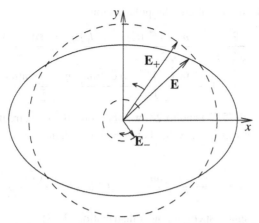

Fig. 5.11. The elliptical polarization as a superposition of two circular modes having opposite helicities for $\omega = \omega_c$, $\psi = 1/\gamma$.

radiation as a sum of either two linear or two circular polarization components:

$$\left\langle \frac{dP}{d\Omega} \right\rangle = \frac{r^2(1 - \mathbf{n} \cdot \boldsymbol{\beta})}{\mu_0 c} \langle |\mathbf{E}|^2 \rangle = \frac{r^2(1 - \mathbf{n} \cdot \boldsymbol{\beta})}{\mu_0 c} [|\mathbf{E}_+|^2 + |\mathbf{E}_-|^2]$$

$$= \frac{r^2(1 - \mathbf{n} \cdot \boldsymbol{\beta})}{2\mu_0 c} \left[\left(\frac{\hat{E}_x + \hat{E}_y}{\sqrt{2}} \right)^2 + \left(\frac{\hat{E}_x - \hat{E}_y)^2}{\sqrt{2}} \right)^2 \right] \qquad (5.29)$$

$$= \frac{r^2(1 - \mathbf{n} \cdot \boldsymbol{\beta})}{2\mu_0 c} [\hat{E}_x^2 + \hat{E}_y^2] = \frac{r^2(1 - \mathbf{n} \cdot \boldsymbol{\beta})}{\mu_0 c} [\langle E_x^2 \rangle + \langle E_y^2 \rangle].$$

We explained the elliptical polarization using a monochromatic wave. However, in most cases the radiation field is described by a continuous spectrum:

$$\tilde{\mathbf{E}}(\omega) = [\tilde{E}_x(\omega), \ \tilde{E}_y(\omega)].$$

Since this is the Fourier transform of a real function $\mathbf{E}(t)$, we have the symmetry relations

$$\tilde{E}_x(-\omega) = \tilde{E}_x(\omega), \qquad \tilde{E}_y(-\omega) = -\tilde{E}_y(\omega). \qquad (5.30)$$

The time-domain presentation is obtained with the inverse Fourier transform:

$$\mathbf{E}(t_p) = \frac{1}{\sqrt{2\pi}} \int_{-\infty}^{\infty} \tilde{\mathbf{E}}(\omega) e^{i\omega t_p} \, d\omega = \frac{1}{\sqrt{2\pi}} \int_{-\infty}^{\infty} [\tilde{E}_x(\omega)\boldsymbol{\eta}_x + \tilde{E}_y(\omega)\boldsymbol{\eta}_y] e^{i\omega t_p} \, d\omega.$$

We filter a narrow frequency band at ω of width $d\omega$. Since the above integration covers positive and negative frequencies, we obtain a contribution at ω and one at $-\omega$. Considering the above symmetry conditions (5.30), we combine the two exponentials to give trigonometric functions and use positive frequencies only:

$$d\mathbf{E}(t_p) = \frac{2}{\sqrt{2\pi}} [\tilde{E}_x(\omega) \cos(\omega t_p) \boldsymbol{\eta}_x + i\tilde{E}_y(\omega) \sin(\omega t_p) \boldsymbol{\eta}_y)] \, d\omega. \qquad (5.31)$$

We split this into two modes of circular polarization:

$$dE(t_p) = \frac{2}{\sqrt{2\pi}} \left[\frac{\tilde{E}_x(\omega) + i\tilde{E}_y(\omega)}{\sqrt{2}} \frac{\cos(\omega t_p)\,\boldsymbol{\eta}_x + i\sin(\omega t_p)\,\boldsymbol{\eta}_y}{\sqrt{2}} \right.$$
$$\left. + \frac{\tilde{E}_x(\omega) - i\tilde{E}_y(\omega)}{\sqrt{2}} \frac{\cos(\omega t_p)\,\boldsymbol{\eta}_x - i\sin(\omega t_p)\,\boldsymbol{\eta}_y}{\sqrt{2}} \right] d\omega.$$

On comparing this with (5.27) and (5.28) we obtain for the two circular-polarization components of the Fourier-transformed field at $\omega > 0$ (note that $\tilde{E}_y$ changes sign for negative frequencies)

$$\tilde{E}_+(\omega) = \frac{\tilde{E}_x(\omega) + i\tilde{E}_y(\omega)}{\sqrt{2}}, \qquad \tilde{E}_-(\omega) = \frac{\tilde{E}_x(\omega) - i\tilde{E}_y(\omega)}{\sqrt{2}}, \qquad (5.32)$$

which gives for the angular spectral power distribution (3.18)

$$\frac{d^2 P}{d\Omega\,d\omega} = \frac{2r^2\omega_0}{2\pi\mu_0 c}[|\tilde{E}_+|^2 + |\tilde{E}_-|^2] = \frac{2r^2\omega_0}{2\pi\mu_0 c}[|\tilde{E}_x|^2 + |\tilde{E}_y|^2]. \qquad (5.33)$$

5.6.2 The linear polarization

The linear polarization of the angular spectral distribution. In the previous discussions we divided the spectral angular power density into two parts. One originates from the horizontal component E_x of the electric field and the other from its vertical component E_y. These parts represent the two modes of linear polarization, which we call the σ-mode and the π-mode. In Fig. 5.3 the normalized spectral angular power densities $F_{s\sigma}$ and $F_{s\pi}$ for these two modes of polarization are plotted against the vertical angle ψ for three different frequencies, $\omega = 0.2\omega_c$, ω_c, and $2\omega_c$ and in Fig. 5.4 they are plotted also for small frequencies. In the median plane the horizontal polarization has a maximum and the vertical one vanishes. With increasing vertical angle ψ the horizontal component decreases monotonically while the vertical component first increases, then reaches a maximum, and thereafter decreases, always being smaller than the horizontal component. This behavior is to be expected from the qualitative picture shown in Fig. 1.6.

So far we have discussed the values of the two components of the polarization. For many experiments the degree of the horizontal polarization is of importance, [43]. It is defined as the difference between the two power components divided by their sum:

$$\frac{\dfrac{d^2 P_\sigma}{d\Omega\,d\omega} - \dfrac{d^2 P_\pi}{d\Omega\,d\omega}}{\dfrac{d^2 P_\sigma}{d\Omega\,d\omega} + \dfrac{d^2 P_\pi}{d\Omega\,d\omega}} = \frac{F_{s\sigma} - F_{s\pi}}{F_{s\sigma} + F_{s\pi}}.$$

This quantity is plotted against the vertical angle ψ in Fig. 5.12 for $\omega = \omega_c$ and in Fig. 5.13 for $\omega \ll \omega_c$. It starts with unity in the median plane where there is perfect horizontal polarization and decreases monotonically with increasing angle ψ. For some applications this degree of polarization has to be weighted by the power radiated at a given angle, which

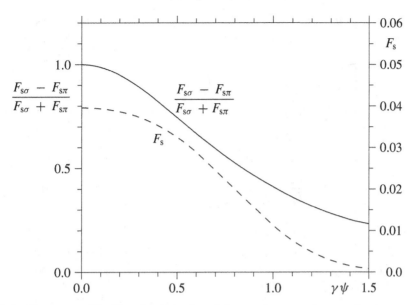

Fig. 5.12. The degree of horizontal polarization of the angular spectral power density versus the vertical angle ψ at $\omega = \omega_c$.

is shown as its normalized value F_s in the figures. At the higher frequency $\omega = \omega_c$, shown in Fig. 5.12, the degree of polarization is relatively large over most of the angular distribution, whereas for the lower frequencies, shown in Fig. 5.13, it is already much reduced where F_s reaches its maximum.

The linear polarization of the spectrum. We can integrate the spectral angular distribution over the angle ψ to obtain the spectral power densities with their normalized polarization components $S_{s\sigma}$ and $S_{s\pi}$. They are shown in Figs. 5.5 and 5.6 in logarithmic and linear representations. At low frequencies the horizontal polarization component of this angle-integrated radiation reaches three quarters of the total. This fraction increases with increasing frequency to reach an asymptotic value of unity. This is clearly shown in Fig. 5.14, where the degree of polarization

$$\frac{\dfrac{dP_\sigma}{d\omega} - \dfrac{dP_\pi}{d\omega}}{\dfrac{dP_\sigma}{d\omega} + \dfrac{dP_\pi}{d\omega}} = \frac{S_{s\sigma} - S_{s\pi}}{S_{s\sigma} + S_{s\pi}}$$

is plotted against frequency. Integrating the polarization components over the vertical angle ψ has some practical applications. In many experiments the vertical angular distribution is not resolved, either because one uses a large sample, which accepts the full angular spread of the radiation, or because the electrons emitting the radiation have themselves an angular spread that smears out the natural angular distribution of the radiation. In both cases the angle-integrated polarization is relevant for polarization-dependent phenomena.

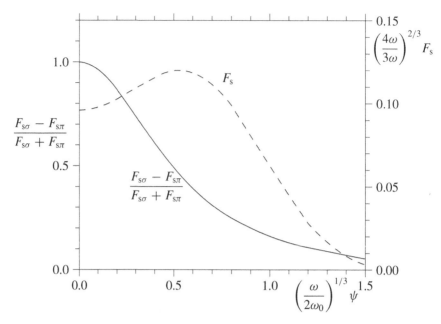

Fig. 5.13. The degree of horizontal polarization of the angular spectral power density versus the vertical angle ψ at low frequencies $\omega \ll \omega_c$.

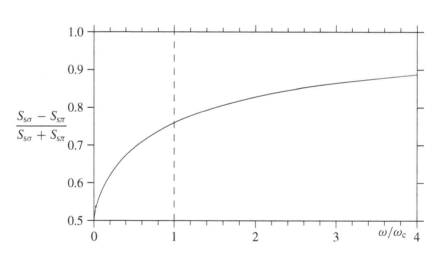

Fig. 5.14. The degree of horizontal polarization of the spectral power density versus frequency.

The frequency-integrated polarization. Polarization-dependent effects usually occur at a well-determined frequency. Integrating the polarization components over frequency is therefore of little practical interest but it can still reveal some properties of the radiation.

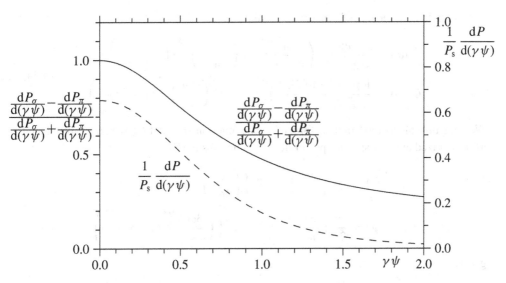

Fig. 5.15. Horizontal polarization degree after integrating over ω.

In Fig. 5.9 the two polarization components of the frequency-integrated radiation are plotted against the vertical angle ψ. The resulting degree of the polarization is given by the equation

$$\frac{\dfrac{\mathrm{d}P_\sigma}{\mathrm{d}(\gamma\psi)} - \dfrac{\mathrm{d}P_\pi}{\mathrm{d}(\gamma\psi)}}{\dfrac{\mathrm{d}P_\sigma}{\mathrm{d}(\gamma\psi)} + \dfrac{\mathrm{d}P_\pi}{\mathrm{d}(\gamma\psi)}} = \frac{1 + 2\gamma^2\psi^2/7}{1 + 12\gamma^2\psi^2/7}$$

and is plotted in Fig. 5.15. As expected, it starts with unity in the median plane, where the vertical polarization component vanishes at all frequencies. For very large angles it approaches the value $\frac{1}{6}$.

The polarization of the total radiation. We integrate the spectral angular power distribution over all frequencies and obtain the fractions radiated into the two modes of polarization, which we discussed before, and the degree of horizontal polarization of the total radiation:

$$P_\sigma = \frac{7}{8}P_s, \qquad P_\pi = \frac{1}{8}P_s, \qquad \frac{P_\sigma - P_\pi}{P_\sigma + P_\pi} = \frac{3}{4}.$$

5.6.3 The elliptical polarization

To understand the elliptical or circular polarization we have to go back to the radiation fields treated in Chapter 4, where we obtained the expression (4.20) for the Fourier-transformed electric-field components of synchrotron radiation:

$$\tilde{E}_x(\omega) = \frac{e\gamma}{\sqrt{2\pi}\,\epsilon_0 c r_p}\left(\frac{3|\omega|}{4\omega_c}\right)^{1/3}\text{Ai}'\left(\left(\frac{3\omega}{4\omega_c}\right)^{2/3}(1+\gamma^2\psi^2)\right)$$
$$\tilde{E}_y(\omega) = \frac{\mathrm{i}e\gamma(\omega/|\omega|)}{\sqrt{2\pi}\,\epsilon_0 c r_p}\left(\frac{3|\omega|}{4\omega_c}\right)^{2/3}\gamma\psi\,\text{Ai}\left(\left(\frac{3\omega}{4\omega_c}\right)^{2/3}(1+\gamma^2\psi^2)\right).$$

(5.34)

We bring this into a more compact form using the functions $F_{s\sigma}$ and $F_{s\pi}$ which we introduced as normalized angular spectral power density functions in (5.8),

$$F_{s\sigma}(\omega,\psi) = \frac{9}{2\pi}\left(\frac{3\omega}{4\omega_c}\right)^{2/3}\text{Ai}'^2\left(\left(\frac{3\omega}{4\omega_c}\right)^{2/3}(1+\gamma^2\psi^2)\right)$$
$$F_{s\pi}(\omega,\psi) = \frac{9}{2\pi}\left(\frac{3\omega}{4\omega_c}\right)^{4/3}\gamma^2\psi^2\,\text{Ai}^2\left(\left(\frac{3\omega}{4\omega_c}\right)^{2/3}(1+\gamma^2\psi^2)\right),$$

giving

$$\tilde{E}_x(\omega) = -\frac{e\gamma}{3\epsilon_0 c r_p}\sqrt{F_{s\sigma}}, \qquad \tilde{E}_y(\omega) = \mathrm{i}\frac{e\gamma}{3\epsilon_0 c r_p}\frac{\omega}{|\omega|}\frac{\psi}{|\psi|}\sqrt{F_{s\pi}}.$$

(5.35)

Since we take here the positive sign of the square roots $\sqrt{F_{s\sigma}}$ and $\sqrt{F_{s\pi}}$, we have to express explicitly the sign of the field as given in (5.34). The derivative Ai$'$ is negative, resulting in a minus sign for $\tilde{E}_x(\omega)$. The vertical field $\tilde{E}_y(\omega)$ changes its sign with frequency ω and with vertical angle ψ.

According to (5.32), we can give the field components in the frequency domain for the two modes of circular polarization:

$$\tilde{E}_+(\omega) = \frac{\tilde{E}_x(\omega)+\mathrm{i}\tilde{E}_y(\omega)}{\sqrt{2}} = -\frac{e\gamma}{3\sqrt{2}\epsilon_0 c r_p}\left(\sqrt{F_{s\sigma}}+\frac{\omega}{|\omega|}\frac{\psi}{|\psi|}\sqrt{F_{s\pi}}\right)$$
$$\tilde{E}_-(\omega) = \frac{\tilde{E}_x(\omega)-\mathrm{i}\tilde{E}_y(\omega)}{\sqrt{2}} = -\frac{e\gamma}{3\sqrt{2}\epsilon_0 c r_p}\left(\sqrt{F_{s\sigma}}-\frac{\omega}{|\omega|}\frac{\psi}{|\psi|}\sqrt{F_{s\pi}}\right).$$

With (3.18) and (5.33) we obtain the angular spectral power distribution:

$$\frac{\mathrm{d}^2 P_+}{\mathrm{d}\Omega\,\mathrm{d}\omega} = \frac{P_s\gamma}{\omega_c}\frac{F_{s\sigma}+F_{s\pi}+2(\psi/|\psi|)\sqrt{F_{s\sigma}F_{s\pi}}}{2} = \frac{P_s\gamma}{\omega_c}F_{s+}$$
$$\frac{\mathrm{d}^2 P_-}{\mathrm{d}\Omega\,\mathrm{d}\omega} = \frac{P_s\gamma}{\omega_c}\frac{F_{s\sigma}+F_{s\pi}-2(\psi/|\psi|)\sqrt{F_{s\sigma}F_{s\pi}}}{2} = \frac{P_s\gamma}{\omega_c}F_{s-}.$$

The two normalized spectral angular power densities F_{s+} and F_{s-} for the two modes of circular polarization can be expressed as combinations of the corresponding quantities $F_{s\sigma}$ and $F_{s\pi}$ of the linear polarization:

$$F_{s+} = \frac{1}{2}\left(F_{s\sigma}+F_{s\pi}+2\frac{\psi}{|\psi|}\sqrt{F_{s\sigma}F_{s\pi}}\right)$$
$$F_{s-} = \frac{1}{2}\left(F_{s\sigma}+F_{s\pi}-2\frac{\psi}{|\psi|}\sqrt{F_{s\sigma}F_{s\pi}}\right).$$

(5.36)

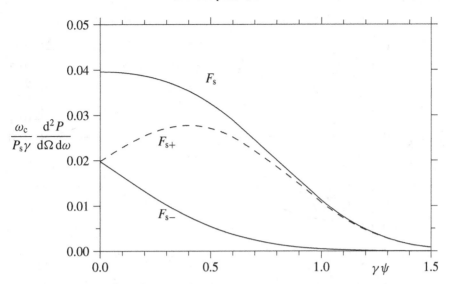

Fig. 5.16. The normalized angular spectral power distributions for the two modes of circular polar-ization and the total radiation versus the vertical angle ψ at $\omega = \omega_c$.

The latter, $F_{s\sigma}$ and $F_{s\pi}$, are expressed by using Airy functions (5.8) or by using modified Bessel functions (5.9) as explained earlier. For the angular spectral power density the sum of the two modes of circular polarization is the same as that of the two linear modes:

$$F_{s+} + F_{s-} = F_{s\sigma} + F_{s\pi} = F_s.$$

In analogy with the linear case we define the degree of circular polarization as

$$\frac{\dfrac{d^2 P_+}{d\Omega\,d\omega} - \dfrac{d^2 P_-}{d\Omega\,d\omega}}{\dfrac{d^2 P_+}{d\Omega\,d\omega} + \dfrac{d^2 P_-}{d\Omega\,d\omega}} = \frac{F_{s+} - F_{s-}}{F_{s+} + F_{s-}} = 2\frac{\psi}{|\psi|}\frac{\sqrt{F_{s\sigma} F_{s\pi}}}{F_{s\sigma} + F_{s\pi}}. \tag{5.37}$$

To illustrate the circular polarization of synchrotron radiation, the two normalized angular spectral power densities F_{s+} and F_{s-} of the two modes, together with the value of the total radiation as well as the degree of circular polarization, are plotted in the next few figures as functions of the vertical angle ψ. First, Figs. 5.16 and 5.17 show these quantities at the critical frequency whereas Figs. 5.18 and 5.19 show the lower part of the spectrum. In the median plane, $\psi = 0$, the two circular modes are equal and the degree of elliptical polarization vanishes. With increasing angle ψ the two modes become more different and the degree of polarization grows monotonically. However, the total angular spectral power density decreases at large angle. In the case of small frequencies this happens only at angles for which the degree of polarization is already large. At low frequencies and very large angles the degree of polarization approaches unity and the radiation becomes circularly polarized. For negative angles the elliptical polarization reverses its sign.

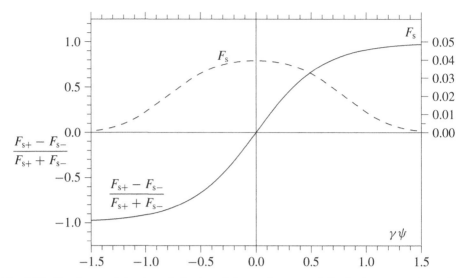

Fig. 5.17. The degree of circular polarization of the angular spectral distribution versus the vertical angle ψ at $\omega = \omega_c$.

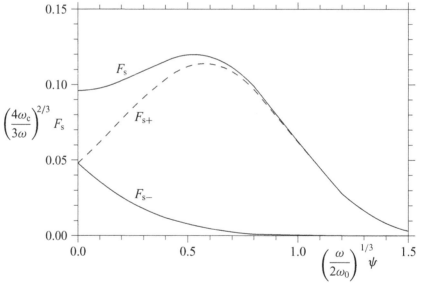

Fig. 5.18. Normalized angular spectral power densities for the two circular polarization modes and the total radiation versus vertical angle ψ at $\omega \ll \omega_c$.

The polarization is a well-determined quantity at a given angle ψ and frequency ω and should be expressed as such. However, it is sometimes useful to integrate the modes of polarization over one of these variables in order to gain some knowledge of the properties of the synchrotron radiation. We have done that already for the linear polarization components and we will do it now for the circular ones.

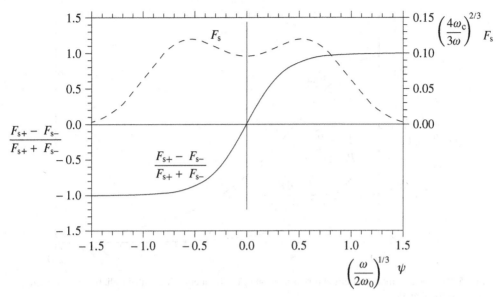

Fig. 5.19. The degree of circular polarization of the angular spectral distribution versus the vertical angle ψ at $\omega \ll \omega_c$.

Since the degree of circular polarization (5.37) is antisymmetric with respect to ψ, there should be no elliptical polarization after integrating it over this angle.

We integrate now over frequency ω to obtain the angular distribution,

$$\frac{dP}{d\Omega} = \frac{d^2 P}{d\psi \, d\phi} = P_s \frac{\gamma}{\omega_c} \int_0^\infty F_s(\omega\psi) \, d\omega,$$

which gives for the two components of the circular polarization

$$\frac{dP_+}{d(\gamma\psi)} = P_s \frac{\pi}{\omega_c} \int_0^\infty \left(F_s + 2\frac{\psi}{|\psi|} \sqrt{F_{s\sigma} F_{s\pi}} \right) d\omega$$

$$\frac{dP_-}{d(\gamma\psi)} = P_s \frac{\pi}{\omega_c} \int_0^\infty \left(F_s - 2\frac{\psi}{|\psi|} \sqrt{F_{s\sigma} F_{s\pi}} \right) d\omega.$$

This involves an integral,

$$\int_0^\infty \sqrt{F_{s\sigma} F_{s\pi}} \, d\omega$$

$$= -\frac{9\gamma\psi}{2\pi} \int_0^\infty \frac{3\omega}{4\omega_c} \text{Ai}\left(\left(\frac{3\omega}{4\omega_c}\right)^{2/3} (1+\gamma^2\psi^2) \right) \text{Ai}'\left(\left(\frac{3\omega}{4\omega_c}\right)^{2/3} (1+\gamma^2\psi^2) \right) d\omega$$

$$= \frac{\omega_c \sqrt{3}}{2\pi} \frac{\gamma\psi}{\pi} \frac{\gamma\psi}{(1+\gamma^2\psi^2)^3}, \tag{5.38}$$

which is obtained from (A.34) in Appendix A using

$$p = \frac{3\omega}{4\omega_c}, \qquad b = 1 + \gamma^2\psi^2.$$

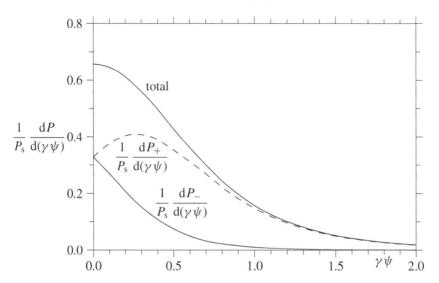

Fig. 5.20. The two modes of circular polarization of the frequency-integrated radiation as a function of the vertical angle.

We obtain for the angular distribution of the frequency-integrated modes of circular polarization

$$\frac{dP_+}{d(\gamma\psi)} = P_s \frac{21}{64} \frac{1}{(1+\gamma^2\psi^2)^{5/2}} \left(1 + \frac{5}{7}\frac{\gamma^2\psi^2}{1+\gamma^2\psi^2} + \frac{64\sqrt{3}}{21\pi}\frac{\gamma\psi}{\sqrt{1+\gamma^2\psi^2}}\right)$$

$$\frac{dP_-}{d(\gamma\psi)} = P_s \frac{21}{64} \frac{1}{(1+\gamma^2\psi^2)^{5/2}} \left(1 + \frac{5}{7}\frac{\gamma^2\psi^2}{1+\gamma^2\psi^2} - \frac{64\sqrt{3}}{21\pi}\frac{\gamma\psi}{\sqrt{1+\gamma^2\psi^2}}\right).$$

They are plotted in Fig. 5.20 as functions of $\gamma\psi$.

5.7 The photon distribution

So far we have treated synchrotron radiation as electromagnetic fields and corresponding power distributions. However, radiation is emitted in the form of quanta, called photons, each having an energy

$$E_\gamma = \hbar\omega \quad \text{with} \quad \hbar = h/(2\pi), \tag{5.39}$$

where ω is the frequency of the electromagnetic wave and $h = 6.6262 \times 10^{-34}$ J s is Planck's constant. We will now introduce an *ad hoc* quantization by taking the classical power distribution in terms of frequency and calculate the photon distribution using the above relation between energy and frequency. One could ask whether this procedure is correct or whether a complete quantum-mechanical calculation of the radiation is necessary. An electron circulating in a storage ring can be regarded as a macroscopic atom. Obviously it is in an energy state having a very large quantum number, in which case the classical treatment is a very good approximation. However, a complete quantum-mechanical treatment of synchrotron radiation can be found in several publications [4, 7, 36].

We investigated before the spectral power density of the radiation which is the power element $dP(\omega)$ of frequency ω emitted into a frequency interval of width $d\omega$ given by (5.15):

$$\frac{dP}{d\omega} = \frac{P_s}{\omega_c} S_s\left(\frac{\omega}{\omega_c}\right). \tag{5.40}$$

We convert this into an energy distribution of the photon flux, which is the number of photons of energy E_γ emitted per second, $\dot{n}(E_\gamma)$, into an energy band of width dE_γ:

$$dP = E_\gamma \, d\dot{n} = E_\gamma \frac{d\dot{n}}{dE_\gamma} dE_\gamma, \qquad \frac{dP}{d\omega} = E_\gamma \frac{d\dot{n}}{dE_\gamma/\hbar} = \hbar \frac{d\dot{n}}{dE_\gamma/E_\gamma}. \tag{5.41}$$

We introduce the critical photon energy

$$E_{\gamma c} = \hbar\omega_c = \frac{3ch\gamma^3}{4\pi|\rho|} = \frac{3m_0c^2\lambda_{\text{Comp}}\gamma^3}{4\pi|\rho|}, \tag{5.42}$$

where

$$\lambda_{\text{Comp}} = h/(m_0c) = 2.426 \times 10^{-12}\,\text{m} \tag{5.43}$$

is the Compton wavelength. Since $E_{\gamma c} > 0$ is positive while the curvature $1/\rho$ can have positive and negative signs, we take its absolute value.

With (5.41) we obtain the photon flux per *relative* energy dE_γ/E_γ or frequency band $d\omega/\omega$, which is directly related to the frequency distribution of the radiated power:

$$\frac{d\dot{n}}{dE_\gamma/E_\gamma} = \frac{d\dot{n}}{d\omega/\omega} = \frac{1}{\hbar}\frac{dP}{d\omega} = \frac{P_s}{E_{\gamma c}}[S_{s\sigma}(E_\gamma) + S_{s\pi}(E_\gamma)]. \tag{5.44}$$

The functions $S_{s\sigma}$ and $S_{s\pi}$, given in (5.16), can be expressed as functions of $E_\gamma/E_{\gamma c}$ instead of ω/ω_c to describe this photon distribution, which now has the same form as the corresponding power distribution with respect to ω shown in Fig. 5.5.

By integrating over the distribution (5.44) we obtain the total number of photons $\dot{n}_s$ radiated per unit time:

$$\dot{n}_s = \frac{P_s}{E_{\gamma c}} \int_0^\infty \frac{S_s(E_\gamma)}{E_\gamma/E_{\gamma c}} \, d(E_\gamma/E_{\gamma c}) = \frac{P_s}{E_{\gamma c}} \frac{15\sqrt{3}}{8}.$$

This integral is obtained from (A.30) and (A.31) given in Appendix A. The average photon energy is simply given by

$$\langle E_\gamma \rangle = \frac{P_s}{\dot{n}_s} = \frac{8\sqrt{3}}{45} E_{\gamma c}.$$

The variance of the photon energy $\langle E_\gamma^2 \rangle$ is of special interest for later applications:

$$\langle E_\gamma^2 \rangle = \frac{1}{\dot{n}_s} \frac{P_s}{E_{\gamma c}} \int_0^\infty E_\gamma^2 \frac{S_s(E_\gamma)}{E_\gamma / E_{\gamma c}} \, \mathrm{d}(E_\gamma / E_{\gamma c}) = \frac{22}{54} E_{\gamma c}^2.$$

This integral is also solved in Appendix A, (A.32) and (A.33).

We summarize the relevant quantities of the emitted photons and give them also for the individual modes of polarization:

$$
\begin{array}{lll}
\dot{n}_\sigma = \dfrac{12\sqrt{3}}{8} \dfrac{P_s}{E_{\gamma c}}, & \dot{n}_\pi = \dfrac{3\sqrt{3}}{8} \dfrac{P_s}{E_{\gamma c}}, & \dot{n}_s = \dfrac{15\sqrt{3}}{8} \dfrac{P_s}{E_{\gamma c}} \\[2ex]
\langle E_{\gamma\sigma} \rangle = \dfrac{35\sqrt{3}}{180} E_{\gamma c}, & \langle E_{\gamma\pi} \rangle = \dfrac{20\sqrt{3}}{180} E_{\gamma c}, & \langle E_\gamma \rangle = \dfrac{32\sqrt{3}}{180} E_{\gamma c} \\[2ex]
\langle E_{\gamma\sigma}^2 \rangle = \dfrac{25}{54} E_{\gamma c}^2, & \langle E_{\gamma\pi}^2 \rangle = \dfrac{10}{54} E_{\gamma c}^2, & \langle E_\gamma^2 \rangle = \dfrac{22}{54} E_{\gamma c}^2.
\end{array}
\tag{5.45}
$$

The numbers of emitted photons given in the top line contain the ratio between the total power P_s and the critical photon energy $E_{\gamma c}$ given in (5.1) and (5.42):

$$\frac{P_s}{E_{\gamma c}} = \frac{8\pi r_0 m_0 c^2 \gamma}{9 h \rho} = \frac{2 e^2 \gamma}{9 \epsilon_0 h \rho} = \frac{4 \omega_0 \gamma \alpha_f}{9}, \tag{5.46}$$

where we used the angular velocity $\omega_0 = c/\rho$ and introduced the fine-structure constant

$$\alpha_f = \frac{e^2}{2 \epsilon_0 c h} = \frac{1}{137.036} = 0.007\,297. \tag{5.47}$$

With this we obtain for the total number of photons $\dot{n}_s$ radiated per second or n_s per revolution

$$\dot{n}_s = \frac{5\sqrt{3}}{6} \alpha_f \omega_0 \gamma, \qquad n_s = \frac{5\pi\sqrt{3}}{3} \alpha_f \gamma = 0.0662 \gamma. \tag{5.48}$$

In cases in which the ring has field-free straight sections the angular frequency ω_0 is replaced by the revolution frequency $\omega_{\mathrm{rev}} < \omega_0$ to obtain the average number of photons per second. On the other hand, the number of photons emitted per revolution stays the same. In a storage ring operating at 1 GeV energy about 132 photons are emitted by each electron per revolution.

With the total number of radiated photons we rewrite the distribution of the photons per relative energy increment $\mathrm{d}E_\gamma / E_\gamma$. For some applications the distribution with respect to an absolute energy band $\mathrm{d}E_\gamma$ is more important. We give both distributions below:

$$\frac{\mathrm{d}\dot{n}}{\mathrm{d}E_\gamma / E_\gamma} = \frac{P_s}{E_{\gamma c}} S_s(E_\gamma) = \frac{8\sqrt{3}\dot{n}_s}{45} [S_{s\sigma}(E_\gamma) + S_{s\pi}(E_\gamma)] = \frac{4\alpha_f \omega_0 \gamma \, S_s(\omega)}{9}$$

$$\frac{\mathrm{d}\dot{n}}{\mathrm{d}E_\gamma} = \frac{P_s}{E_{\gamma c}^2} \frac{S_s(E_\gamma)}{E_\gamma / E_{\gamma c}} = \frac{8\sqrt{3}\dot{n}_s}{45 E_{\gamma c}} \frac{[S_{s\sigma}(E_\gamma) + S_{s\pi}(E_\gamma)]}{E_\gamma / E_{\gamma c}} = \frac{4\alpha_f \omega_0 \gamma \, S_s(\omega)}{9 E_{\gamma c} \omega / \omega_c}.$$

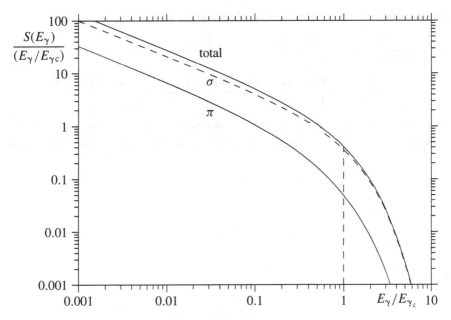

Fig. 5.21. The normalized photon spectrum.

As mentioned before, the first distribution has the same form as the spectral power distribution shown in Fig. 5.5. The second distribution is plotted in Fig. 5.21, which shows that there is a large enhancement at low photon energies compared with the distribution in terms of frequency.

For experiments in which the full vertical extension of the radiation is accepted, the number of photons radiated per horizontal angle element $d\phi$ and relative frequency band $d\omega/\omega$ is of interest. For one electron we have

$$\frac{d^2n}{d\phi\,d\omega/\omega} = \frac{4\alpha_f\gamma\,S_s(\omega/\omega_c)}{9}$$

and, for a current I representing I/e electrons passing per time unit through a point on the orbit, the resulting photon flux per angle and relative frequency band is

$$\frac{d^2\dot{n}}{d\phi\,d\omega/\omega} = \frac{4\alpha_f\gamma\,I\,S_s(\omega/\omega_c)}{9e}. \tag{5.49}$$

We can also give the angular energy distribution of the photon flux from the angular spectral power distribution (5.7) using the relations (5.41),

$$\frac{d^2\dot{n}}{d\Omega\,dE_\gamma/E_\gamma} = \frac{d^2P}{d\Omega\,\hbar\,d\omega}, \tag{5.50}$$

giving this distribution per relative or absolute energy increment,

$$\frac{\mathrm{d}^2\dot{n}}{\mathrm{d}\Omega\,\mathrm{d}E_\gamma/E_\gamma} = \frac{P_s\gamma}{E_{\gamma c}}F_s(E_\gamma,\psi) = \frac{8\sqrt{3}\dot{n}_s\gamma}{45}[F_{s\sigma}(E_\gamma,\psi)+F_{s\pi}(E_\gamma,\psi)]$$

$$\frac{\mathrm{d}^2\dot{n}}{\mathrm{d}\Omega\,\mathrm{d}E_\gamma} = \frac{P_s\gamma}{E_{\gamma c}^2}\frac{F_s(E_\gamma,\psi)}{E_\gamma/E_{\gamma c}} = \frac{8\sqrt{3}\dot{n}_s}{45E_{\gamma c}}\frac{[F_\sigma(E_\gamma,\psi)+F_\pi(E_\gamma,\psi)]}{E_\gamma/E_{\gamma c}},$$

(5.51)

with the normalized angular spectral distribution functions for the two modes of polarization $F_{s\sigma}$ and $F_{s\pi}$ given by (5.8) but expressed in terms of $E_\gamma/E_{\gamma c}$ instead of in terms of ω/ω_c.

Part III

Undulator radiation

6

A qualitative treatment

6.1 Introduction

An undulator is a spatially periodic magnetic structure designed to produce quasi-monochromatic synchrotron radiation from relativistic particles. There are several types of such devices and we start here with a plane harmonic undulator having a magnetic field of the form

$$B(z) = B_y = B_0 \cos(2\pi c/\lambda_u) = B_0 \cos(k_u z) \tag{6.1}$$

as illustrated in Fig. 6.1. The period length of this field is λ_u and $k_u = 2\pi/\lambda_u$ is the corresponding wave number. If the field B_0 is relatively weak, the trajectory of an ultra-relativistic particle going along the axis of the undulator is, to a good approximation, of the form

$$x(z) = a \cos(k_u z), \qquad \frac{dx}{ds} = -a k_u \sin(k_u z) = -\psi_0 \sin(k_u z),$$

which will be shown later in detail. This trajectory is characterized by the amplitude a and the maximum deflection angle $\hat{x}' = \psi_0$ with respect to the z-axis. The properties of the undulator radiation depend strongly on the magnitude of the deflection angle ψ_0 compared with the natural opening angle $1/\gamma$ of the emitted radiation. This ratio is quantified by the undulator parameter K_u:

$$K_u = \frac{\psi_0}{1/\gamma} = \gamma \psi_0. \tag{6.2}$$

Depending on the value of K_u, we have two possible cases.

- For $K_u < 1$ the angle ψ_0 of the particle trajectory is smaller than the approximate natural opening angle $1/\gamma$ of the emitted radiation. As a result, the charge traversing the undulator emits a smoothly modulated field, leading to so-called *weak-undulator* radiation, which has a simple pattern and quasi-monochromatic properties.
- For $K_u > 1$ the deflection angle ψ_0 is larger than the natural opening angle $1/\gamma$ of the radiation. As a consequence the emitted fields are strongly modulated, leading to so-called *strong-undulator* radiation with a more complicated pattern and a spectrum containing many harmonics.

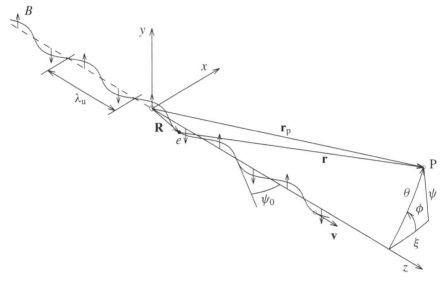

Fig. 6.1. The geometry of undulator radiation.

We treat first the plane harmonic undulator, which is widely used, and abbreviate its radiation sometimes by UR. It has a median symmetry plane with a perpendicular magnetic field, in which the nominal particle trajectory lies. This field is periodic with period length λ_u, given by (6.1), of total length

$$L_\mathrm{u} = N_\mathrm{u}\lambda_\mathrm{u} \qquad (6.3)$$

and contains N_u periods. There exist other undulators that have no simple period, such as the modulated undulator, or have no median plane, such as the helical undulator. Some of these will be discussed later.

Many properties of undulator radiation can be understood from a qualitative treatment. This also helps us to understand the basic physics, to estimate the importance of the parameters involved, to judge the validity of certain approximations, and to obtain some basic results. In the following we use various pictures to describe and understand undulator radiation.

6.2 The interference

The frequency emitted by a particle going through an undulator can be obtained by considering the interference between the parts of the radiation created at successive periods. A charge traversing the undulator with velocity βc is moving along a sinusoidal trajectory (Fig. 6.2). For a strong undulator its drift velocity $\langle\beta\rangle c = \beta^* c$, i.e. its average velocity along the axis, is slightly smaller than the instantaneous particle velocity βc. However, we consider here weak undulators and neglect this difference.

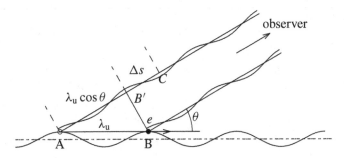

Fig. 6.2. Constructive interference of the fields emitted at different periods.

We investigate first the radiation observed in the forward direction, $\theta = 0$. The wave emitted in the first period is propagating with velocity c while the charge is moving with the drift velocity βc along the axis. The radiation arrives one period λ_u downstream at a time $t_\gamma = \lambda_u/c$ while the charge takes a little longer, $t_e = \lambda_u/(\beta c)$. The radiation emitted in the second period is delayed with respect to the first wave by $\Delta t = t_e - t_\gamma$. The same delay occurs between the emissions at later successive periods. An observer at $\theta = 0$ therefore sees a periodic field emitted in each period with successive delays $t_e - t_\gamma$. These field contributions will add up for a wavelength equal to this delay multiplied by the speed of light. We therefore have constructive interference for the wavelength and frequency:

$$\lambda_1(0) = \lambda_{10} = c\,\Delta t = \frac{\lambda_u}{\beta}(1 - \beta), \qquad \omega_{10} = \frac{2\pi c}{\lambda_u}\frac{\beta}{1 - \beta}.$$

For an observer at an angle θ the situation is different since the projection of the particle velocity on this new direction is relevant for the observed delays. In the direction of angle θ contributions from successive periods are delayed by $\lambda_u/(\beta c) - \lambda_u \cos\theta/c$. We therefore have constructive interference for a wavelength

$$\lambda_1(\theta) = \frac{\lambda_u}{\beta}(1 - \beta\cos\theta), \qquad \omega_1 = \frac{2\pi\beta c}{\lambda_u}\frac{1}{1 - \beta\cos\theta}. \tag{6.4}$$

Making the ultra-relativistic approximation $\beta \approx 1$, $\gamma \gg 1$, we have

$$\omega_1 = \frac{2\pi c}{\lambda_u}\frac{2\gamma^2}{1 + \gamma^2\theta^2} = \frac{\omega_{10}}{1 + \gamma^2\theta^2}. \tag{6.5}$$

At a given angle θ there is a frequency $\omega_1(\theta)$, which we will call the 'proper frequency.' Since in a strong undulator also higher harmonics are emitted, we use the subscript '1' for the lowest harmonic and 'm' for the higher harmonics. On the axis, $\theta = 0$, the proper frequency is $\omega_1(0) = \omega_{10}$. It is $2\gamma^2$ times larger than that of the particle motion, $\Omega_u = 2\pi c/\lambda_u$. This allows us to obtain radiation of rather short wavelength from an undulator having a realizable period length of a few centimeters. For a limited number of periods the frequency ω observed at the angle θ has a distribution around the proper frequency ω_1 that becomes narrower with increasing number of periods.

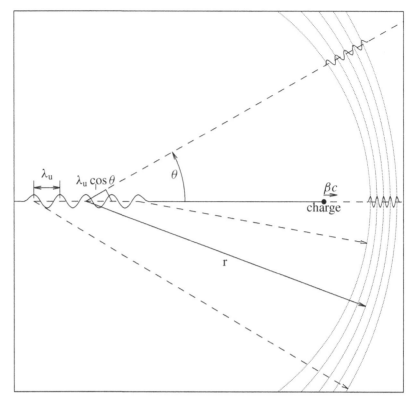

Fig. 6.3. An illustration of undulator radiation using the wave fronts of the fields emitted at different periods, $\beta = 0.8$.

6.3 The undulator radiation as a wave front

Undulator radiation can be illustrated by drawing the wave fronts created in each period. This is shown in Fig. 6.3 for $\beta = 0.8$. A more relativistic particle velocity would be closer to practical examples but the resulting small angles would be difficult to draw.

As we saw before in the interference picture, a particle going through the undulator creates in each period a wave with intervals spaced by $\lambda_u/(\beta c)$ in time and λ_u in longitudinal position. The particle passing the center of the undulator at $t' = 0$ emits a spherical wave that has reached a radius r_1 at the time $t = r_1/c$. The next period is reached by the particle at the time $t' = \lambda_u/(\beta c)$ and it emits there a spherical wave that has reached a radius $r_2 = c(t - t') = ct - \lambda_u/\beta$ at the same observation time t as when the first wave has reached a radius r_1. Since the origin of the second wave is displaced by λ_u from the first one, the two wave fronts are separated at time t by $\lambda_u/\beta - \lambda_u = (1 - \beta)\lambda_u/\beta \approx \lambda_u/(2\gamma^2)$ in the forward direction, and about $\lambda_u(1 + \gamma^2\theta^2)/(2\gamma^2)$ in the direction of the angle θ. This is shown in Fig. 6.3, where the emitted wave fronts are compressed strongly in the forward direction but less so at larger angles θ.

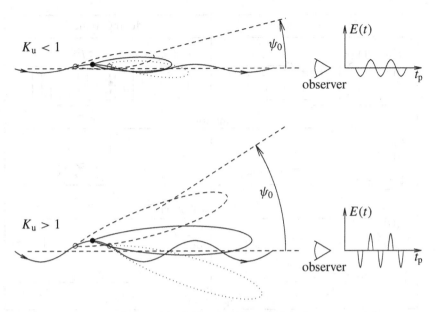

Fig. 6.4. The modulation of the field emitted in a weak and a strong undulator.

6.4 The modulation of the emitted field

Another way to understand the radiation emitted in a weak and a strong undulator is illustrated in Fig. 6.4. Owing to the transverse acceleration, a charge going through an undulator emits radiation with an instantaneous natural opening angle of about $1/\gamma$. In a weak undulator the deflection angle ψ_0 is small or comparable to this natural opening angle. As a consequence an observer sees just a small angular modulation of the incoming radiation, as indicated in the top part of Fig. 6.4. The observed proper frequency ω_1 is the Doppler-shifted frequency Ω_u of the particle motion:

$$\omega_1 \approx \frac{\Omega_u}{1 - \beta \cos \theta} \approx \Omega_u \frac{2\gamma^2}{1 + \gamma^2 \theta^2}.$$

For a strong undulator the angular modulation ψ_0 of the radiation is much larger than the natural opening angle $1/\gamma$. Instead of a smooth modulation the observer receives the field as a periodic train of short pulses, as indicated in the lower part of Fig. 6.4. The spectrum of this field contains many harmonics of the basic frequency, [44].

6.5 The weak undulator in the laboratory and moving frames

Probably the best understanding of undulator radiation can be obtained by considering a system moving with the drift velocity βc along the z-axis. In this frame the particle motion

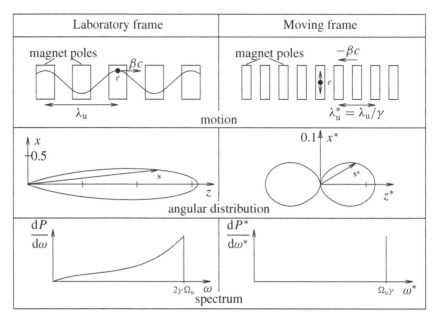

Fig. 6.5. The weak undulator in the laboratory and moving frames.

is not relativistic for a weak undulator and consists of a simple harmonic oscillation that emits dipole radiation. On going back into the laboratory frame, the frequency of this field is Doppler shifted.

The situation in the two frames is shown in Fig. 6.5. The angle of the trajectory is very small, $\psi_0 \ll 1/\gamma$, and the drift velocity is approximately βc. In the laboratory frame the particle follows a sinusoidal trajectory with frequency $\Omega_u = 2\pi\beta c/\lambda_u$. In the moving frame this motion becomes a simple harmonic oscillation. Its frequency is increased by the Lorentz factor γ corresponding to the frame motion, because the period length of the undulator seen by the particle becomes Lorentz contracted, $\lambda_u^* = \lambda_u/\gamma$. The motion of the particle represents an oscillating charge that creates the well-known dipole radiation. It is emitted mainly perpendicular to the x^*-axis but with a large opening angle. Transformed back into the laboratory frame, this radiation becomes confined mainly into a cone around the z-axis with opening angle $1/\gamma$. The spectrum of the radiation in the moving frame has a single frequency, $\Omega_u^* = \Omega_u\gamma$. The transformation of the field into the laboratory frame creates a Doppler shift, resulting in a higher frequency ω_1, which depends on the observation angle θ. For the ultra-relativistic case it can be approximated by

$$\omega_1 = \frac{\Omega_u^*}{\gamma(1 - \beta\cos\theta)} \approx \frac{2\gamma^2\Omega_u}{1 + \gamma^2\theta^2}. \tag{6.6}$$

In the laboratory frame the spectrum is therefore no longer monochromatic but has the above relation between the proper frequency ω_1 and the angle θ of observation.

6.6 The strong undulator in the laboratory and moving frames

In the case of a strong undulator, $K_u \geq 1$, the angle of the trajectory is larger, $\psi_0 \geq 1/\gamma$, and the drift velocity of the particle is smaller than its instantaneous velocity:

$$\beta^* c < \beta c, \qquad \gamma^* = 1/\sqrt{1 - \beta^{*2}} < \gamma.$$

The moving frame has a Lorentz factor γ^* that is smaller than that of the particle. The period length λ_u of the undulator is contracted to $\lambda_u^* = \lambda_u \gamma^*$. Since the absolute value of the particle momentum is constant in the laboratory system, its components in the transverse and longitudinal directions are modulated by the large deflection angle. As a consequence the motion in the moving frame is no longer a linear oscillation but follows a 'figure eight' as indicated in Fig. 6.6. As an approximation we can decompose this non-linear motion into an oscillation along the x^*-axis with odd harmonics of the frequency $\Omega_u^* \approx \Omega_u \gamma^*$ emitting radiation mainly along the z-axis, and an oscillation along the z^*-axis having even harmonics of this frequency with the radiation emitted perpendicular to the axis. The transformation back into the laboratory frame results again in radiation emitted mainly in the forward direction, but now only the odd harmonics are concentrated along the axis with opening angle $1/\gamma^*$, while the even harmonics are distributed in a ring around it. The spectrum in the moving system consists of lines of both odd and even harmonics of the frequency Ω_u^*. Observed in the laboratory frame these frequencies are Doppler shifted resulting in the

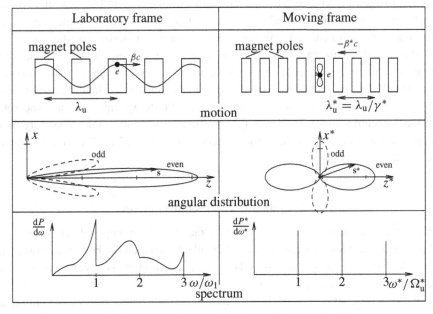

Fig. 6.6. The strong undulator in the laboratory and moving frames.

proper frequencies

$$\omega_m = m\omega_1 = m\frac{\Omega_u^*}{\gamma^*1 - \beta^*\cos\theta} \approx m\frac{2\gamma^{*2}\Omega_u}{1 + \gamma^{*2}\theta^2}. \tag{6.7}$$

The overall spectrum is broad, but at a given angle θ it consists of a series of lines $m\omega_1$. In the forward direction $\theta = 0$, only odd harmonics are observed.

6.7 The helical undulator

Helical undulators provide radiation with circular polarization. The electron trajectory in this device has the form of a helix around the longitudinal z-axis, given in Cartesian coordinates by

$$\mathbf{R} = [x, y, z] = [a\cos(k_u z), a\sin(k_u z), z].$$

This represents a superposition of two plane undulator trajectories, rotated by an angle of $\pi/2$ and shifted by a quarter of an undulator period. The emitted radiation has the same superposition properties and consists of two waves, one with horizontal and the other with vertical linear polarization, with a phase shift of $\pi/2$ with respect to each other. This results in a wave with circular polarization. Circular polarization can in fact be obtained with two separate plane undulators rotated and shifted with respect to each other [45, 46].

Going into a frame that moves with the electron drift velocity along the z-direction can help us to understand helical-undulator radiation in a different way. In this frame the electron moves on a circular orbit with a normalized velocity $\beta \ll 1$ for a weak undulator, and with $\beta \approx 1$ for a strong undulator. In the first case the electron radiates over a large solid angle. The waves emitted along the axis have circular polarization whereas the ones emitted in the orbit plane have linear horizontal polarization. Transforming this back to the laboratory plane results in circular polarization along the axis and linear polarization on a cone with half opening angle $1/\gamma$. For a strong helical undulator the circular electron motion in the moving frame is relativistic and represents a little storage ring emitting synchrotron radiation concentrated close to the orbit plane. Transformed back to the laboratory frame, this radiation will be concentrated around a cone of half angle $1/\gamma^*$ with little intensity along the axis.

6.8 Undulators and related devices

Undulators as sources of quasi-monochromatic radiation were proposed [47, 48] and realized [49] early on. Most of the properties of undulator radiation can be found in the review article [50], which is followed closely here. This radiation is also treated in books and articles [8, 9, 17, 25, 51–56].

So far we have considered undulators having a periodic magnetic field that emit quasi-monochromatic radiation of high intensity, due to the superposition of the radiation from each period. For most applications the narrow bandwidth is of importance and its frequency can be tuned to the desired value by varying the strength of the magnetic field, i.e. the undulator parameter K, which changes the drift velocity $\beta^* c$. The bandwidth obtained is rarely sufficiently narrow to be used directly, but it helps to reduce the heat load on the monochromator, which is usually needed. For some experiments monochromaticity is not as important as intensity. In this case the overlap of the radiation emitted in each undulator is important. Devices that are optimized for this aspect usually have a high value for the magnetic field and the undulator parameter K_u. These devices are called *wigglers*. The distinction between wigglers and undulators is not sharp and is determined more by the application of the radiation than by its nature.

Undulators and wigglers produce in general no overall deflection or displacement of the electron beam. This allows us to adjust the strength of the magnetic field with negligible effects on the electron beam in the rest of the ring. For this reason they are often called *insertion devices* that can be inserted into a straight section of a storage ring. They also include short assemblies of magnets with very high fields but no overall deflection. Such devices are supposed to provide radiation with a high critical frequency and are called *wavelength shifters*. Apart from the mentioned books on synchrotron radiation, there is a rich literature on insertion devices, of which we mention here only two review articles [57, 58].

Undulators and multipole wigglers are usually realized with permanent magnets since the short period length leaves little space for the coils needed in electromagnets. The field is adjusted by mechanical variation of the gap between poles of magnets. The very high fields in a wavelength shifter are realized with superconducting magnets.

Insertion devices produce no overall deflection and displacement and are located in straight sections. Usually the radiation emitted along the axis of the insertion device is used. As a consequence, also radiation from the adjacent ring magnets upstream and downstream of the straight section enters the experiment. This contamination not only gives additional intensity but also leads to interference among the three sources and can alter the spectrum. In the case of long undulators the intensity of the monochromatic peaks dominates and the effect of the ring magnets is not important. A similar situation exists for high-field wigglers, which emit a very high critical frequency. However, there are other situations in which the interplay among the sources has to be considered.

7

The plane weak undulator

7.1 The trajectory

7.1.1 The equation of motion

We consider a plane magnetic undulator with a harmonic field of period length λ_u, corresponding wave number $k_u = \lambda_u/(2\pi)$, number N_u of periods, and total length $L_u = N_u\lambda_u$, as shown in Fig. 7.1. The strength of the undulator field is assumed to be weak in the sense that the maximum angle of the trajectory is small compared with the opening angle of the emitted radiation, $\psi_0 \ll 1/\gamma$. The ratio between the two is expressed in terms of the undulator parameter K_u defined in the introduction, (6.2),

$$K_u = \gamma\psi_0,$$

which will be assumed to be smaller than unity.

   We investigate the trajectory and the related parameters of the particle going through this undulator. We start with the general case, approximate first for a weak undulator, $K_u < 1$, then for observation from a large distance, $r_p \gg L_u$, and finally for ultra-relativistic particles, $\gamma \gg 1$.

   For the field in the median plane $y = 0$ we take (6.1)

$$\mathbf{B}(x, 0, z) = B_0[0, \cos(k_u z), 0]. \tag{7.1}$$

This determines also the field outside the plane in a current- and iron-free region, but we will not consider this here. The undulator has a length $L_u = N_u\lambda_u$ containing N_u periods. However, a sharp termination of this periodic field or its derivative at $z = \pm L_u/2$ leads to unrealistic high frequencies in the calculated spectrum. Since the radiation emitted in the relatively long, periodic main part of the undulator dominates, we may ignore the end field and its effects for the time being. Undulators are usually inserted into a straight section of a storage ring. For this reason they are, together with wiggler magnets, called *insertion devices*. To minimize the disturbance of the electron orbit in the storage ring one usually demands that the overall transverse deflection and displacement due to the undulator field vanish:

$$x(-L_u/2) = x(L_u/2) \quad \text{and} \quad \frac{dx}{dz}(-L_u/2) = \frac{dx}{dz}(L_u/2) = 0.$$

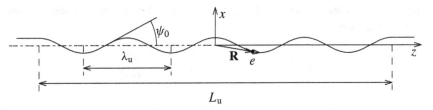

Fig. 7.1. The trajectory in a plane harmonic undulator.

The second condition can be expressed with a vanishing field integral. By integrating twice over the field we could also satisfy the first condition. However, there are several reasons that favor either a symmetric or an antisymmetric field with respect to the longitudinal coordinate z. With this we can satisfy both conditions,

$$B_y(-z) = B_y(z) \quad \text{or} \quad B_y(-z) = -B_y(z) \quad \text{and} \quad \int_{-L_u/2}^{L_u/2} B_y(z)\,dz = 0.$$

Both choices of symmetry have been realized in practice. Apart from an unimportant phase difference, the radiation from both solutions has the same properties. With the field (7.1) we have here already chosen the symmetric case.

We calculate the nominal trajectory of a charged particle going through a weak undulator with $K_u < 1$. It follows on average the z-axis, as indicated in Fig. 7.1. Since we have only a magnetic deflecting field the absolute value of the particle velocity stays constant, $v = \beta c$. We assume that the particle traverses the origin $z = 0$ at time $t' = 0$. The chosen symmetry condition results in a vanishing transverse velocity at this moment, leading to the initial conditions

$$z(0) = 0, \qquad \dot{x}(0) = 0, \qquad \dot{z}(0) = \beta c \qquad (7.2)$$

with $\dot{x} = dx/dt'$. The Lorentz force on the particle in the magnetic undulator field,

$$\mathbf{F} = e[\mathbf{v} \times \mathbf{B}] = m_0 \gamma [\ddot{x}, \ddot{y}, \ddot{z}] = e B_0 [-\cos(k_u z)\,\dot{z}, 0, \cos(k_u z)\,\dot{x}],$$

leads to the system of two differential equations

$$\ddot{x} = \frac{d^2 x}{dt'^2} = -\frac{e B_0}{m_0 \gamma} \cos(k_u z)\,\dot{z}, \qquad \ddot{z} = \frac{d^2 z}{dt'^2} = \frac{e B_0}{m_0 \gamma} \cos(k_u z)\,\dot{x}. \qquad (7.3)$$

The first one can be written as

$$d\dot{x} = -\frac{e B_0}{m_0 \gamma} \cos(k_u z)\,dz$$

and integrated to give

$$\dot{x} = -\frac{e B_0}{m_0 \gamma k_u} \sin(k_u z).$$

satisfying the initial conditions (7.2). The z-component of the velocity is obtained from the conservation of energy:

$$\dot{x}^2 + \dot{z}^2 = \beta^2 c^2, \qquad \dot{z} = \beta c \sqrt{1 - \frac{\dot{x}^2}{\beta^2 c^2}}. \tag{7.4}$$

7.1.2 The approximation for a weak undulator

We assume that the angle of the trajectory is much smaller than the natural opening angle of the radiation, $K_u < 1$, and approximate the above expression for the longitudinal velocity $\dot{z} \approx \beta c$. The derivative of x with respect to z is

$$x'(z) = \frac{dx}{dz} = \frac{\dot{x}}{\dot{z}} = -\frac{e B_0}{m_0 c \beta \gamma k_u} \sin(k_u z).$$

We introduced before the undulator parameter K_u as the ratio between the trajectory and the opening angle $1/\gamma$ of the radiation. We give now a more exact definition,

$$K_u = \beta \gamma \hat{x}' = \beta \gamma \psi_0 = \frac{e B_0}{m_0 c k_u}, \tag{7.5}$$

and use it to express the trajectory in the weak undulator,

$$x'(z) = -\frac{K_u}{\beta \gamma} \sin(k_u z), \qquad x(z) = \frac{K_u}{\beta \gamma k_u} \cos(k_u z) = a \cos(k_u z)$$

with the maximum angle and excursion

$$\hat{x}' = \frac{K_u}{\beta \gamma} \approx \psi_0 < \frac{1}{\gamma}, \qquad \hat{x} = a = \frac{K_u}{\beta \gamma k_u}. \tag{7.6}$$

From the longitudinal motion $\dot{z} = \beta c$, $z = \beta c t'$, we obtain the trajectory as a function of the time t', which we write in vector form with Cartesian components

$$\mathbf{R}(t') = \left[\frac{K_u}{\beta \gamma k_u} \cos(\Omega_u t'), \ 0, \ \beta c t' \right]$$

$$\boldsymbol{\beta}(t') = \left[-\frac{K_u}{\gamma} \sin(\Omega_u t'), \ 0, \ \beta \right] \tag{7.7}$$

$$\dot{\boldsymbol{\beta}}(t') = \left[-\frac{K_u c k_u \beta}{\gamma} \cos(\Omega_u t'), \ 0, \ 0 \right]$$

using the frequency of the periodic particle motion in the laboratory frame,

$$\Omega_u = k_u \beta c. \tag{7.8}$$

In this approximation $K_u < 1$, the z-component of the velocity is nearly equal to the full speed of the particle, neglecting the influence of the transverse motion which is only of second order in K_u. On the basis of this trajectory we calculate some expressions needed later for the radiation field.

The vectors $\mathbf{r}(t')$ from the particle to the observer and $\mathbf{r}_p$ from the center of the undulator to the observer are, according to Fig. 6.1,

$$\mathbf{r}(t') = \mathbf{r}_p - \mathbf{R}(t'), \qquad \mathbf{r}_p = r_p[\sin\theta\cos\phi, \ \sin\theta\sin\phi, \ \cos\theta],$$

which gives with the first equation in (7.7):

$$\mathbf{r}(t') = r_p\left[\sin\theta\cos\phi - \frac{K_u}{r_p\beta\gamma k_u}\cos(\Omega_u t'), \ \sin\theta\sin\phi, \ \cos\theta - \frac{\beta c t'}{r_p}\right]. \tag{7.9}$$

7.1.3 The observation from a large distance

As a next approximation we observe the radiation from a distance r_p large compared with the length of the undulator L_u, resulting in

$$\frac{|\beta c t'|}{r_p} \leq \frac{L_u}{2r_p} \ll 1, \qquad \frac{K_u}{r_p\beta\gamma k_u} = \frac{L_u K_u}{2\pi r_p\beta\gamma N_u} \ll 1. \tag{7.10}$$

We calculate first the absolute value of the distance r from (7.9) which is needed in order to obtain the relation between the time scales. Since this depends on the small difference $r - \beta c t'$, we have to consider the above quantities in (7.10):

$$r(t') = r_p - \beta c t'\cos\theta - \frac{K_u\sin\theta\cos\phi}{\beta\gamma k_u}\cos(\Omega_u t'). \tag{7.11}$$

The relation between the emission and observation times becomes

$$t = t' + \frac{r(t')}{c}, \qquad t_p = t - \frac{r_p}{c} = t' + \frac{r(t') - r_p}{c}.$$

Since the time t consists to a large extent of an irrelevant constant traveling time r_p/c for light to pass between the center of the undulator and the observer, we use the reduced observation time $t_p = t - r_p/c$ relative to the arrival time of the radiation originating from the center of the undulator as we did previously for ordinary synchrotron radiation (4.13). We obtain

$$t_p = t'(1 - \beta\cos\theta) - \frac{K_u\sin\theta\cos\phi}{\beta c\gamma k_u}\cos(\Omega_u t'). \tag{7.12}$$

This consists of a linear relation and an oscillating term. To estimate the importance of the latter, we multiply the time by the frequency of the radiation, $\omega_1 = k_u\beta(1 - \beta\cos\theta)$, which is expected from the qualitative arguments in (6.4). This results in a phase modulation of

$$\omega_1 t_p = \Omega_u t' - \frac{K_u\sin\theta\cos\phi}{\gamma(1 - \beta\cos\theta)}\cos(\Omega_u t').$$

To obtain its magnitude we take a typical opening angle of the radiation (1.1),

$$\sin\theta \approx 1/\gamma, \qquad \cos\theta \approx \beta$$

and obtain for the factor

$$\frac{K_u \sin \theta}{\gamma(1 - \beta \cos \theta)} \approx K_u < 1.$$

This quantity is smaller than unity and results in a phase modulation of small amplitude, which we will neglect. We are left with a linear relation between the time scales for a weak plane undulator:

$$t_p = t'(1 - \beta \cos \theta).$$

The distance r, as it appears in the denominator of the Liénard–Wiechert equation, can be determined with less precision since its variation leads only to a smooth change in amplitude of the radiation. We neglect the corresponding terms of the quantities (7.11) and are left with a constant for this distance and the unit vector pointing from the particle to the observer:

$$r = r_p, \qquad \mathbf{n} = \frac{\mathbf{r}(t')}{r(t')} \approx \frac{\mathbf{r}_p}{r_p} = [\sin \theta \cos \phi, \ \sin \theta \sin \phi, \ \cos \theta]. \tag{7.13}$$

With these ingredients we obtain the relation between the emission and observation times, the denominator and the triple vector product appearing in the Liénard–Wiechert equation:

$$t_p = t'(1 - \beta \cos \theta)$$

$$1 - \mathbf{n} \cdot \boldsymbol{\beta} = 1 - \beta \cos \theta \tag{7.14}$$

$$[\mathbf{n} \times [(\mathbf{n} - \boldsymbol{\beta}) \times \dot{\boldsymbol{\beta}}]] = \frac{\beta c k_u K_u \cos(\Omega_u t')}{\gamma}$$

$$[1 - \beta \cos \theta - \sin^2 \theta \cos^2 \phi, \ -\sin^2 \theta \sin \phi \cos \phi, \ \sin \theta \cos \phi(\beta - \cos \theta)].$$

7.1.4 The ultra-relativistic approximation

For an ultra-relativistic particle $\gamma \gg 1$ we have only to consider small angles $\theta \ll 1$ and can make the usual approximations, leading to

$$t_p = \frac{1 + \gamma^2 \theta^2}{2\gamma^2} t'$$

$$1 - \mathbf{n} \cdot \boldsymbol{\beta} = \frac{1 + \gamma^2 \theta^2}{2\gamma^2} \tag{7.15}$$

$$[\mathbf{n} \times [(\mathbf{n} - \boldsymbol{\beta}) \times \dot{\boldsymbol{\beta}}]] = \frac{c k_u K_u \cos(\Omega_u t')}{2\gamma^3} [1 - \gamma^2 \theta^2 \cos(2\phi), \ -\gamma^2 \theta^2 \sin(2\phi), \ 0].$$

The particle motion in a weak undulator has a periodic velocity and corresponds to this class treated in Chapter 2. It has a drift velocity $\langle \boldsymbol{\beta} \rangle = \boldsymbol{\beta}^* = \beta[0, \ 0, \ 1]$ along the axis and a time period $T_{rev} = 2\pi / \Omega_u$. For the observer at an angle θ the time period of the radiation

(2.42) and the corresponding frequency are

$$T_p = \frac{1 + \gamma^2\theta^2}{2\gamma^2} T_{\text{rev}}, \qquad \omega_1 = \frac{2\gamma^2}{1 + \gamma^2\theta^2} \Omega_u. \qquad (7.16)$$

We could calculate the emitted field from (2.44) in the frequency domain. We will do that later for the strong undulator, but here we use the more transparent Liénard–Wiechert equations.

7.1.5 The particle motion in the moving system

It is interesting to calculate the particle motion in the undulator also in a frame S^* moving with the drift velocity of the particle along the z-direction. Applying the Lorentz transformation

$$x^* = x, \qquad z^* = \gamma(z - \beta ct'), \qquad ct^* = \gamma(ct' - \beta z) \qquad (7.17)$$

to (7.7) gives

$$x^*(t') = \frac{K_u}{\beta\gamma k_u} \cos(\Omega_u ct') = a\cos(\Omega_u ct'), \qquad z^*(t') = 0.$$

We express t' with the time t^* of the moving frame,

$$ct^* = \gamma ct'(1 - \beta^2) = ct'/\gamma,$$

and obtain in this frame only a motion along the x^*-axis:

$$x^*(t^*) = \frac{K_u}{\beta\gamma k_u} \cos(\gamma\Omega_u t^*) = \frac{K_u}{\beta\gamma k_u} \cos(\Omega_u^* t^*). \qquad (7.18)$$

In the moving frame the particle executes a harmonic oscillation with the same amplitude a but a frequency $\Omega_u^* = \gamma\Omega_u$ that is increased by the Lorentz factor. We found this result already in the qualitative discussion of undulator radiation.

7.2 The radiation field

7.2.1 The field calculated from the Liénard–Wiechert equation

We calculate the radiation field emitted in a weak undulator from the second term of the Liénard–Wiechert expression (2.17) in Chapter 2:

$$\mathbf{E}(t) = \frac{e}{4\pi c\epsilon_0} \left\{ \frac{[\mathbf{n} \times [(\mathbf{n} - \boldsymbol{\beta}) \times \dot{\boldsymbol{\beta}}]]}{r(1 - \mathbf{n} \cdot \boldsymbol{\beta})^3} \right\}_{\text{ret}}, \qquad \mathbf{B} = \frac{[\mathbf{n} \times \mathbf{E}]}{c}. \qquad (7.19)$$

Using equation (7.14) and expressing the emission time and frequency in terms of those of the observer,

$$\Omega_u t' = \Omega_u \frac{t_p}{1 - \beta\cos\theta} = \omega_1 t_p \quad \text{with} \quad \omega_1 = \frac{\Omega_u}{1 - \beta\cos\theta},$$

we obtain the radiation fields as functions of the observation time t_p,

$$
\mathbf{E}(t_p) = \frac{e\Omega_u K_u}{4\pi c\epsilon_0 \gamma r_p} \cos(\omega_1 t_p)
$$
$$
\times \frac{[1 - \beta\cos\theta - \sin^2\theta\cos^2\phi, \ -\sin^2\theta\sin\phi\cos\phi, \ \sin\theta\cos\phi(\beta - \cos\theta)]}{(1 - \beta\cos\theta)^3}
$$

$$(7.20)$$

$$
\mathbf{B}(t_p) = \frac{e\Omega_u K_u}{4\pi\epsilon_0 c^2 \gamma r_p} \cos(\omega_1 t_p)
$$
$$
\times \frac{[\beta\sin^2\theta\sin\phi\cos\phi, \ \cos\theta - \beta(1 - \sin^2\theta\sin^2\phi), \ -(1 - \beta\cos\theta)\sin\theta\sin\phi]}{(1 - \beta\cos\theta)^3}
$$

and, in the ultra-relativistic approximation,

$$
\mathbf{E}(t_p) = \frac{e\Omega_u K_u \gamma^3}{\pi\epsilon_0 c r_p} \frac{[1 - \gamma^2\theta^2\cos(2\phi), \ -\gamma^2\theta^2\sin(2\phi), \ 0]}{(1 + \gamma^2\theta^2)^3} \cos(\omega_1 t_p)
$$
$$
\mathbf{B}(t_p) = \frac{e\Omega_u K_u \gamma^3}{\pi\epsilon_0 c^2 r_p} \frac{[\gamma^2\theta^2\sin(2\phi), \ 1 - \gamma^2\theta^2\cos(2\phi), \ 0]}{(1 + \gamma^2\theta^2)^3} \cos(\omega_1 t_p).
$$

$$(7.21)$$

Since the radiation is concentrated within a small angle θ, the z-component of the field becomes negligible.

7.2.2 The undulator field as Lorentz-transformed dipole radiation

It is illustrative to calculate the weak undulator radiation as a Lorentz transformation of the field emitted by an oscillating dipole. In the previous section we calculated the motion of a charge traversing the undulator in the moving frame, (7.18), and obtained in the weak-field approximation $K_u \ll 1$

$$
x^*(t^*) = a\cos(\Omega_u^* t^*), \qquad z^*(t^*) = 0 \tag{7.22}
$$

with

$$
a = \frac{K_u}{\beta\gamma k_u}, \qquad \Omega_u^* = \Omega_u \gamma.
$$

This motion (7.22) represents a harmonic oscillation along the x^*-axis with frequency $\Omega_u^* = \Omega_u \gamma$, amplitude a, and dipole moment

$$
\mathbf{M}^*(t^*) = ae[1, \ 0, \ 0]\cos(\Omega_u^* t^*).
$$

In Chapter 2 we calculated the field emitted by such a dipole oscillating along the z-axis; see Fig. 2.11. We adapt it to our present case, with the dipole moment along the x-axis, and mark the coordinates and angles by asterisks to indicate that they correspond to the moving system S^*, Fig. 7.2. Taking now only the far fields $\mathbf{E}^*$ and $\mathbf{B}^* = [\mathbf{n}^* \times \mathbf{E}^*]/c$ in (2.29) and

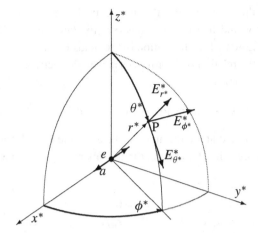

Fig. 7.2. Coordinates for the fields of an oscillating dipole.

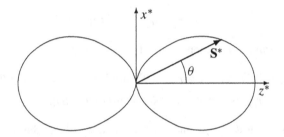

Fig. 7.3. The angular power distribution of the dipole radiation.

expressing them in terms of the coordinates of Fig. 7.2 which are rotated relative to those in Fig. 2.11, we have

$$\mathbf{E}^*(t^*) = [E_x^*, \ E_y^*, \ E_z^*] = \frac{ea\Omega_u^{*2}}{4\pi\epsilon_0 c^2 r^*}\cos(k_u^* r^* - \Omega_u^* t^*)$$
$$\times [1 - \sin^2\theta^*\cos^2\phi^*, \ -\sin^2\theta^*\sin\phi^*\cos\phi^*, \ -\sin\theta^*\cos\theta^*\cos\phi^*]$$
$$(7.23)$$

$$\mathbf{B}^*(t^*) = \frac{ea\Omega_u^{*2}}{4\pi\epsilon_0 c^3 r^*}\cos(k_u^* r^* - \Omega_u^* t^*)\,[0, \ \cos\theta^*, \ -\sin\theta^*\sin\phi^*]. \qquad (7.24)$$

The properties of this radiation field were discussed in Chapter 2. It has cylindrical symmetry around the x^*-axis and is emitted mainly around the (y, z)-plane but with a large opening angle. This is illustrated in Fig. 7.3, which shows the distribution of the Poynting vector.

We have obtained the trajectory (7.22) in the moving frame from the motion in the laboratory system,

$$x(t') = a\cos(\Omega_u t'), \ z(t') = \beta c t'.$$

We then calculated the fields $\mathbf{E}^*$ and $\mathbf{B}^*$ emitted in the moving frame by the oscillating dipole moment. We would like now to transform the fields into the laboratory frame by applying an inverse Lorentz transformation to the fields $\mathbf{E}^*$ and $\mathbf{B}^*$.

We first establish the relation of coordinates and angles between the two frames and use the Lorentz transformation

$$x^* = x, \qquad z^* = \gamma(z - \beta ct), \qquad ct^* = \gamma(ct - \beta z). \tag{7.25}$$

For the radiation emitted at the origin of space and time in both frames and observed at the distance r_p^* and time t^* in S^* and at r_p, t in the laboratory frame we have the coordinates

	S^*		S	
emission	$z^* = 0$	$t^* = 0$	$z = 0$	$t' = 0$
observation	$z^* = r_p^* \cos \theta^*$	$t^* = r_p^*/c$	$z = r_p \cos \theta$	$t = r_p/c$

On expressing t^* with a Lorentz transformation, we find the relation between the distances r_p^* and r_p from the center to the observer in the two frames:

$$ct^* = r_p^* = r_p\gamma(1 - \beta \cos \theta). \tag{7.26}$$

The Lorentz transform for the components of the vector r_p^* is

$$\mathbf{r}_p^* = [x^*, y^*, z^*] = r_p^*[\sin \theta^* \sin \phi^*, \sin \theta^* \sin \phi^*, \cos \theta^*]$$
$$= r_p[\cos \theta \sin \phi, \sin \theta \sin \phi, \gamma(\cos \theta - \beta)].$$

From this, and using $\tan \phi = y/x$ and $\cos \theta = z/r_p$, we obtain the relations between the angles:

$$\phi^* = \phi, \qquad \cos \theta^* = \frac{\cos \theta - \beta}{1 - \beta \cos \theta}, \qquad \sin \theta^* = \frac{\sin \theta}{\gamma(1 - \beta \cos \theta)}. \tag{7.27}$$

Next we treat the phase $(\mathbf{k}_u^* \mathbf{r}^* - \Omega_u^* t^*)$ of the emitted radiation which is invariant under a Lorentz transformation,

$$(\mathbf{k}_u^* \mathbf{r}^* - \Omega_u^* t^*) = (\mathbf{k}_1 \mathbf{r} - \omega_1 t) = \xi, \tag{7.28}$$

with the wave vector

$$\mathbf{k}_u^* = k_u^*[\sin \theta^* \cos \phi^*, \sin \theta^* \sin \phi^*, \cos \theta^*], \qquad k_u^* = \gamma k_u$$

and a corresponding expression for $\mathbf{k}_u$. The left-hand side of (7.28) can be written as

$$\xi = k_u(\sin \theta^* \cos \phi^* x^* + \sin \theta^* \sin \phi^* y^* + \cos \theta^* z^*) - \Omega_u^* t^*.$$

With the angle relation (7.27) and the Lorentz transformation (7.25) this becomes

$$\xi = \frac{k_u^*(\sin \theta \cos \phi \, x + \sin \theta \sin \phi \, y + \cos \theta \, z) - \Omega_u^* t}{\gamma(1 - \beta \cos \theta)}.$$

The Lorentz invariance (7.28) can be satisfied with the relations

$$k_1 = \frac{k_u^*}{\gamma(1 - \beta\cos\theta)} = \frac{k_u}{(1 - \beta\cos\theta)}, \qquad \omega_1 = \frac{\Omega_u^*}{\gamma(1 - \beta\cos\theta)} = \frac{\Omega_u}{(1 - \beta\cos\theta)}.$$

In the moving frame the observer moves towards the oscillating charge and thus the distance r^* between the two changes. This change is relatively small during the duration of the oscillation and we neglect it, as we did in the laboratory frame for $L_u \ll r_p$, and replace r^* by r_p^*. Its transformation has been obtained in (7.26), but can also be calculated by taking the vector $\mathbf{r}_p$ and transforming the components according to (7.25),

$$\mathbf{r}_p^* = r_p(\sin\theta\cos\phi, \ \sin\theta\sin\phi, \ \gamma(\cos\theta - \beta)),$$

where we used the relation $r_p = ct'$. The absolute value of $\mathbf{r}_p^*$ becomes

$$r_p^* = r_p\gamma(1 - \beta\cos\theta).$$

For the relation between the electric and magnetic fields in the two frames we use the inverse Lorentz transformation

$$E_x = \gamma(E_x^* + \beta c B_y^*), \qquad E_y = \gamma(E_y^* - \beta c B_x^*), \qquad E_z = E_z^*$$
$$B_x = \gamma(B_x^* - \beta E_y^*/c), \qquad B_y = \gamma(B_y^* + \beta E_x^*/c), \qquad B_z = B_z^*$$

to calculate the field in the laboratory frame from (7.23) and (7.24). We obtain a propagating wave:

$$\mathbf{E}(t) = \frac{e\Omega_u K_u}{4\pi\epsilon_0 c r_p\gamma}\cos(k_1 r_p - \omega_1 t)$$
$$\times \frac{[1 - \beta\cos\theta - \sin^2\theta\cos^2\phi, \ -\sin^2\theta\cos\phi\sin\phi, \ (\beta - \cos\theta)\sin\theta\cos\phi]}{(1 - \beta\cos\theta)^3}$$

$$\mathbf{B}(t) = \frac{e\Omega_u K_u}{4\pi\epsilon_0 c^2\gamma r_p}\cos(k_1 r_p - \omega_1 t)$$
$$\times \frac{[\beta\sin^2\theta\sin\phi\cos\phi, \ \cos\theta - \beta(1 - \sin^2\theta\sin^2\phi), \ -(1 - \beta\cos\theta)\sin\theta\sin\phi]}{(1 - \beta\cos\theta)^3}.$$

On expressing this in terms of the reduced observer time t_p, we obtain oscillating fields that are identical to those in (7.20) and (7.21), which we derived from the Liénard–Wiechert expression.

7.2.3 The undulator radiation in the frequency domain

The undulator radiation field $\mathbf{E}(t_p)$ was derived before in the time domain. We take the Fourier transformation of this field:

$$\tilde{\mathbf{E}}(\omega) = \frac{1}{\sqrt{2\pi}}\int_{-\infty}^{\infty} \mathbf{E}(t_p)e^{-i\omega t_p}\,dt_p.$$

The only time-dependent part in the expression (7.20) for the electric field is the factor $\cos(\omega_1 t_p)$, and we can restrict ourself to the Fourier transform for this term:

$$\mathcal{F}(\cos(\omega_1 t_p)) = \frac{1}{\sqrt{2\pi}} \int_{N_u\pi/\omega_1}^{N_u\pi/\omega_1} \cos(\omega_1 t_p)\, e^{-i\omega t_p}\, dt_p.$$

We take $u = \omega_1 t_p$ as the new integration variable which goes from $-\pi N_u$ to πN_u during the passage of the particle through the undulator with N_u periods. This gives

$$\mathcal{F}(\cos(\omega_1 t_p)) = \frac{1}{\sqrt{2\pi}\,\omega_1} \int_{-\pi N_u}^{\pi N_u} \cos(u)\cos(u\omega/\omega_1)\, du$$

$$= \frac{1}{\sqrt{2\pi}} \frac{\pi N_u}{\omega_1} \left(\frac{\sin((\omega/\omega_1 - 1)\pi N_u)}{(\omega/\omega_1 - 1)\pi N_u} + \frac{\sin((\omega/\omega_1 + 1)\pi N_u)}{(\omega/\omega_1 + 1)\pi N_u} \right)$$

$$= \frac{1}{\sqrt{2\pi}} \frac{\pi N_u}{\omega_1} \frac{\sin((\Delta\omega/\omega_1)\pi N_u)}{(\Delta\omega/\omega_1)\pi N_u} \left(1 + \frac{\Delta\omega/\omega_1}{2 + \Delta\omega/\omega_1} \right), \tag{7.29}$$

where we introduced the relative deviation of the observed frequency ω from ω_1:

$$\Delta\omega = \omega - \omega_1.$$

The function $\sin z/z$ appearing above has the first zero at $z = \pi$, i.e. at $\Delta\omega/\omega_1 = 1/N_u$, which is small for a large number of periods. We assume now that $N_u \gg 1$, in which case the second term in the brackets of (7.29) can be neglected relative to unity and we obtain approximately

$$\mathcal{F}(\cos(\omega_1 t_p)) = \frac{1}{\sqrt{2\pi}} \frac{\pi N_u}{\omega_1} \frac{\sin\left(\dfrac{\Delta\omega}{\omega_1}\pi N_u \right)}{\dfrac{\Delta\omega}{\omega_1}\pi N_u},$$

which is shown in Fig. 7.4. The Fourier transformation of the electric field (7.21) is therefore

$$\tilde{E}_\perp(\omega) = \frac{e\Omega_u K_u \gamma^3}{\pi\epsilon_0 c r_p} \frac{[1 - \gamma^2\theta^2\cos(2\phi),\ -\gamma^2\theta^2\sin(2\phi)]}{(1 + \gamma^2\theta^2)^3} \frac{\pi N_u}{\sqrt{2\pi}\omega_1} \frac{\sin\left(\dfrac{\Delta\omega}{\omega_1}\pi N_u \right)}{\dfrac{\Delta\omega}{\omega_1}\pi N_u}.$$

At a given angle θ, the radiation field has a basic frequency ω_1 with a distribution of the form $\sin z/z$ around it. It has a typical width $\Delta\omega/\omega_1 = 1/N_u$ where the field vanishes. For a large number of periods the field becomes more and more concentrated around ω_1.

7.2.4 A discussion of the weak-undulator radiation field

The electric field emitted in a weak plane undulator with many periods and observed from a large distance can be written in the ultra-relativistic approximation in the time and frequency

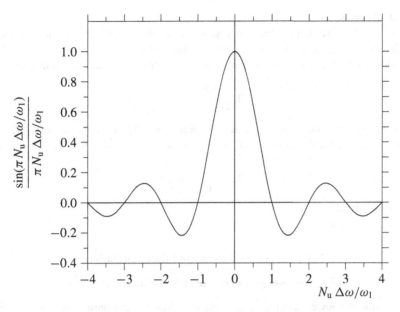

Fig. 7.4. The frequency distribution of the field around ω_1.

domains as

$$
\tilde{\mathbf{E}}_\perp(t_p) = \hat{E}_u \frac{[1 - \gamma^2\theta^2\cos(2\phi), -\gamma^2\theta^2\sin(2\phi)]}{(1 + \gamma^2\theta^2)^3} \cos(\omega_1 t_p)
$$

$$
\tilde{\mathbf{E}}_\perp(\omega) = \hat{E}_u \frac{[1 - \gamma^2\theta^2\cos(2\phi), -\gamma^2\theta^2\sin(2\phi)]}{(1 + \gamma^2\theta^2)^3} \sqrt{\frac{\pi}{2} \frac{N_u}{\omega_1}} \frac{\sin\left(\dfrac{\Delta\omega}{\omega_1}\pi N_u\right)}{\dfrac{\Delta\omega}{\omega_1}\pi N_u}
$$

(7.30)

with the on-axis amplitude used before in (3.11),

$$
\hat{E}_u = c\hat{B}_u = \frac{e\Omega_u K_u \gamma^3}{\pi\epsilon_0 c r_p} = \frac{ek_u K_u \gamma^3}{\pi\epsilon_0 r_p} = \frac{4r_0 c B_0 \gamma^3}{r_p}.
$$

(7.31)

It has the following main properties.

- The magnetic field in the time or frequency domain is obtained from the relation

$$
\mathbf{B} = \frac{[\mathbf{n} \times \mathbf{E}]}{c}.
$$

- The horizontal electric- and the vertical magnetic-field components E_x and B_y represent the horizontal polarization or σ-mode. They have a maximum on-axis $\theta = 0$ with amplitudes $\hat{E}_u$ and $\hat{E}_u/c$, respectively.
- The vertical electric- and the horizontal magnetic-field components represent the vertical polarization or π-mode. They vanish not only on the axis but also in the horizontal ($\phi = 0$) and vertical

($\phi = \pi/2$) planes.

- The horizontal and vertical electric-field components are both in phase, which indicates that the polarization is everywhere linear without any circular component.
- The fields observed as functions of time t_p consist of harmonic oscillations with frequency

$$\omega_1 = \Omega_u \frac{2\gamma^2}{1 + \gamma^2\theta^2} = \frac{\omega_{10}}{1 + \gamma^2\theta^2},$$

having a maximum value of ω_{10} on the axis. At an angle $\theta = 1/\gamma$ the frequency is half this value.
- The wave from an undulator of length $L_u = N_u\lambda_u$, observed at an angle θ, lasts for a time T_p,

$$T_p = \frac{L_u}{c} \frac{1 + \gamma^2\theta^2}{2\gamma^2},$$

and contains N_u periods.

7.3 Properties of weak-undulator radiation

7.3.1 The energy and power radiated in an undulator

The instantaneous power emitted by one electron traversing a harmonic undulator is, according to the general expression (3.13),

$$P(t') = \frac{2r_0 c^3 e^2 E_e^2 B_0^2 \cos^2(\Omega_u t')}{3(m_0 c^2)^3}$$

with $E_e = m_0 c^2 \gamma$ being the energy of the particle. We are interested not in the fast variation with time of the power but rather in its average value. We use the undulator parameters

$$L_u = N_u\lambda_u, \qquad K_u = \frac{eB_0}{m_0 c k_u} \quad \text{and} \quad \langle B^2 \rangle = \frac{B_0^2}{T_0} \int_0^{T_0} \cos^2(\Omega_u t') \, dt' = \frac{1}{2} B_0^2$$

to express the *emitted power averaged over one period* and the energy radiated in one traversal through the undulator by one electron,

$$\langle P \rangle = P_u = \frac{P_s}{2}, \qquad U_u = \frac{P_u L_u}{c},$$

giving

$$P_u = \frac{r_0 c^3 e^2 E_e^2 B_0^2}{3(m_0 c^2)^3} = \frac{r_0 c m_0 c^2 \gamma^2 k_u^2 K_u^2}{3} = \frac{e^2 c \gamma^2 k_u^2 K_u^2}{12\pi \epsilon_0}$$

$$U_u = \frac{r_0 c^2 e^2 E_e^2 B_0^2 L_u}{3(m_0 c^2)^3} = \frac{r_0 m_0 c^2 \gamma^2 k_u^2 K_u^2 L_u}{3} = \frac{e^2 \gamma^2 k_u K_u^2 N_u}{6\epsilon_0}.$$

(7.32)

So far we have calculated the power or energy emitted by a single particle of charge e. We can also give it for a beam of many particles n_e, representing a current $I = e\dot{n}_e$ with $\dot{n}_e$ being the number of electrons entering an undulator per unit time. In most practical cases there is no systematic time relation between the fields emitted by individual particles and the total power radiated by the current is just the sum of the individual power contributions

$$P_{uI} = \dot{n}_e U_u = I U_u / e,$$

giving

$$P_{uI} = \frac{r_0 c^2 e E_e^2 I B_0^2 L_u}{3(m_0 c^2)^3} = \frac{r_0 m_0 c^2 \gamma^2 I k_u^2 K_u^2 N_u \lambda_u}{3e} = \frac{e \gamma^2 I k_u K_u^2 N_u}{6 \epsilon_0}. \tag{7.33}$$

In Chapter 15 we will discuss cases of coherent radiation in which groups of particles radiate in phase for certain frequencies, leading to an enhancement of the emitted power. We exclude this rather rare condition here.

7.3.2 The angular spectral power distribution

The angular spectral energy density of undulator radiation can be obtained from the expression (3.17),

$$\frac{d^2 U}{d\Omega \, d\omega} = \frac{2 r_p^2 |\tilde{\mathbf{E}}(\omega)|^2}{\mu_0 c}.$$

Since $\theta \ll 1$, the solid angle $d\Omega$ can be approximated in the ultra-relativistic case by

$$d\Omega = \sin\theta \, d\theta \, d\phi \approx \theta \, d\theta \, d\phi.$$

The average power distribution is

$$\frac{d^2 P_u}{d\Omega \, d\omega} = \frac{d^2 U}{d\Omega \, d\omega} \frac{c}{L_u} = \frac{2 r_p^2 |\tilde{\mathbf{E}}(\omega)|^2}{\mu_0 L_u}. \tag{7.34}$$

From (7.30) we obtain for the spectral power density of undulator radiation

$$\frac{d^2 P_u}{d\Omega \, d\omega} = \frac{r_0 c m_0 c^2 k_u^2 K_u^2 \gamma^4}{\pi}$$

$$\times \frac{[(1 - \gamma^2 \theta^2 \cos(2\phi))^2 + (\gamma^2 \theta^2 \sin(2\phi))^2]}{(1 + \gamma^2 \theta^2)^5} \frac{N_u}{\omega_1} \left(\frac{\sin(\pi N_u \, \Delta\omega / \omega_1)}{\pi N_u \, \Delta\omega / \omega_1} \right)^2, \tag{7.35}$$

where we use again the classical electron radius $r_0 = e^2 / (4\pi \epsilon_0 m_0 c^2)$ introduced before in (3.8). The two terms in the square brackets correspond to the horizontal and vertical electric fields, i.e. the σ and π polarization modes.

We give the angular spectral power distribution in compact form by expressing the factor in front by the total power (7.32),

$$\frac{r_0 c m_0 c^2 k_u^2 K_u^2 \gamma^4}{\pi} = P_u \frac{3 \gamma^2}{\pi},$$

and we use three normalized angular-distribution functions, $F_{u\sigma}(\theta, \phi)$, $F_{u\pi}(\theta, \phi)$, and $F_u(\theta, \phi)$,

$$\boxed{\frac{d^2 P_u}{d\Omega \, d\omega} = P_u \gamma^2 (F_{u\sigma}(\theta, \phi) + F_{u\pi}(\theta, \phi)) f_N(\Delta\omega)} \tag{7.36}$$

with

$$F_{u\sigma}(\theta, \phi) = \frac{3}{\pi} \frac{(1 - \gamma^2 \theta^2 \cos(2\phi))^2}{(1 + \gamma^2 \theta^2)^5}, \qquad F_{u\pi}(\theta, \phi) = \frac{3}{\pi} \frac{(\gamma^2 \theta^2 \sin(2\phi))^2}{(1 + \gamma^2 \theta^2)^5}$$

$$F_u(\theta, \phi) = F_{u\sigma} + F_{u\sigma} = \frac{3}{\pi} \frac{[1 - 2\gamma^2 \theta^2 \cos(2\phi) + \gamma^4 \theta^4]}{(1 + \gamma^2 \theta^2)^5} \tag{7.37}$$

and the spectral function $f_N(\Delta\omega)$,

$$f_N(\Delta\omega) = \frac{N_u}{\omega_1} \left(\frac{\sin(\pi N_u \, \Delta\omega/\omega_1)}{\pi N_u \, \Delta\omega/\omega_1} \right)^2, \qquad \omega_1 = \frac{2\gamma^2 \Omega_u}{1 + \gamma^2 \theta^2}, \qquad \Delta\omega = \omega - \omega_1$$

$$f_N(\Delta\omega) \rightarrow \delta(\omega - \omega_1) \quad \text{for} \quad N_u \rightarrow \infty, \qquad \int_{-\infty}^{\infty} f_N(\Delta\omega) \, d\omega = 1, \tag{7.38}$$

which is plotted in Fig. 7.5.

This expression for the angular spectral power density contains the total power P_u as a factor. On replacing it by the total energy U_u we obtain the spectral angular energy

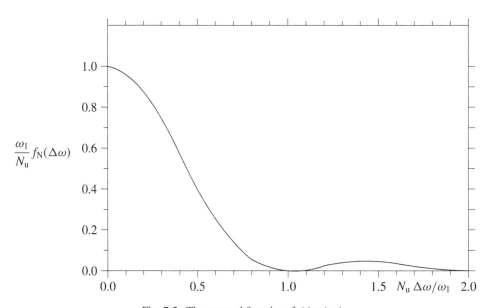

Fig. 7.5. The spectral function $f_N(\Delta\omega/\omega_1)$.

distribution. The functions $F_{u\sigma}$ and $F_{u\pi}$ represent basically the angular distributions. They depend only on the product $\gamma^2\theta^2$, not on γ or θ individually. We will now discuss the dependences of undulator radiation on the angle and frequency separately.

7.3.3 The angular power distribution

To obtain the angular distribution, we integrate (7.36) over the frequency deviation $\Delta\omega = \omega - \omega_1$. For a large number of periods, $N_u \gg 1$, this is reduced to an integration of the spectral function $f_N(\Delta\omega)$:

$$\frac{dP_u}{d\Omega} = P_u\gamma^2[F_{u\sigma}(\theta, \phi) + F_{u\pi}(\theta, \phi)] \int_{-\infty}^{\infty} f_N(\Delta\omega)\, d\omega.$$

According to (7.38) the above integral gives unity, resulting in the angular power distribution

$$
\begin{aligned}
\frac{dP_u}{d\Omega} &= P_u\gamma^2[F_{u\sigma}(\theta, \phi) + F_{u\pi}(\theta, \phi)] \\
&= P_u\frac{3\gamma^2}{\pi}\frac{[(1 - \gamma^2\theta^2\cos(2\phi))^2 + (\gamma^2\theta^2\sin(2\phi))^2]}{(1 + \gamma^2\theta^2)^5} \\
&= P_u\frac{3\gamma^2}{\pi}\frac{[1 - 2\gamma^2\theta^2\cos(2\phi) + \gamma^4\theta^4]}{(1 + \gamma^2\theta^2)^5}.
\end{aligned}
\tag{7.39}
$$

We summarize the properties of the angular power distribution.

- In the first two lines of the above expression we keep the two modes of polarization separated. The π-mode vanishes at $\phi = 0$ and $\phi = \pi/2$, i.e. in the median plane and the perpendicular (y, z)-plane. This can be understood from the picture of undulator radiation being Lorentz-transformed fields emitted by an oscillating dipole, as illustrated in Fig. 7.2. In the two planes corresponding to $\phi^* = 0$ and $\phi^* = \pi/2$ the electric field emitted by the dipole has no component in the y^*-direction that could give rise to the vertical polarization of the undulator radiation. The π-mode of the power distribution is proportional to $\sin^2(2\phi) = (1 - \cos(4\phi))/2$ and has therefore a four-fold axial symmetry.
- The σ-mode of the polarization vanishes in the two directions $\phi = 0$, $\theta = 1/\gamma$ and $\phi = \pi$, $\theta = 1/\gamma$. These two values correspond to the $\pm x^*$-directions of the dipole axis in the moving system, along which no radiation is emitted.
- It is interesting to note that the sum $F_u(\theta, \phi)$ of the two normalized modes of polarization is identical to the expression (3.14) for the instantaneous distribution of the radiation due to a transverse acceleration shown in Fig. 3.5. Since the maximum angle, $\psi_0 < 1/\gamma$, of the trajectory in a weak undulator is smaller than the natural radiation opening angle, this instantaneous distribution can be observed.

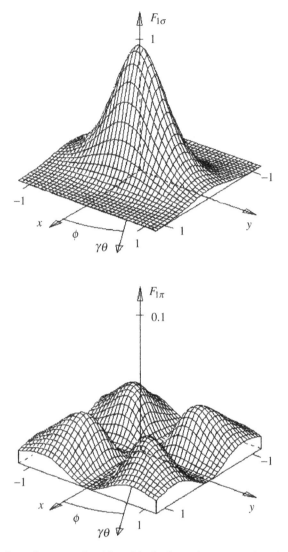

Fig. 7.6. Normalized angular power densities of the horizontal (upper) and vertical (lower) modes of polarization for weak-undulator radiation.

To illustrate the angular distribution we plot in Fig. 7.6 the angular distribution for the two modes of polarization and in Fig. 7.7 two cuts along the planes $\phi = 0$ and $\phi = \pi/2$. The latter represent the σ-mode only since the π-mode vanishes here. Figure 7.7 is therefore identical to Fig. 3.4, where we plotted the instantaneous power distributions in the same planes.

Instead of the spherical coordinates ϕ and θ we use now the angles ξ and ψ, lying in the (x, z)- and (y, z)-planes, respectively, as shown in Fig. 6.1. The two sets of coordinates

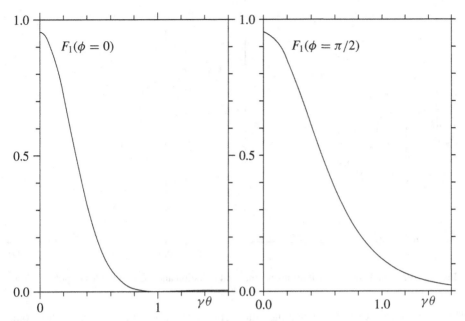

Fig. 7.7. Cuts through the power distribution for $\gamma \gg 1$.

are related by

$$\psi = \theta \sin\phi, \qquad \xi = \theta \cos\phi, \qquad d\Omega = \theta \, d\phi \, d\theta = d\xi \, d\psi.$$

The angular distribution expressed in these angles is given by the two functions

$$F_{u\sigma} = \frac{3}{\pi} \frac{(1 - \gamma^2\xi^2 + \gamma^2\psi^2)^2}{(1 + \gamma^2\xi^2 + \gamma^2\psi^2)^5}, \qquad F_{u\pi} = \frac{3}{\pi} \frac{(2\gamma\xi\gamma\psi)^2}{(1 + \gamma^2\xi^2 + \gamma^2\psi^2)^5},$$

and the total angular distribution becomes

$$\frac{dP}{d\Omega} = P_u \frac{3\gamma^2}{\pi} \frac{1 - 2\gamma^2\xi^2 + 2\gamma^2\psi^2 + \gamma^4\xi^4 + 2\gamma^2\xi^2\gamma^2\psi^2 + \gamma^4\psi^4}{(1 + \gamma^2\xi^2 + \gamma^2\psi^2)^5}.$$

Integrating over ξ or ψ gives the projections of the distribution onto the horizontal (x, z)- and vertical (y, z)-planes,

$$\frac{dP}{d\xi} = P_u \frac{3\gamma}{32} \left(\frac{12 - 16\gamma^2\xi^2 + 7\gamma^4\xi^4}{(1 + \gamma^2\xi^2)^{9/2}} + \frac{5\gamma^2\xi^2}{(1 + \gamma^2\xi^2)^{7/2}} \right)$$

$$\frac{dP}{d\psi} = P_u \frac{3\gamma}{32} \left(\frac{7}{(1 + \gamma^2\psi^2)^{5/2}} + \frac{5\gamma^2\psi^2}{(1 + \gamma^2\psi^2)^{7/2}} \right),$$

where we have again separated the two modes of polarization inside the large brackets. It is interesting to note that the projection on the (y, z)-plane $dP/d\psi$ has the same expression as the vertical distribution of frequency-integrated synchrotron radiation from long magnets.

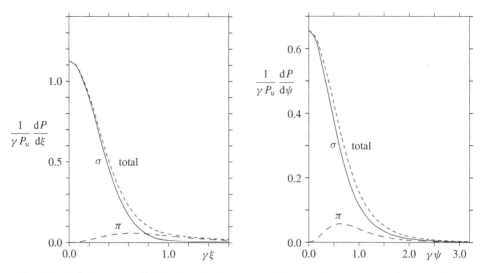

Fig. 7.8. Projected power distributions for the total radiation and the two modes of polarization.

This is, of course, expected, since the latter represents an integration over the horizontal angle, which is equivalent to a projection onto the vertical (y, z)-plane. The two projected distributions are shown in Fig. 7.8.

We calculate the variances of the two distributions:

$$\langle \xi_\sigma^2 \rangle = \frac{3}{14\gamma^2}, \qquad \langle \xi_\pi^2 \rangle = \frac{3}{2\gamma^2}, \qquad \langle \xi^2 \rangle = \frac{3}{8\gamma^2}$$

and

$$\langle \psi_\sigma^2 \rangle = \frac{1}{2\gamma^2}, \qquad \langle \psi_\pi^2 \rangle = \frac{3}{2\gamma^2}, \qquad \langle \psi^2 \rangle = \frac{5}{8\gamma^2}.$$

Summing up the variances of the angles ξ and ψ results in the variance and RMS value of the angle θ:

$$\langle \xi^2 \rangle + \langle \psi^2 \rangle = \langle \theta^2 \rangle = \frac{1}{\gamma^2}, \qquad \theta_{\text{RMS}} = \frac{1}{\gamma}.$$

We consider now the angular distribution of the radiation at a given particular frequency ω. We start with the general angular frequency distribution (7.36),

$$\frac{\mathrm{d}^2 P_\mathrm{u}}{\mathrm{d}\Omega \, \mathrm{d}\omega} = P_\mathrm{u} \gamma^2 (F_{\mathrm{u}\sigma}(\phi, \theta) + F_{\mathrm{u}\pi}(\phi, \theta)) f_\mathrm{N}(\Delta\omega)$$

and replace the frequency ω_1 in $f_\mathrm{N}(\Delta\omega)$ by the angle θ,

$$\omega_1 = \frac{\omega_{10}}{1 + \gamma^2 \theta^2}, \tag{7.40}$$

giving

$$f_N(\theta, \omega) = \frac{(1 + \gamma^2\theta^2)N_u}{\omega_{10}} \left(\frac{\sin((\omega(1 + \gamma^2\theta^2) - \omega_{10})\pi N_u/\omega_{10})}{(\omega(1 + \gamma^2\theta^2) - \omega_{10})\pi N_u/\omega_{10}} \right)^2.$$

For $\omega < \omega_{10}$ and $N_u \gg 1$, this angular distribution of the monochromatized radiation forms a cone around the axis with half opening angle θ_0 and a conical region of width $\Delta\theta_0$:

$$\theta_0 = \sqrt{\frac{\omega_{10} - \omega}{\omega}} \frac{1}{\gamma}, \qquad \Delta\theta_0 \approx \frac{1}{\gamma^2\theta N_u} \frac{\omega_{10}}{\omega}.$$

Of special interest is the angular distribution for the frequency $\omega = \omega_{10}$, which is often selected by a monochromator. We can obtain this power distribution directly with the help of (7.36); however, we will later also need the electric field. In (7.30), we set $\omega = \omega_{10}$ and express ω_1 in terms of θ;

$$\tilde{\mathbf{E}}_\perp(\omega_{10}) = \frac{e\gamma^3\Omega_u K_u}{\sqrt{2\pi}\,\pi\,\epsilon_0 c r_p}$$

$$\times \frac{[1 - \gamma^2\theta^2\cos(2\phi), \; -\gamma^2\theta^2\sin(2\phi)]}{(1 + \gamma^2\theta^2)^3} \frac{\pi N_u}{\omega_1} \frac{\sin(\gamma^2\theta^2\pi N_u)}{\gamma^2\theta^2\pi N_u},$$

where the negligible z-component is omitted and the remaining two components are marked with a subscript '$\perp$.' The last term of the above expression is large only for $\gamma\theta < 1/\sqrt{N_u} \ll 1$ and we can neglect $\gamma^2\theta^2$ terms relative to unity in the rest of the expression. The resulting field has only a horizontal component:

$$|\tilde{\mathbf{E}}(\omega)| = \tilde{E}_x(\omega) = \frac{e\gamma^3\Omega_u K_u}{\sqrt{2\pi}\,\pi\,\epsilon_0 c r_p} \frac{\pi N_u}{\omega_{10}} \frac{\sin(\gamma^2\theta^2\pi N_u)}{\gamma^2\theta^2\pi N_u}. \tag{7.41}$$

We calculate the spectral angular power density at $\omega = \omega_{10}$,

$$\frac{d^2 P_u}{d\Omega\, d\omega} = P_u \gamma^2 \frac{3}{\pi} \frac{N_u}{\omega_{10}} \left(\frac{\sin(\gamma^2\theta^2\pi N_u)}{\gamma^2\theta^2\pi N_u} \right)^2, \tag{7.42}$$

and plot the resulting distribution in Fig. 7.9. It has a diverging variance $\langle\theta^2\rangle \to \infty$. The unphysical termination of the undulator field at $\pm L_u/2$ creates high frequencies in the spectrum. As a consequence the selected frequency ω_{10} is still present at relatively large angles. In many cases only the central part of the above distribution is of interest, and we approximate it by an exponential, having at the origin the same value and the same first non-vanishing derivative,

$$\left(\frac{\sin(\gamma^2\theta^2\pi N_u)}{\gamma^2\theta^2\pi N_u} \right)^2 \approx e^{-(\gamma^2\theta^2\pi N_u)^2/3}, \tag{7.43}$$

giving

$$\frac{d^2 P_u}{d\Omega\, d\omega} = P_u \gamma^2 \frac{3}{\pi} \frac{N_u}{\omega_{10}} e^{-(\gamma^2\theta^2\pi N_u)^2/3}.$$

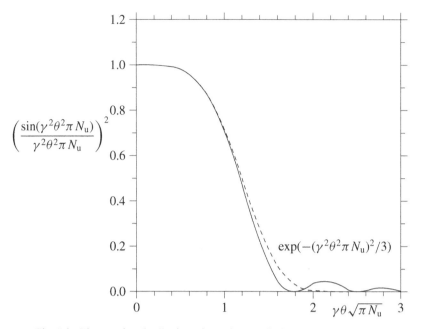

Fig. 7.9. The angular distribution of undulator radiation filtered at $\omega = \omega_{10}$.

with the RMS value

$$\theta_{\text{RMS}} = \frac{\sqrt[4]{3\pi}}{\pi} \frac{1}{\gamma \sqrt{N_{\text{u}}}} = \frac{0.5577}{\gamma \sqrt{N_{\text{u}}}}, \tag{7.44}$$

which can be used as an approximation.

7.3.4 The spectral power distribution

We calculate the spectrum of the undulator radiation with respect to the proper frequency ω_1, which consists actually only of a conversion of the observation angle into frequency. For a large period number, $N_{\text{u}} \gg 1$, we observe at a given angle θ a narrow band around ω_1 given by

$$\omega \approx \omega_1 = \frac{\omega_{10}}{1 + \gamma^2 \theta^2} = \frac{2 k_{\text{u}} c \gamma^2}{1 + \gamma^2 \theta^2}$$

$$\tag{7.45}$$

$$d\omega_1 = -\frac{2\omega_{10}\gamma^2}{(1 + \gamma^2\theta^2)^2}\theta \, d\theta = -2\omega_{10}\gamma^2 \left(\frac{\omega_1}{\omega_{10}}\right)^2 \theta \, d\theta.$$

Increasing θ reduces the frequency, which is expressed by the minus sign in the relation between the differentials. This is of no importance since it will just interchange the limits of integration. With the above relation we can convert the angular distribution into a power density with respect to the nominal frequency ω_1 at an angle θ:

$$\frac{dP}{d\omega_1} = \frac{1}{2\omega_{10}\gamma^2}\left(\frac{\omega_{10}}{\omega_1}\right)^2 \frac{dP}{\theta\,d\theta} = \frac{1}{2\omega_{10}\gamma^2}\left(\frac{\omega_{10}}{\omega_1}\right)^2 \int_0^{2\pi}\frac{dP}{d\Omega}\,d\phi$$

$$= P_u\frac{1}{2\omega_{10}}\left(\frac{\omega_{10}}{\omega_1}\right)^2 \int_0^{2\pi} F_u(\theta,\phi)\,d\phi.$$

We integrate the angular power distribution functions over the azimuthal angle ϕ,

$$\int_{-\pi}^{\pi} F_{u\sigma}\,d\phi = \frac{3(2+\gamma^4\theta^4)}{(1+\gamma^2\theta^2)^5}, \qquad \int_{-\pi}^{\pi} F_{u\pi}\,d\phi = \frac{3\gamma^4\theta^4}{(1+\gamma^2\theta^2)^5},$$

and obtain for the power distribution in θ

$$\frac{dP}{\theta\,d\theta} = 3\gamma^2 P_u\frac{[(2+\gamma^4\theta^4)+(\gamma^4\theta^4)]}{(1+\gamma^2\theta^2)^5}. \tag{7.46}$$

Using (7.45), this is converted into a distribution with respect to ω_1:

$$\frac{dP}{d\omega_1} = \frac{3P_u}{\omega_{10}}\frac{\omega_1}{\omega_{10}}\left(1 - 2\frac{\omega_1}{\omega_{10}} + 2\left(\frac{\omega_1}{\omega_{10}}\right)^2\right). \tag{7.47}$$

We would like to obtain the distribution with respect to the frequency ω, which is a little different from the above for a finite number of periods:

$$\frac{dP}{d\omega} = \int_0^{\omega_{10}} \frac{d^2P}{d\omega_1\,d\omega}\,d\omega_1 = \int_{-\infty}^{\infty}\frac{dP}{d\omega_1} f_N(\Delta\omega)\,d\omega_1$$

$$= \frac{3P_u}{\omega_{10}}\int_0^{\omega_{10}}\frac{\omega_1}{\omega_{10}}\left(1 - 2\frac{\omega_1}{\omega_{10}} + 2\left(\frac{\omega_1}{\omega_{10}}\right)^2\right) f_N(\Delta\omega)\,d\omega_1. \tag{7.48}$$

We assume now that we have a very large number of undulator periods and approximate the spectral function by the δ-function;

$$f_N(\Delta\omega) \rightarrow \delta(\omega - \omega_1) \quad \text{for} \quad N_u \rightarrow \infty,$$

whereupon the integral (7.48) vanishes except if $\omega = \omega_1$. This results in power spectra of the two modes of polarization and the total radiation:

$$\frac{dP_{u\sigma}}{d\omega} = \frac{3P_u}{\omega_{10}}\frac{\omega}{\omega_{10}}\left(\frac{1}{2} - \frac{\omega}{\omega_{10}} + \frac{3}{2}\left(\frac{\omega}{\omega_{10}}\right)^2\right)$$

$$\frac{dP_{u\pi}}{d\omega} = \frac{3P_u}{\omega_{10}}\frac{\omega}{\omega_{10}}\left(\frac{1}{2} - \frac{\omega}{\omega_{10}} + \frac{1}{2}\left(\frac{\omega}{\omega_{10}}\right)^2\right) \tag{7.49}$$

$$\frac{dP_u}{d\omega} = \frac{3P_u}{\omega_{10}}\frac{\omega}{\omega_{10}}\left(1 - 2\frac{\omega}{\omega_{10}} + 2\left(\frac{\omega}{\omega_{10}}\right)^2\right).$$

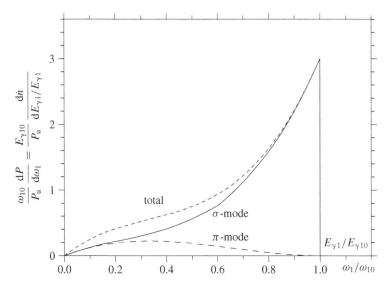

Fig. 7.10. The spectral power density or photon distribution per relative energy band of weak undulator radiation with $N_u \gg 1$.

This spectrum is shown in Fig. 7.10 for the two modes of polarization and the total radiation. It is interesting to note that, at very small frequencies, $\omega \ll \omega_{10}$, we have

$$\frac{dP_{u\sigma}}{d\omega_1} = \frac{dP_{u\pi}}{d\omega_1} \approx \frac{3P_u}{2\omega_{10}} \frac{\omega_1}{\omega_{10}},$$

and the two components of the polarization have the same intensity and increase proportionally to ω. It should be noted that this spectrum is obtained by converting angles into frequency. The full spectrum is therefore observed only if we accept the total solid angle of the radiation. Replacing the spectral function $f_N(\Delta\omega)$ by the δ-function is for $N_u \gg 1$ a good approximation for the whole spectrum except around $\omega \approx \omega_{10}$. There, we have to replace (7.49) by the proper convolution (7.48), as shown in the example in Fig. 7.11, where the sharp drop is rounded off.

For a smaller number of periods the more general expression (7.29) for the spectral function has to be used. In this case the end fields of the undulators can have a significant influence on the spectrum and should be included in a numerical computation.

7.4 The photon distribution

7.4.1 The number and energy of photons

We can express the spectrum as the power or the photon flux per frequency band. Each emitted photon of frequency ω has a certain energy

$$E_\gamma = \hbar\omega \tag{7.50}$$

with $\hbar = h/(2\pi)$ and $h = 6.6262 \times 10^{-34}$ being Planck's constant. If $d\dot{n}$ photons are emitted with energy E_γ in a band dE_γ, the power carried by them is $dP = E_\gamma\,d\dot{n}$. This relation was used before in (5.44) to convert a spectral power distribution of synchrotron radiation into one of the photon flux and is applied now to undulators:

$$\frac{d\dot{n}}{d\Omega} = \frac{1}{E_\gamma}\frac{dP}{d\Omega} = \frac{1}{\hbar\omega}\frac{dP}{d\Omega}$$

$$\frac{d\dot{n}}{d\omega_1} = \frac{1}{E_{\gamma 1}}\frac{dP}{d\omega_1} = \frac{1}{\hbar\omega_1}\frac{dP}{d\omega_1} \quad \text{or} \quad \frac{d\dot{n}}{d\omega_1/\omega_1} = \frac{1}{\hbar}\frac{dP}{d\omega_1}, \tag{7.51}$$

$$\frac{d^2\dot{n}}{d\Omega\,d\omega/\omega} = \frac{1}{\hbar}\frac{d^2P}{d\Omega\,d\omega}.$$

The bottom equation gives the number of photons radiated per unit time into a solid angle $d\Omega$ and a relative frequency $d\omega/\omega$ or photon-energy increment $d\omega/\omega = dE_\gamma/E_\gamma$.

From the power spectrum (7.48) with respect to the proper frequency ω_1 we obtain the photon flux per *relative* proper frequency or energy band,

$$\frac{d\dot{n}}{d\omega_1/\omega_1} = \frac{d\dot{n}}{dE_{\gamma 1}/E_{\gamma 1}} = \frac{3P_u}{E_{\gamma 10}}\frac{E_{\gamma 1}}{E_{\gamma 10}}\left(1 - 2\frac{E_{\gamma 1}}{E_{\gamma 10}} + 2\left(\frac{E_{\gamma 1}}{E_{\gamma 10}}\right)^2\right), \tag{7.52}$$

with the proper and maximum photon energies

$$E_{\gamma 1} = \hbar\omega_1, \qquad E_{\gamma 10} = \hbar\omega_{10}. \tag{7.53}$$

This distribution is shown in Figs. 7.10 and 7.11 and can be given in terms of the photon energy or frequency $E_\gamma/E_{\gamma 10} = \omega/\omega_{10}$.

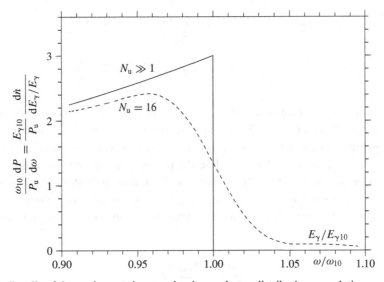

Fig. 7.11. Details of the total spectral power density or photon distribution per relative energy band of weak-undulator radiation for a finite number N_u of periods.

Integrating these spectra (7.53) over all photon energies gives the total number of photons emitted per unit time, called the photon flux, by an electron traversing the undulator. We also calculate the average value of the photon energy, its variances for the two modes of polarization, and the total radiation emitted by one electron:

$$
\begin{aligned}
\dot{n}_{u\sigma} &= \frac{3}{2}\frac{P_u}{E_{\gamma 10}}, & \dot{n}_{u\pi} &= \frac{1}{2}\frac{P_u}{E_{\gamma 10}}, & \dot{n}_u &= 2\frac{P_u}{E_{\gamma 10}} \\
\langle E_{\gamma\sigma}\rangle &= \frac{7}{12}E_{\gamma 10}, & \langle E_{\gamma\pi}\rangle &= \frac{3}{12}E_{\gamma 10}, & \langle E_\gamma\rangle &= \frac{6}{12}E_{\gamma 10} \\
\langle E_{\gamma\sigma}^2\rangle &= \frac{26}{60}E_{\gamma 10}^2, & \langle E_{\gamma\pi}^2\rangle &= \frac{6}{60}E_{\gamma 10}^2, & \langle E_\gamma^2\rangle &= \frac{21}{60}E_{\gamma 10}^2.
\end{aligned}
\tag{7.54}
$$

We refer here to the photon energy E_γ but could also give the expressions in terms of the frequency $\omega = E_\gamma/\hbar$. Of the total number of photons emitted, three quarters correspond to the horizontal polarization and one quarter to the vertical one. The average photon energy of the total radiation is just half of the peak energy.

We now express the photon flux $\dot{n}_u$ in terms of basic parameters and give also the number of photons n_u radiated by the electron during one traversal and the flux due to a beam current I, which represents I/e electrons entering the undulator per unit time:

$$
\begin{aligned}
\dot{n}_u &= \frac{2P_u}{E_{\gamma 10}} = \frac{P_u}{\hbar c k_u \gamma^2} = \frac{r_0 m_0 c^2 k_u K_u^2}{3\hbar} = \frac{\alpha_f c k_u K_u^2}{3} \\
n_u &= \frac{\dot{n}_u L_u}{c} = \frac{2U_u}{E_{\gamma 10}} = \frac{2\pi r_0 m_0 c^2 K_u^2 N_u}{3\hbar c} = \frac{2\pi \alpha_f K_u^2 N_u}{3} \\
\dot{n}_{uI} &= \frac{n_u I}{e} = \frac{2U_u I}{e E_{\gamma 10}} = \frac{2\pi r_0 m_0 c^2 I K_u^2 N_u}{3e\hbar c} = \frac{2\pi \alpha_f I K_u^2 N_u}{3e}.
\end{aligned}
\tag{7.55}
$$

It is interesting to compare this with the number of photons emitted in ordinary synchrotron radiation presented in (5.48), expressed in a similar way. The two cases have different γ-dependences. The radiated power in ordinary synchrotron radiation for a fixed bending radius is proportional to γ^4 while the photon energies are proportional to γ^3. For a weak undulator we have a fixed magnetic field, making both the total power and the photon energies proportional to γ^2 and resulting in a number of photons that is independent of γ. The number of photons emitted by one electron during one undulator period is

$$
\frac{n_u}{N_u} = \frac{2\pi \alpha_f K_u^2}{3}.
$$

This is much smaller than unity, which is of interest for discussing interference for a small number of photons.

7.4.2 The photon spectrum

For an undulator with many periods, $N_u \gg 1$, we can substitute $\omega \approx \omega_1$ into (7.52) and obtain the photon flux per relative observed frequency $d\omega/\omega$ or energy band dE_γ/E_γ. We now give the results also for the two modes of polarization and use the total photon flux $\dot{n}_u = 2P_u/E_{\gamma 10}$ to express the spectrum

$$\frac{d\dot{n}_\sigma/\dot{n}_u}{d\omega/\omega} = \frac{d\dot{n}_\sigma/\dot{n}_u}{dE_\gamma/E_\gamma} = \frac{3}{2}\frac{\omega}{\omega_{10}}\left(\frac{1}{2} - \frac{\omega}{\omega_{10}} + \frac{3}{2}\left(\frac{\omega}{\omega_{10}}\right)^2\right)$$

$$\frac{d\dot{n}_\pi/\dot{n}_u}{d\omega/\omega} = \frac{d\dot{n}_\pi/\dot{n}_u}{dE_\gamma/E_\gamma} = \frac{3}{2}\frac{\omega}{\omega_{10}}\left(\frac{1}{2} - \frac{\omega}{\omega_{10}} + \frac{1}{2}\left(\frac{\omega}{\omega_{10}}\right)^2\right) \qquad (7.56)$$

$$\frac{d\dot{n}/\dot{n}_u}{d\omega/\omega} = \frac{d\dot{n}/\dot{n}_u}{dE_\gamma/E_\gamma} = \frac{3}{2}\frac{\omega}{\omega_{10}}\left(1 - 2\frac{\omega}{\omega_{10}} + 2\left(\frac{\omega}{\omega_{10}}\right)^2\right),$$

which is shown in Figs. 7.10 and 7.11.

It is sometimes interesting to give the photon distribution with respect to the absolute frequency or photon-energy increment $dE_\gamma = \hbar\, d\omega$:

$$\frac{d\dot{n}/\dot{n}_u}{d\omega_1} = \frac{1}{\omega_1}\frac{d\dot{n}/\dot{n}_u}{d\omega_1/\omega_1} = \frac{3}{2\omega_{10}}\left(1 - 2\frac{\omega_1}{\omega_{10}} + 2\left(\frac{\omega_1}{\omega_{10}}\right)^2\right),$$

which is shown in Fig. 7.12. This distribution looks very different, being large at low frequencies because, for the same power injected into an energy band, more photons are necessary at low than at large energy. The total distribution is symmetric around $\omega/\omega_{10} = 0.5$, which is consistent with the average photon energy being half the maximum value.

7.4.3 The angular spectral photon distribution

We obtain the photon-flux distribution in terms of angle and relative frequency or energy from (7.37) using the relations (7.51),

$$\frac{d^2\dot{n}}{d\Omega\, d\omega/\omega} = P_u\frac{\gamma^2}{\hbar}F_u(\theta, \phi)f_N(\Delta\omega) = \dot{n}_u\frac{\omega_{10}}{2}F_u(\theta, \phi)f_N(\Delta\omega), \qquad (7.57)$$

or with respect to an increment in absolute frequency $d\omega$ or energy dE_γ

$$\frac{d^2\dot{n}}{d\Omega\, d\omega} = \frac{\dot{n}_u\gamma^2}{2}\frac{[F_{u\sigma}(\theta, \phi) + F_{u\pi}(\theta, \phi)]}{\omega/\omega_{10}}f_N(\Delta\omega). \qquad (7.58)$$

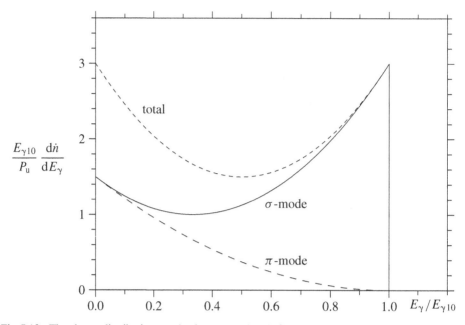

Fig. 7.12. The photon distribution per absolute energy band of weak-undulator radiation with $N_u \gg 1$.

7.4.4 The undulator radiation on the axis

Undulator radiation is mostly used close to the axis and to the maximum frequency $\theta = 0$, $\omega = \omega_{10}$, which gives, from (7.37) and (7.38),

$$F_u(0) = F_{u\sigma}(0) = \frac{3}{\pi}, \qquad f_N(\Delta\omega) = \frac{N_u}{\omega_{10}} \left(\frac{\sin(\pi N_u \, \Delta\omega / \omega_{10})}{\pi N_u \, \Delta\omega / \omega_{10}} \right)^2, \qquad \Delta\omega = \omega - \omega_{10}.$$

We obtain for the on-axis angular power and photon flux or photon distribution radiated by a single electron or a current I

$$\frac{dP}{d\Omega} = P_u \frac{3\gamma^2}{\pi}, \qquad \frac{d\dot{n}}{d\Omega} = \frac{1}{\hbar\omega_{10}} \frac{dP}{d\Omega} = \frac{P_u}{\hbar\omega_{10}} \frac{3\gamma^2}{\pi} = \frac{\alpha_f c \gamma^2 k_u K_u^2}{2\pi}$$

$$\frac{dn}{d\Omega} = \frac{d\dot{n}}{d\Omega} \frac{N_u \lambda_u}{c} = \frac{3\dot{n}_u}{2} = \alpha_f \gamma^2 K_u^2 N_u, \qquad \frac{d\dot{n}_I}{d\Omega} = \frac{\alpha_f \gamma^2 I K_u^2 N_u}{e},$$

where we used the relations (7.51) and $n_u = \dot{n}_u N_u \lambda_u / c$.

With the relation (7.45) we convert the angular distribution into one with respect to the proper frequency:

$$\frac{dP}{d\omega_1} = P_u \frac{3}{\omega_{10}}, \qquad \frac{d\dot{n}}{d\omega_1/\omega_1} = \frac{1}{\hbar} \frac{dP}{d\omega_1} = \frac{\alpha_f c k_u K_u^2}{2}$$

$$\frac{dn}{d\omega_1/\omega_1} = \pi \alpha_f K_u^2 N_u, \qquad \frac{d\dot{n}_I}{d\omega_1/\omega_1} = \frac{\pi \alpha_f I K_u^2 N_u}{e}.$$

The angular spectral distributions on the axis is

$$\frac{d^2 P}{d\Omega \, d\omega} = P_u \frac{3\gamma^2 N_u}{\pi \omega_{10}} \left(\frac{\sin(\pi N_u \, \Delta\omega/\omega_{10})}{\pi N_u \, \Delta\omega/\omega_{10}} \right)^2$$

$$\frac{d^2 \dot{n}}{d\Omega \, d\omega/\omega} = \frac{\alpha_f c \gamma^2 k_u K_u^2 N_u}{2\pi} \left(\frac{\sin(\pi N_u \, \Delta\omega/\omega_{10})}{\pi N_u \, \Delta\omega/\omega_{10}} \right)^2 .$$

If we also select the frequency $\omega = \omega_{10}$, we obtain the photon flux due to one electron traversing the undulator:

$$\frac{d^2 \dot{n}}{d\Omega \, d\omega/\omega} = \frac{\alpha_f c \gamma^2 k_u K_u^2 N_u}{2\pi} . \qquad (7.59)$$

For practical applications the photon flux emitted by a current I is important:

$$\boxed{\frac{d^2 n_I}{d\Omega \, d\omega/\omega} = \frac{I}{e} \frac{d^2 n}{d\Omega \, d\omega/\omega} = \frac{\alpha_f \gamma^2 I K_u^2 N_u^2}{e} .}$$

8

The plane strong undulator

8.1 The trajectory

8.1.1 The trajectory in the laboratory frame

We consider again a plane, harmonic undulator with period length $\lambda_u = k_u/2\pi$, as shown in Fig. 7.1. However, we allow the maximum deflecting angle ψ_0 to be larger than the natural radiation opening angle of about $1/\gamma$, resulting in a related undulator parameter (7.5),

$$K_u = \frac{eB_0}{m_0 c k_u},$$
(8.1)

which can be larger than unity. We follow here the method used in [50].

To calculate the trajectory of the particle going along the axis through the undulator, we generalize the calculation performed before for a weak-field device. It is determined by the Lorentz force, leading to a pair of differential equations (7.3):

$$\ddot{x} = \frac{d^2 x}{dt'^2} = -\frac{eB_0}{m_0 \gamma} \cos(k_u z)\, \dot{z}, \qquad \ddot{z} = \frac{d^2 z}{dt'^2} = \frac{eB_0}{m_0 \gamma} \cos(k_u z)\, \dot{x}.$$
(8.2)

The first one can be integrated,

$$\dot{x} = -\frac{eB_0}{m_0 \gamma k_u} \sin(k_u z) = -\frac{cK_u}{\gamma} \sin(k_u z),$$

satisfying the initial conditions (7.2) for our choice of an undulator field that is symmetric in z. Using the constancy of the velocity,

$$\dot{x}^2 + \dot{z}^2 = \beta^2 c^2,$$

we obtain for the z-component

$$\dot{z} = \beta c \sqrt{1 - \frac{K_u^2}{\beta^2 \gamma^2} \sin^2(k_u z)}.$$
(8.3)

From the ratio between $\dot{x}$ and $\dot{z}$ we obtain the derivative of x with respect to z:

$$x' = \frac{dx}{dz} = \frac{\dot{x}}{\dot{z}} = -\frac{K_\mathrm{u} \sin(k_\mathrm{u} z)}{\beta\gamma\sqrt{1 - \dfrac{K_\mathrm{u}^2}{\beta^2\gamma^2}\sin^2(k_\mathrm{u} z)}}. \tag{8.4}$$

The trajectory $x(z)$ is obtained from the above expression, which leads to an elliptical integral. This exact solution is of little practical interest. We see already from the above equation that, for $|K_\mathrm{u}/\beta\gamma| \geq 1$, the trajectory has points with a vertical slope and the particle might be trapped within a single undulator period.

We discuss now the approximations to be made. First we assume that the maximum angle of the trajectory stays small relative to unity:

$$\hat{x}' = \tan\psi_0 = \frac{K_\mathrm{u}}{\beta\gamma\sqrt{1 - \dfrac{K_\mathrm{u}^2}{\beta^2\gamma^2}}} \ll 1, \quad \text{giving} \quad \hat{x}' \approx \psi_0 \approx \frac{K_\mathrm{u}}{\beta\gamma} \ll 1.$$

This condition is fulfilled in all practical cases. Synchrotron radiation from electrons is only of interest in the ultra-relativistic case, for which ψ_0 is small even for relatively large values of the undulator parameter K_u. We will later consider also radiation from protons, which are not extremely relativistic. However, due to the large mass in the denominator of (8.1), the parameter K_u is in this case small for fields that can be obtained in practice.

The square of the term $K_\mathrm{u}/(\beta\gamma)$ is usually very small but we can not neglect it here. In calculating the relation between the emission and the observation time it will be compared with the normalized difference in velocity between the radiation and the particle, $1 - \beta \approx 1/(2\gamma^2)$, which is of the same order as $(K_\mathrm{u}/\gamma)^2$. For other parameters, such as the triple vector product, we can make further approximations. Therefore, we include here terms up to order $(K_\mathrm{u}/\beta\gamma)^2$ at first and see later which terms can be neglected. As a second approximation we assume that the radiation is observed from a large distance r_p, much bigger than the length L_u of the undulator,

$$\frac{L_\mathrm{u}}{r_\mathrm{p}} = \frac{2\pi N_\mathrm{u}}{r_\mathrm{p} k_\mathrm{u}} \ll 1,$$

and make the corresponding approximations later in the derivations.

With this we obtain the transverse motion up to square terms in $K_\mathrm{u}/(\beta\gamma)$,

$$x'(z) = -\frac{K_\mathrm{u}}{\beta\gamma}\sin(k_\mathrm{u} z), \qquad x(z) = \frac{K_\mathrm{u}}{\beta\gamma k_\mathrm{u}}\cos(k_\mathrm{u} z) = \hat{x}\cos(k_\mathrm{u} z), \tag{8.5}$$

where we chose the integration constant to satisfy the initial conditions (7.2) in order to obtain a symmetric trajectory with respect to z having vanishing average transverse excursion $\langle x \rangle = 0$. This is an arbitrary choice since we could equally well consider the antisymmetric trajectory. The amplitude of the trajectory is

$$\hat{x} = a = \frac{K_\mathrm{u}}{\beta\gamma k_\mathrm{u}}.$$

We would like to know the particle motion as a function of time t' and apply our approximation to the expression (8.3):

$$\dot{z} = \beta c \sqrt{1 - \frac{K_u^2}{\beta^2 \gamma^2} \sin^2(k_u z)} \approx \beta c \left(1 - \frac{K_u^2}{2\beta^2 \gamma^2} \sin^2(k_u z)\right).$$

We obtain $z(t')$ by integrating the above equation, satisfying our initial condition (7.2). This integral is listed in standard tables, giving, for $K_u^2/(2\beta^2\gamma^2) < 1$,

$$\beta c t' = \int_0^z \frac{1}{1 - \frac{K_u^2}{2\beta^2 \gamma^2} \sin^2(k_u z)} \, dz$$

$$= \frac{1}{k_u \sqrt{1 - \frac{K_u^2}{2\beta^2 \gamma^2}}} \arctan\left(\sqrt{1 - \frac{K_u^2}{2\beta^2 \gamma^2}} \tan(k_u z)\right)$$

or

$$k_u z = \arctan\left(\frac{1}{\sqrt{1 - \frac{K_u^2}{2\beta^2 \gamma^2}}} \tan\left(\sqrt{1 - \frac{K_u^2}{2\beta^2 \gamma^2}} k_u \beta c t'\right)\right).$$

We take the derivative with respect to t', obtaining in the first step

$$\dot{z}(t') = \frac{\left(1 - \frac{K_u^2}{\beta^2 \gamma^2}\right)\beta c}{1 - \frac{K_u^2}{2\beta^2 \gamma^2} \cos^2\left(\sqrt{1 - \frac{K_u^2}{2\beta^2 \gamma^2}} k_u \beta c t'\right)},$$

and develop this with respect to $K_u/(\sqrt{2}\beta\gamma)$, neglecting terms higher than the quadratic ones,

$$\dot{z}(t') \approx \beta c \left(1 - \frac{K_u^2}{2\beta^2 \gamma^2} \sin^2\left(\sqrt{1 - \frac{K_u^2}{2\beta^2 \gamma^2}} k_u \beta c t'\right)\right)$$

$$\approx \beta c \left(1 - \frac{K_u^2}{4\beta^2 \gamma^2} + \frac{K_u^2}{4\beta^2 \gamma^2} \cos\left(\left(1 - \frac{K_u^2}{4\beta^2 \gamma^2}\right) 2 k_u \beta c t'\right)\right), \quad (8.6)$$

where we used $\sin^2 x = (1 - \cos(2x))/2$ and approximated the square root.

The motion in the z-direction consists of an average drift velocity

$$\langle \dot{z} \rangle = \beta c \left(1 - \frac{K_u^2}{4\beta^2 \gamma^2}\right) = \beta^* c \quad (8.7)$$

and a modulation with frequency $2\Omega_u$ and velocity amplitude $\hat{\dot{z}}$:

$$\Omega_u = \left(1 - \frac{K_u^2}{4\beta^2\gamma^2}\right)k_u\beta c = k_u\beta^* c, \qquad \hat{\dot{z}} = \frac{K_u^2}{4\beta^2\gamma^2}\beta c.$$

From (8.7) it is evident that, by neglecting terms of order $(K_u/(\beta\gamma k_u))^2$, we would miss an important property of the motion in a strong undulator. We use here the normalized drift velocity β^* of the particle motion and also the corresponding Lorentz factor γ^*,

$$\beta^* = \beta\left(1 - \frac{K_u^2}{4\beta^2\gamma^2}\right), \qquad \gamma^* = \frac{1}{\sqrt{1 - \beta^{*2}}} \approx \frac{\gamma}{\sqrt{1 + K_u^2/2}}, \tag{8.8}$$

which we introduced briefly in the qualitative treatment presented in Chapter 6. As a further parameter we introduce K_u^*, which is related to K_u by

$$K_u^* = \frac{K_u}{\sqrt{1 + K_u^2/2}} \approx \psi_0\gamma^*, \qquad K_u = \frac{K_u^*}{\sqrt{1 - K_u^{*2}/2}}, \tag{8.9}$$

which can take a range of values $0 \le K_u^* \le \sqrt{2}$. This allows us not only to make some equations more compact but also to express them in forms similar to the ones for the weak undulator. In the qualitative treatment, we investigated the undulator radiation in a moving system with normalized velocity β^* and Lorentz factor γ^*. The transformation back to the laboratory system gives $1/\gamma^*$ for the typical opening angle of the radiation. The physical meaning of the reduced undulator parameter K_u^* is still the ratio between the trajectory angle ψ_0 and the natural opening angle $1/\gamma^*$ of the strong-undulator radiation.

We can write the longitudinal velocity and position components as functions of t'

$$\dot{z}(t') = \beta^* c + \frac{cK_u^2}{4\beta\gamma^2}\cos(2k_u\beta^* ct'), \qquad z(t') = \beta^* ct' + \frac{K_u^2}{8\beta^2\gamma^2 k_u}\sin(2k_u\beta^* ct').$$

With this we can express also the transverse motion (8.5) as a function of time:

$$\dot{x}(t') = -\frac{cK_u}{\gamma}\sin(k_u z) \approx -\frac{cK_u}{\gamma}\sin(\Omega_u t'), \qquad x(t') = \frac{K_u}{\beta\gamma k_u}\cos(\Omega_u t'),$$

where we neglected again terms of higher than second order in $K_u/(\beta\gamma)$. We summarize the particle motion in a plane strong harmonic undulator:

$$\dot{x}(t') = -\frac{cK_u}{\gamma}\sin(\Omega_u t'), \qquad \dot{z}(t') = \beta^* c + \frac{cK_u^2}{4\beta\gamma^2}\cos(2\Omega_u t')$$

$$x(t') = \frac{K_u}{\beta\gamma k_u}\cos(\Omega_u t'), \qquad z(t') = \beta^* ct' + \frac{K_u^2}{8\beta^2\gamma^2 k_u}\sin(2\Omega_u t'). \tag{8.10}$$

8.1.2 The trajectory in the moving frame

Many aspects of the particle motion and the radiation emitted in an undulator can be understood by investigating them in a frame F^* that moves with the drift velocity $\beta^* c$ along the z-axis. Using the Lorentz transformation

$$x^* = x, \qquad z^* = \gamma^*(z - \beta^* ct'), \qquad ct^* = \gamma^*(ct' - \beta^* z) \tag{8.11}$$

we obtain

$$x^*(t') = a\cos(\Omega_u ct'), \qquad z^*(t') = \frac{aK_u\sin(2\Omega_u t')}{8\beta\sqrt{1 + K_u^2/2}} = \frac{aK_u^*}{8\beta}\sin(2\Omega_u t') \tag{8.12}$$

with

$$a = \frac{K_u}{\beta\gamma k_u} = \frac{K_u^*}{\beta\gamma^* k_u}.$$

To obtain the complete motion in the moving frame we should express the coordinates as functions of the corresponding time t^* instead of the time t' of the laboratory frame. However, we concentrate here on the form of the trajectory and eliminate the time dependence in (8.12) to obtain a relation between the components of the motion:

$$z^* = \frac{aK_u^*}{4\beta}\frac{x^*}{a}\sqrt{1 - \frac{x^{*2}}{a^2}}. \tag{8.13}$$

This trajectory $z^*(x^*)$ in the moving frame is shown on the left-hand side of Fig. 8.1, normalized by $\beta\gamma k_u$. It has the form of a 'figure eight' which is very thin for small values of K_u but becomes wider for larger values. The maximum excursion in x^* is reached at $z^* = 0$ and the largest horizontal excursion is obtained at $x^* = a/\sqrt{2}$ with values

$$x_{max}^* = a, \qquad z_{max}^* = \frac{aK_u^*}{8\beta}. \tag{8.14}$$

At the origin the trajectory has an angle ψ_0^* with respect to the x^*-axis given by

$$\tan\psi_0^* = \left|\frac{dz^*}{dx^*}\right| = \frac{K_u^*}{4\beta} = \frac{2z_{max}^*}{a},$$

which grows with K_u to reach a maximum of $\sqrt{2}/(4\beta)$ for $K_u \to \infty$. On the right-hand side of Fig. 8.1 the trajectory is normalized with the maximum transverse amplitude a to show how the trajectory becomes wider with increasing K_u.

We discuss the nature of the approximation made. The expressions (8.10) for the trajectories $x(t')$ and $y(t')$ in the laboratory frame were accurate up to second order in $K_u/(\beta\gamma)$. If this term is neglected there is no z^*-motion in the moving frame and an important part of the particle trajectory and the resulting radiation from strong undulators would be missed. The expression (8.13) for the trajectory in the moving frame is now accurate up to first order in $K_u/(\beta\gamma)$. Because there is a β in the denominator in the equation for $z^*(t')$ one could think that this quantity becomes large for small velocities. Since we assumed at the beginning $K_u/(\beta\gamma) \ll 1$ we also have $K_u/\beta \ll \gamma$, which is of the order of unity for small values of β.

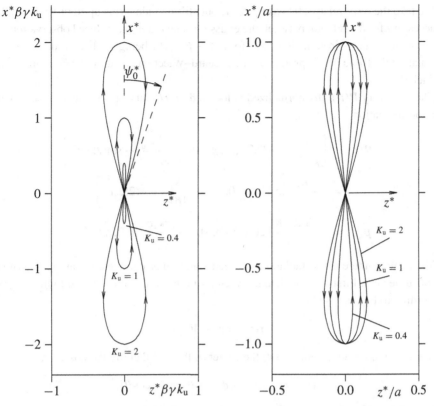

Fig. 8.1. Particle motion in a system moving with the drift velocity through a strong undulator with parameter $K_u = 0.4$, 1.0, and 2.0. Left: actual scale; right: normalized with the amplitude a.

8.1.3 The relevant motion in a strong undulator

We calculated the motion in the laboratory frame of a particle moving with velocity $v = \beta c$ along the axis through an undulator with period length λ_u in the laboratory frame, and introduced the parameters

$$k_u = \frac{2\pi}{\lambda_u}, \qquad K_u = \frac{eB_0}{m_0 c k_u}$$

$$\hat{x} = a = \frac{K_u}{\beta\gamma k_u}, \qquad \hat{x}' = \frac{K_u}{\beta\gamma} = ak_u \approx \frac{K_u}{\gamma} \approx \psi_0$$

$$\beta^* = \beta\left(1 - \frac{K_u^2}{4\beta^2\gamma^2}\right) \approx \beta, \qquad \gamma^* = \frac{1}{\sqrt{1-\beta^{*2}}} \approx \frac{\gamma}{\sqrt{1+K_u^2/2}} \qquad (8.15)$$

$$\Omega_u = k_u\beta^* c \approx k_u c, \qquad K_u^* = \frac{K_u}{\sqrt{1+K_u^2/2}},$$

which are used to express the trajectory $\mathbf{R}(t') = [x, y, z]$ and velocity $\mathbf{v}(t') = [\dot{x}, \dot{y}, \dot{z}]$ given in (8.10).

We take the ultra-relativistic case $\gamma \gg 1$, but still include terms up to order $1/\gamma^2$. This is necessary for the relation between the emission time t' and the reduced observation time t_p, for which the main term is of the form $1 - \beta \cos\theta$, being itself of order $1/\gamma^2$. For calculating the triple vector product in the Liénard–Wiechert equation, fewer terms need to be included.

The trajectory $\mathbf{R}(t')$, the normalized velocity $\boldsymbol{\beta} = \mathbf{v}/c$, and its derivative $\dot{\boldsymbol{\beta}}$ are, within our approximation,

$$\mathbf{R}(t') = \left[\frac{K_u}{\beta\gamma k_u}\cos(\Omega_u t'), \; 0, \; \beta^* ct' + \frac{K_u^2}{8\beta^2\gamma^2 k_u}\sin(2\Omega_u t') \right]$$

$$\boldsymbol{\beta}(t') = \left[-\frac{K_u}{\gamma}\sin(\Omega_u t'), \; 0, \; \beta^* + \frac{K_u^2}{4\beta\gamma^2}\cos(2\Omega_u t') \right]$$

$$\dot{\boldsymbol{\beta}}(t') = \left[-\frac{\beta c k_u K_u}{\gamma}\cos(\Omega_u t'), \; 0, \; -\frac{K_u^2 k_u c}{2\gamma^2}\sin(2\Omega_u t') \right].$$

To calculate the emitted radiation we need some other parameters that are determined by the trajectory. First we calculate the vector $\mathbf{r}$ pointing from the charge to the observer. According to Fig. 6.1 we have

$$\mathbf{r}(t') = \mathbf{r}_p - \mathbf{R}(t'),$$

where $\mathbf{r}_p$ is the vector pointing from the center of the undulator to the observer,

$$\mathbf{r}_p = r_p[\sin\theta\cos\phi, \; \sin\theta\sin\phi, \; \cos\theta],$$

and r_p is its absolute value. We need r for the relation between the two time scales, $ct_p = ct' + r - r_p$, where its difference from r_p appears. For this reason we have to carry terms L_u/r_p along. Furthermore, $r(t')$ changes with the speed of the particle βc and is, in the above relation, compared with ct' changing with the speed of light. For this reason we take terms of order $1/\gamma^2$ along but neglect higher powers.

We obtain for the vector $\mathbf{r}(t')$ and its absolute value

$$\frac{\mathbf{r}(t')}{r_p} = \left[\sin\theta\cos\phi - \frac{K_u\cos(\Omega_u t')}{r_p\beta\gamma k_u}, \; \sin\theta\sin\phi, \; \cos\theta - \frac{\beta^* ct'}{r_p} - \frac{K_u^2\sin(2\Omega t')}{8r_p\gamma^2 k_u} \right]$$

(8.16)

$$\frac{r(t')}{r_p} = \left[1 - \frac{\cos\theta\,\beta^* ct'}{r_p} - \frac{K_u\sin\theta\cos\phi\cos(\Omega_u t')}{r_p\beta\gamma k_u} - \frac{K_u^2\cos\theta\sin(2\Omega_u t')}{8r_p\gamma^2 k_u} \right]. \quad (8.17)$$

From this we obtain the relation between the time t' of emission and the time $t = t' + r(t')/c$ or $t_p = t - r_p/c$ of the observation:

$$t_p(t') = t'(1 - \beta^*\cos\theta) - \frac{K_u\sin\theta\cos\phi\cos(\Omega_u t')}{\beta c\gamma k_u} - \frac{K_u^2\cos\theta\sin(2\Omega_u t')}{8c\gamma^2 k_u}. \quad (8.18)$$

Using the expressions for Ω_u and K_u^* and applying the ultra-relativistic approximations $\gamma \gg 1, \theta \ll 1$, we have

$$1 - \beta^* \cos\theta = \frac{1 + \gamma^{*2}\theta^2}{2\gamma^{*2}}$$

and

$$
\begin{aligned}
t_p(t') &= \frac{1 + \gamma^{*2}\theta^2}{2\gamma^{*2}\Omega_u}\left(\Omega_u t' - \frac{2K_u^*\gamma^*\theta\cos\phi\cos(\Omega_u t')}{1 + \gamma^{*2}\theta^2} - \frac{K_u^{*2}\sin(2\Omega_u t')}{4(1 + \gamma^{*2}\theta^2)}\right) \\
&= \frac{1}{\omega_1}(\Omega_u t' - b_u\cos(\Omega_u t') - a_u\sin(2\Omega_u t'))
\end{aligned}
\tag{8.19}
$$

with

$$\omega_1 = \frac{2\gamma^{*2}\Omega_u}{1 + \gamma^{*2}\theta^2}, \qquad a_u = \frac{K_u^{*2}}{4(1 + \gamma^{*2}\theta^2)}, \qquad b_u = \frac{2K_u^*\gamma^*\theta\cos\phi}{1 + \gamma^{*2}\theta^2}. \tag{8.20}$$

All three terms in the above equation (8.19) are of the same order with respect to $1/\gamma$, whereas all omitted terms contain an extra factor of either $1/\gamma$ or L_u/r_p. The frequency ω_1 was encountered before in the qualitative treatment of strong-undulator radiation (6.7). At a given angle the emitted radiation contains this frequency and harmonics m of it. It should be pointed out that, with the above definition, the observation time t_p does not vanish for $t' = 0$. This is caused by the fact that the particle is at the longitudinal coordinate $z(0) = 0$ at $t' = 0$ but has a transverse excursion $x(0) = a$, which contributes to its distance r from the observer.

For the denominator $1 - \mathbf{n}\cdot\boldsymbol{\beta}$ in the Liénard–Wiechert expression we need the unit vector $\mathbf{n} = \mathbf{r}/r$ pointing from the particle to the observer. We calculate it from (8.16) and (8.17), neglecting terms of order L_u/r_p since we are not comparing it with another distance:

$$\mathbf{n} = [\sin\theta\cos\phi, \ \sin\theta\sin\phi, \ \cos\theta] \approx [\theta\cos\phi, \ \phi\sin\phi, \ 1 - \theta^2/2].$$

However, we retain terms of order $1/\gamma^2$ in the expression for $\boldsymbol{\beta}$ since we compare the product $\mathbf{n}\cdot\boldsymbol{\beta}$ with unity:

$$\frac{dt_p}{dt'} = 1 - \mathbf{n}\cdot\boldsymbol{\beta} = \frac{\Omega_u}{\omega_1}(1 + b_u\sin(\Omega_u t') - 2a_u\cos(2\Omega_u t')). \tag{8.21}$$

Next we calculate the triple vector product appearing in the Fourier transformed Liénard–Wiechert expression for the radiation field (2.44). Since its largest term is of order $1/\gamma$ we neglect higher terms in $\boldsymbol{\beta}$ and $\dot{\boldsymbol{\beta}}$. We also assume that the radiation is observed from a large distance and neglect terms like L_u/r_p:

$$
\begin{aligned}
\boldsymbol{\beta}(t') &= \beta^*[-ak_u\sin(\Omega_u t'), \ 0, \ 1] \\
\dot{\boldsymbol{\beta}}(t') &= \beta^{*2}[-ack_u^2\cos(\Omega_u t'), \ 0, \ 0].
\end{aligned}
\tag{8.22}
$$

For the triple vector product appearing in the expression (2.44) we find

$$[\mathbf{n} \times [\mathbf{n} \times \boldsymbol{\beta}]] = \left[\frac{K_u^*}{\gamma^*}(1 - \sin^2\theta\cos^2\phi)\sin(\Omega_u t') + \beta^*\cos\theta\sin\theta\cos\phi, \right.$$
$$-\frac{K_u^*}{\gamma^*}\sin^2\theta\cos\phi\sin\phi\sin(\Omega_u t') + \beta^*\cos\theta\sin\theta\sin\phi,$$
$$\left. -\frac{K_u^*}{\gamma^*}\cos\theta\sin\theta\cos\phi\sin(\Omega_u t') - \beta^*\sin^2\theta \right]$$

and, in the ultra-relativistic approximation,

$$[\mathbf{n} \times [\mathbf{n} \times \boldsymbol{\beta}]] = \frac{1}{\gamma^*}[\gamma^*\theta\cos\phi + K_u^*\sin(\Omega_u t'), \ \gamma^*\theta\sin\phi, \ 0]. \tag{8.23}$$

To make the connection to a weak-field undulator we can approximate the above expressions for $K_u < 1$. We do this in two steps. First we omit terms containing K_u^2 but keep the parameters β^*, γ^*, and K_u^*. This leads to approximations that are slightly more accurate than the weak-undulator expressions derived in Chapter 7. In this form the expressions are first-order developments in the parameter K_u^*. In a second step we replace the parameters β^*, γ^*, and K_u^* by the original ones β, γ, and K_u to end up with the expressions for weak undulators.

8.2 The radiation from a plane strong undulator

8.2.1 The radiation field

We calculate now the radiation from a strong undulator with a large number of periods N_u in the frequency domain and follow closely the methods used in [50]. While traversing this undulator, the radiating charge has a periodic velocity with period $T_0 = 2\pi/\Omega_u$. We investigated this situation in Chapter 2 and found in (2.42) that the radiation is approximately periodic with period

$$T_p = T_0 \frac{1 + \gamma^{*2}\theta^2}{2\gamma^{*2}} = \frac{2\pi}{\omega_1}. \tag{8.24}$$

Here we use the reduced observation time $t_p = t' + (r - r_p)/c$ and expand the radiation field $\mathbf{E}(t_p)$ into a Fourier series, with coefficients $\mathbf{E}_m$ given in Chapter 2 by

$$\mathbf{E}(t_p) = \sum_{m=-\infty}^{\infty} \mathbf{E}_m e^{im\omega_1 t_p} \quad \text{with} \quad \mathbf{E}_m = \frac{1}{T_p}\int_0^{T_p} \mathbf{E}(t_p)e^{-im\omega_1 t_p}\, dt_p. \tag{8.25}$$

Changing the integration variable from t to t' and integrating by parts gives, according to (2.44),

$$\mathbf{E}_m = \frac{iem\omega_1^2}{8\pi^2\epsilon_0 cr_p}\int_0^{T_0} [\mathbf{n} \times [\mathbf{n} \times \boldsymbol{\beta}]] \times e^{-im\omega_1(t' + (r(t') - r_p)/c)}\, dt'. \tag{8.26}$$

The expressions for the triple vector product (8.23) and the relation between the two time scales (8.19) with abbreviations (8.20) are inserted into (8.26), giving

$$\mathbf{E}_m = \frac{iem\omega_1^2}{8\pi^2\epsilon_0 cr_p} \int_0^{T_0} \frac{[\gamma^*\theta\cos\phi + K_u^*\sin(\Omega_u t'),\ \gamma^*\theta\sin\phi,\ 0]}{\gamma^*}$$
$$\times e^{-im(\Omega_u t' - b_u\cos(\Omega_u t') - a_u\sin(2\Omega_u t'))}\, dt'.$$

To solve the integral we expand the exponential of the trigonometric functions into series of Bessel functions J_n according to (B.21),

$$e^{iz\sin\xi} = \sum_{n=-\infty}^{\infty} J_n(z)e^{in\xi},$$

which gives for our two cases

$$e^{imb_u\cos(\Omega_u t')} = e^{imb_u\sin(\Omega_u t' + \pi/2)} = \sum_{n=-\infty}^{\infty} J_n(mb_u)e^{in\pi/2}e^{in\Omega_u t'}$$

$$e^{ima_u\sin(2\Omega_u t')} = e^{ima_u\sin(-2\Omega_u t' - \pi)} = \sum_{l=-\infty}^{\infty} J_l(ma_u)e^{-il\pi}e^{-il2\Omega_u t'}.$$

(8.27)

The exponent of the second equation seems unnecessarily complicated but leads to a simpler expression later on.

The Fourier component of the electric field for the mth harmonic becomes

$$\mathbf{E}_m = \frac{ime\omega_1^2}{8\pi^2\epsilon_0 cr_p\gamma^*} \int_0^{T_0} [\gamma^*\theta\cos\phi + K_u^*\sin(\Omega_u t'),\ \gamma^*\theta\sin\phi,\ 0]$$
$$\times \sum_{n=-\infty}^{\infty}\sum_{l=-\infty}^{\infty} e^{i\pi(n-2l)/2} J_n(mb_u)J_l(ma_u)e^{i\Omega_u t'(-m+n-2l)}\, dt'.$$

(8.28)

We have two integrals to solve:

$$I_1 = \int_0^{T_0} e^{i\Omega_u t'(-m+n-2l)}\, dt = \begin{cases} 0 & \text{if } -m+n-2l \neq 0 \\ T_0 & \text{if } -m+n-2l = 0 \end{cases}$$

$$I_2 = \int_0^{T_0} \sin(\Omega_u t')e^{i\Omega_u t'(-m+n-2l)}\, dt'$$
$$= \int_0^{T_0} \sin(\Omega_u t')[\cos(\Omega_u t'(m-n+2l)) - i\sin(\Omega_u t'(m-n+2l))]\, dt'.$$

The first integral in the lower line of I_2 vanishes and the second one gives

$$I_2 = i\frac{T_0}{2}\left(\frac{\sin(2\pi(m-n+2l+1))}{2\pi(m-n+2l+1)} - \frac{\sin(2\pi(m-n+2l-1))}{2\pi(m-n+2l-1)}\right).$$

Each of the two terms vanishes unless the argument of the sine function is zero, in which case $\sin x/x \to 1$ and $|I_2| = T_0/2$ for each case. This gives the conditions for the summing indices to obtain non-vanishing values for the integrals,

$$
\begin{aligned}
&\text{in } I_1 & m = n - 2l &\rightarrow n = m + 2l \rightarrow I_1 = T_0 \\
&\text{term 1 in } I_2 & m = n - 2l - 1 &\rightarrow n = m + 2l + 1 \rightarrow I_2 = iT_0/2 \\
&\text{term 2 in } I_2 & m = n - 2l + 1 &\rightarrow n = m + 2l - 1 \rightarrow I_2 = -iT_0/2,
\end{aligned}
$$

which allows us later to eliminate the sum over n. We split the vector $[\mathbf{n} \times [\mathbf{n} \times \boldsymbol{\beta}]]$ into two parts and obtain from (8.28) the Fourier component $\mathbf{E}_m$:

$$
\mathbf{E}_m = \frac{ime\omega_1^2}{8\pi^2\epsilon_0 c r_p \gamma^*}([\gamma^*\theta\cos\phi,\ \gamma^*\theta\sin\phi, 0]I_1 + [K_u^*\sin(\Omega_u t'),\ 0, 0]I_2)
$$

$$
\times \sum_{n=-\infty}^{\infty}\sum_{l=-\infty}^{\infty} e^{i\pi(n-2l)/2} J_n(mb_u)J_l(ma_u). \tag{8.29}
$$

With the above selection rule for the sum index n, we obtain the Fourier component $\mathbf{E}_m$ using $e^{im\pi/2} = i^m$, omitting the vanishing z-component:

$$
\mathbf{E}_{\perp m} = \frac{i^{(1+m)}me\omega_1^2 T_0}{8\pi^2\epsilon_0 c r_p \gamma^*}\left([\gamma^*\theta\cos\phi,\ \gamma^*\theta\sin\phi]\sum_{l=-\infty}^{\infty} J_{m+2l}(mb_u)J_l(ma_u)\right.
$$

$$
\left. + \frac{1}{2}[-K_u^*,\ 0]\sum_{l=-\infty}^{\infty} J_l(ma_u)[J_{m+2l+1}(mb_u) + J_{m+2l-1}(mb_u)]\right).
$$

To make the expression more compact we introduce abbreviations for the sums:

$$
\Sigma_{m1} = \sum_{l=-\infty}^{\infty} J_l(ma_u)J_{m+2l}(mb_u)
$$

$$
\Sigma_{m2} = \sum_{l=-\infty}^{\infty} J_l(ma_u)[J_{m+2l+1}(mb_u) + J_{m+2l-1}(mb_u)] \tag{8.30}
$$

$$
= \sum_{l=-\infty}^{\infty} \frac{2(m+2l)}{mb_u} J_l(ma_u)J_{m+2l}(mb_u).
$$

The alternate expression for Σ_{m2} is obtained from the relation (B.2):

$$
J_{n-1}(z) + J_{n+1}(z) = \frac{2n}{z} J_n(z).
$$

With this and the expression (8.24) we obtain the Fourier component of the electric field emitted in strong undulators:

$$
\mathbf{E}_{\perp m} = \frac{i^{(1+m)}mek_u}{\pi\epsilon_0 r_p}\frac{\gamma^{*3}}{(1+\gamma^{*2}\theta^2)^2}[\gamma^*\theta\cos\phi\,\Sigma_{m1} - \tfrac{1}{2}K_u^*\Sigma_{m2},\ \gamma^*\theta\sin\phi\,\Sigma_{m1}]. \tag{8.31}
$$

The electric field can now be expressed as a Fourier series of frequencies $m\omega_1$:

$$\mathbf{E}_\perp(t_p) = \sum_{m=-\infty}^{\infty} \mathbf{E}_m e^{im\omega_1 t_p} = \frac{ek_u}{\pi\epsilon_0 r_p} \frac{\gamma^{*3}}{(1+\gamma^{*2}\theta^2)^2}$$

$$\times \sum_{m=-\infty}^{\infty} i^{(m+1)} m e^{im\omega_1 t_p} [\gamma^*\theta \cos\phi \Sigma_{m1} - \tfrac{1}{2}K_u^*\Sigma_{m2}, \ \gamma^*\theta \sin\phi \Sigma_{m1}]. \quad (8.32)$$

Since this series is presented in complex notation, it contains positive and negative frequencies $\pm m\omega_1$. To obtain a purely real presentation we collect the two terms at each harmonic and obtain a series expressed in trigonometric functions with positive harmonics, $m > 0$ only. The values of the sums (8.30) for negative values can be obtained with the help of the symmetry relations of Bessel functions (B.5) and (B.6),

$$J_{-m}(z) = (-1)^m J_m(z) \quad \text{and} \quad J_m(-z) = (-1)^m J_m(z), \quad (8.33)$$

giving the relation between the sums (8.30) for positive and negative values of m:

$$\Sigma_{-m1} = \Sigma_{m1} \quad \text{and} \quad \Sigma_{-m2} = \Sigma_{m2}.$$

We obtain the radiation field in the time domain expressed as a Fourier series with the components

$$\mathbf{E}_{\perp m}(t_p) = \frac{-ek_u\gamma^{*3}}{\pi\epsilon_0 r_p(1+\gamma^{*2}\theta^2)^2} \sin(m\omega_1 t_p + m\pi/2)$$

$$\times [2\gamma^*\theta \cos\phi \ \Sigma_{m1} - K_u^*\Sigma_{m2}, \ 2\gamma^*\theta \sin\phi \ \Sigma_{m1}] \quad (8.34)$$

$$\mathbf{E}_\perp(t_p) = \sum_{m=1}^{\infty} \mathbf{E}_{\perp m}(t_p),$$

where we have absorbed the factor of two into the square bracket. The two components of the vector in the sum Σ_{m2} give the horizontal and vertical polarizations of the radiation. For a given harmonic m the two are in phase, indicating that we have linear polarization without any circular component as in the case of a weak undulator.

We express the oscillatory term separately for even and odd harmonics m:

$$\sin(m\omega_1 t_p + m\pi/2) = \begin{cases} (-1)^{(m-1)/2} \cos(m\omega_1 t_p) & \text{for } m \text{ odd} \\ (-1)^{m/2} \sin(m\omega_1 t_p) & \text{for } m \text{ even.} \end{cases}$$

The fields of odd and even harmonics are out of phase. This is expected from the qualitative treatment of this radiation illustrated in Fig. 6.6, and from the particle motion in the moving frame shown in Fig. 8.1. The transverse acceleration responsible for the odd harmonics has a maximum when the longitudinal acceleration vanishes.

It is instructive to check the field (8.34) on the axis, $\theta = 0$, and at the central time $t_p = 0$. According to the time relation (8.18) this field was created at the time $t' = 0$ when the charge was in the center of the undulator in the magnetic field $B_0 = K_u m_0 c k_u / e$. From (3.12) or (7.31) we know that the field must be

$$\hat{E}_u = \frac{e^2 B_0 \gamma^3}{\pi \epsilon_0 m_0 c r_p} = \frac{e k_u K_u \gamma^3}{\pi \epsilon_0 r_p}.$$

To calculate the field also from (8.34), we evaluate first the square bracket, which reduces in the forward direction to $K^* \Sigma_{m2}$. Since $b_u = 0$ for $\theta = 0$, the Bessel functions of this argument vanish except if they are of order zero. Therefore Σ_{m2} reduces to two terms with $m + 2l + 1 = 0$ and $m + 2l - 1 = 0$, giving $J_0(m b_u) = 1$. This results in

$$\Sigma_{m2} = (-1)^{(m-1)/2} \big(J_{(m-1)/2}(m a_u) - J_{(m+1)/2}(m a_u) \big), \tag{8.35}$$

where we used the symmetry relation (8.33). The time-dependent part of (8.34) reduces for $t_p = 0$ to $\sin(m\pi/2) = (-1)^{(m-1)/2}$. At $t_p = 0$ we obtain for the field on the axis

$$\mathbf{E}_\perp(0) = \frac{e k_u K_u \gamma^3}{\pi \epsilon_0 r_p} \left\{ \frac{\sum_{m=1}^{\infty} m \big(J_{(m-1)/2}(m a_u) - J_{(m+1)/2}(m a_u) \big)}{\big(1 + K_u^2/2\big)^2} \right\}.$$

It can be shown numerically that the expression in the curly brackets is unity and the field agrees with the general expression (3.12).

Going back to the general case (8.34), we now take the large but finite number $N_u \gg 1$ of periods into account, and calculate the Fourier transform of each harmonic component m. The time span $N_u T_0 = 2\pi N_u / \Omega_u$ of emission translates into the observation time

$$N_u T_p = N_u T_0 \frac{1 + \gamma^{*2}\theta^2}{2\gamma^{*2}} = \frac{2\pi N_u}{\omega_1},$$

both containing N_u periods.

The time-domain signal (8.34) consists of oscillations at different harmonics $m > 0$, each lasting N_u periods. To obtain their Fourier transform we separate out the time-dependent term

$$\mathcal{F}[\sin(m\omega_1 t_p + m\pi/2)] = \frac{1}{\sqrt{2\pi}} \int_{-\pi N_u/\omega_1}^{\pi N_u/\omega_1} [\sin(m\omega_1 t_p + m\pi/2)] e^{-i\omega t_p} \, dt_p$$

$$= -\frac{i^{(m+1)}}{\sqrt{2\pi}} \frac{\pi N_u}{\omega_1} \frac{\sin\left(\dfrac{\Delta\omega}{\omega_1}\pi N_u\right)}{\dfrac{\Delta\omega}{\omega_1}\pi N_u},$$

where we used the abbreviation

$$\Delta\omega_m = \omega - m\omega_1.$$

With this we obtain from the expression (8.34) the components $\mathbf{E}_m$ of the strong-undulator field harmonics in the frequency domain:

$$\tilde{E}_{\perp m}(\omega) = \frac{i^{1+m} m e k_u \gamma^{*3}}{\pi \sqrt{2\pi} \epsilon_0 r_p} \frac{[2\gamma^*\theta \cos\phi \, \Sigma_{m1} - K_u^* \Sigma_{m2}, \, 2\gamma^*\theta \sin\phi \, \Sigma_{m1}]}{(1 + \gamma^{*2}\theta^2)^2}$$

$$\times \frac{\pi N_u}{\omega_1} \frac{\sin\left(\dfrac{\Delta\omega}{\omega_1} \pi N_u\right)}{\dfrac{\Delta\omega}{\omega_1} \pi N_u}. \tag{8.36}$$

It should be noted that this is a complex Fourier transformation, containing positive and negative frequencies ω, but only positive harmonics m.

8.3 Properties of strong-undulator radiation

8.3.1 The angular spectral power distribution

The averaged total power and the total energy emitted by a charge e in an undulator are given by the general expressions (7.32),

$$P_u = \frac{2r_0 c^3 e^2 \langle B^2 \rangle E^2}{3(m_0 c^2)^3} = \frac{r_0 c m_0 c^2 k_u^2 K_u^2 \gamma^2}{3}$$

$$U_u = \frac{r_0 m_0 c^2 \gamma^2 k_u^2 K_u^2 L_u}{3} = \frac{e^2 \gamma^2 k_u K_u^2 N_u}{6\epsilon_0}, \tag{8.37}$$

which we will use to express the various distributions of the radiation in a compact form.

We assume that the number of undulator periods is sufficiently large that, at a given angle, the contributions from different harmonics are separated in frequency and there is no overlap between them. This allows us to calculate the power P_{um} for each harmonic m separately and to obtain the total power as a sum, $P = \sum P_{um}$. If the number of periods were not large enough we might receive at the same angle radiation of the same frequency, but belonging to different harmonics, which can produce interference effects. We exclude this possibility here.

We start with the angular spectral power distribution of each harmonic m, obtained with the relation (7.34), and obtain the total radiation as a sum over the harmonic contributions:

$$\frac{d^2 P_m}{d\Omega \, d\omega} = \frac{2r_p^2 |\tilde{E}_m(\omega)|^2}{\mu_0 L_u}, \qquad \frac{d^2 P}{d\Omega \, d\omega} = \sum_{m=1}^{\infty} \frac{d^2 P_m}{d\Omega \, d\omega}.$$

From the expression for the radiation field (8.36) we obtain the angular spectral power density for each harmonic $m > 0$,

$$\boxed{\frac{d^2 P_m}{d\Omega \, d\omega} = \frac{d P_m}{d\Omega} f_N(\Delta\omega_m) = P_u \gamma^{*2} [F_{m\sigma}(\theta, \phi) + F_{m\pi}(\theta, \phi)] f_N(\Delta\omega_m),} \tag{8.38}$$

with the normalized angular distribution functions

$$
F_{m\sigma}(\theta, \phi) = \frac{3m^2}{\pi\left(1 + K_u^2/2\right)^2 K_u^{*2}} \frac{(2\Sigma_{m1}\gamma^*\theta\cos\phi - \Sigma_{m2}K_u^*)^2}{\left(1 + \gamma^{*2}\theta^2\right)^3}
$$

$$
F_{m\pi}(\theta, \phi) = \frac{3m^2}{\pi\left(1 + K_u^2/2\right)^2 K_u^{*2}} \frac{(2\Sigma_{m1}\gamma^*\theta\sin\phi)^2}{(1 + \gamma^{*2}\theta^2)^3}
$$

$$(8.39)$$

containing the sums

$$
\Sigma_{m1} = \sum_{l=-\infty}^{\infty} J_l(ma_u) J_{m+2l}(mb_u)
$$

$$
\Sigma_{m2} = \sum_{l=-\infty}^{\infty} J_l(ma_u)(J_{m+2l+1}(mb_u) + J_{m+2l-1}(mb_u)),
$$

$$(8.40)$$

which involve the parameters (8.19)

$$
a_u = \frac{K_u^{*2}}{4(1 + \gamma^{*2}\theta^2)}, \qquad b_u = \frac{2K_u^*\gamma^*\theta\cos\phi}{1 + \gamma^{*2}\theta^2}.
$$

$$(8.41)$$

The spectral function $f_N(\Delta\omega_m)$ has the same form as (7.38) but is generalized to include higher harmonics $\omega_m = m\omega_1$:

$$
f_N(\Delta\omega_m) = \frac{N_u}{\omega_1}\left(\frac{\sin\left(\frac{\Delta\omega_m}{\omega_1}\pi N_u\right)}{\frac{\Delta\omega_m}{\omega_1}\pi N_u}\right)^2, \qquad \omega_m = m\omega_1 = m\frac{2\gamma^{*2}\Omega_u}{1 + \gamma^{*2}\theta^2}
$$

$$
\frac{\Delta\omega_m}{\omega_1} = \frac{\omega - m\omega_1}{\omega_1}, \qquad \int_{-\infty}^{\infty} f_N(\Delta\omega_m)\,d\omega = 1.
$$

$$(8.42)$$

8.3.2 The angular power distribution

Integrating this over the frequency ω gives the angular power distribution of each harmonic m:

$$
\boxed{
\begin{aligned}
\frac{dP_m}{d\Omega} &= P_u\gamma^{*2}[F_{m\sigma}(\theta, \phi) + F_{m\pi}(\theta, \phi)] \\
&= P_u\gamma^{*2}\frac{3m^2[(2\Sigma_{m1}\gamma^*\theta\cos\phi - \Sigma_{m2}K_u^*)^2 + (2\Sigma_{m1}\gamma^*\theta\sin\phi)^2]}{\pi\left(1 + K_u^2/2\right)^2 K_u^{*2}(1 + \gamma^{*2}\theta^2)^3}.
\end{aligned}
}
$$

$$(8.43)$$

To discuss the above representation of strong-undulator radiation we start by giving with Fig. 8.2 an overall graphical representation of the angular power density. A value of

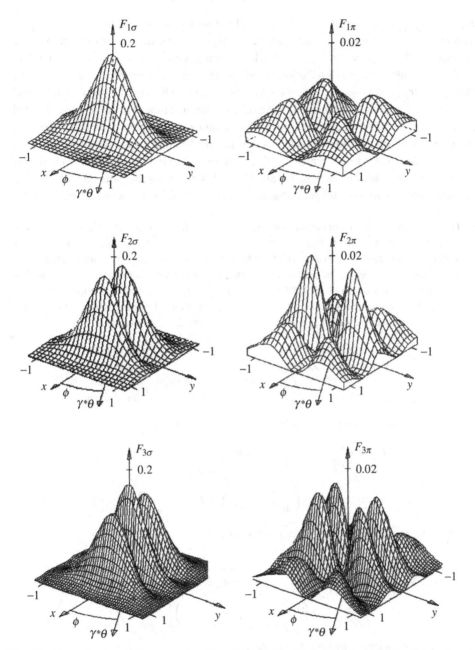

Fig. 8.2. Normalized angular power densities of the horizontal (left) and vertical (right) modes of polarization for the first three harmonics of the radiation from an undulator with $K_u = \sqrt{2}$ ($K_u^* = 1$).

$K_u = \sqrt{2}$ ($K_u^* = 1$) was chosen, which is still not very large but produces clearly higher harmonics. The fundamental mode, shown at the top, has the same features as that for weak undulator radiation shown in Fig. 7.6. The horizontal mode of polarization has a maximum on the axis with a smooth distribution around it. The vertical polarization vanishes in the two symmetry planes and has four maxima located close to the two diagonals. The second harmonic (center) has no intensity on the axis. Its horizontal mode has two maxima located in the horizontal plane, whereas the vertical polarization has no intensity in this plane but a group of three maxima below and above it. The pattern becomes even more complicated for the third harmonic (bottom), which has three maxima lying in the horizontal plane for the σ-mode and four maxima, each below and above this plane, for the π-mode. We illustrate the properties of strong-undulator radiation further by considering special cases.

- The angular power density in the horizontal (x, z)-plane. In the horizontal plane $\phi = 0$ the vertical polarization vanishes. This is a general property of the radiation emitted by a charge moving on a trajectory lying in the (x, z)-plane, which we found earlier for ordinary synchrotron radiation.
- The angular power density in the vertical (y, z)-plane. In the vertical plane $\phi = \pm\pi/2$ the first term inside the brackets in (8.43) for the horizontal polarization vanishes. Furthermore, since also $b_u = 0$, all Bessel functions of this argument vanish except those of zeroth order, as we saw before (8.35). This reduces the sum Σ_{m1} to one term, and the sum Σ_{m2} to two terms:

$$\Sigma_{m1} = 0 \qquad\qquad\qquad\qquad\qquad \text{for } m \text{ odd}$$
$$\Sigma_{m1} = (-1)^{m/2} J_{m/2}(ma_u) \qquad\qquad \text{for } m \text{ even}$$
$$\Sigma_{m2} = (-1)^{(m-1)/2}\big(J_{(m-1)/2}(ma_u) - J_{(m+1)/2}(ma_u) \big) \quad \text{for } m \text{ odd}$$
$$\Sigma_{m2} = 0 \qquad\qquad\qquad\qquad\qquad \text{for } m \text{ even.}$$

We obtain for the angular power distribution in the (y, z)-plane

$$\frac{dP_m}{d\Omega} = \frac{dP_{m\sigma}}{d\Omega} = P_u \frac{3m^2\gamma^{*2}}{\pi\left(1 + K_u^2/2\right)^2(1 + \gamma^{*2}\theta^2)^3}$$
$$\times \left[J_{(m-1)/2}\left(\frac{mK_u^{*2}}{4(1 + \gamma^{*2}\theta^2)}\right) - J_{(m+1)/2}\left(\frac{mK_u^{*2}}{4(1 + \gamma^{*2}\theta^2)}\right) \right]^2 \qquad (8.44)$$

for m odd, containing only horizontal polarization, and

$$\frac{dP_m}{d\Omega} = \frac{dP_{m\pi}}{d\Omega} = P_u \frac{3m^2\gamma^{*2}}{\pi\left(1 + K_u^2/2\right)^2 K_u^{*2}(1 + \gamma^{*2}\theta^2)^3}$$
$$\times \gamma^{*2}\theta^2 \left[J_{m/2}\left(\frac{mK_u^{*2}}{4(1 + \gamma^{*2}\theta^2)}\right) \right]^2$$

for m even, containing only vertical polarization.

Later we will discuss in some detail the properties of the radiation emitted on the axis, $\theta = 0$. Furthermore, we will give some approximations of the expression (8.43) by developing it into powers of K_u^*.

8.3.3 The spectral density of the radiation

At a given angle θ the radiation from a strong undulator has a spectrum containing a set of narrow bands centered around the harmonic frequencies $m\omega_1$. It is given by the spectral function $f_N(\Delta\omega_m)$ (8.42), which we approximate now by a δ-function. Integrating over it gives, at a given angle θ, a set of single-frequency lines $\omega = \omega_m = m\omega_1$. We extend the procedure (7.45) to include higher harmonics and convert the angular power density (8.43) into a spectral distribution:

$$\omega \approx \omega_m = m\omega_1 = \frac{m\omega_{10}}{1+\gamma^{*2}\theta^2} = \frac{2mk_u c\gamma^{*2}}{1+\gamma^{*2}\theta^2}$$

$$\mathrm{d}\omega_m = -\frac{2m\omega_{10}\gamma^{*2}}{(1+\gamma^{*2}\theta^2)^2}\theta\,\mathrm{d}\theta = -2m\omega_{10}\gamma^{*2}\left(\frac{\omega_1}{\omega_{10}}\right)^2\theta\,\mathrm{d}\theta \qquad (8.45)$$

$$\frac{\mathrm{d}P_m}{\mathrm{d}\omega_m} = \frac{1+K_u^2/2}{2m\omega_{10}\gamma^2}\left(\frac{\omega_{10}}{\omega_1}\right)^2\int_0^{2\pi}\frac{\mathrm{d}P_m}{\mathrm{d}\Omega}\,\mathrm{d}\phi.$$

The factor $1 + K_u^2/2$ reflects the fact that, in a strong undulator, the increment $\theta\,\Delta\theta$ corresponds to a smaller frequency increment $\Delta\omega_1$ than it does in a weak one. This results in a larger power for the same frequency bin. The above integration over ϕ has to be carried out numerically. The resulting spectral power density is shown in Fig. 8.3 with $K_u = \sqrt{2}$ ($K_u^* = 1$) for the first three harmonics, each for the total and the two modes of polarization. The two odd harmonics have sharp peaks at the highest frequencies because their angular distributions are concentrated around small angles where $\omega \approx m\omega_{10}$. The even harmonics have no intensity on the axis where the highest frequency is emitted. For this reason their spectra vanish at the top in a smooth way. The third harmonic has a bump below the top frequency caused by the two secondary maxima of its angular distribution shown in Fig. 8.2.

8.3.4 The power contained in each harmonic

Integrating the spectral distribution (8.45) over the frequency, or the angular distribution (8.43) over the solid angle, gives the total power contained in each harmonic. This operation was carried out numerically and the result is shown in Fig. 8.4, where the P_{um} in each of the first five harmonics, divided by the total power P_u, is plotted against the undulator parameter K_u. With increasing undulator strength more and more harmonics contribute. In this normalized presentation the fundamental is unity at the origin but decreases rapidly with increasing K_u, for which the higher harmonics become important. We also plot the sum of the five harmonics, which approximates the total power quite well at low values of K_u, but for $K = \sqrt{2}$ ($K^* = 1$) misses already about 20% of the power which is emitted into harmonics $m > 5$.

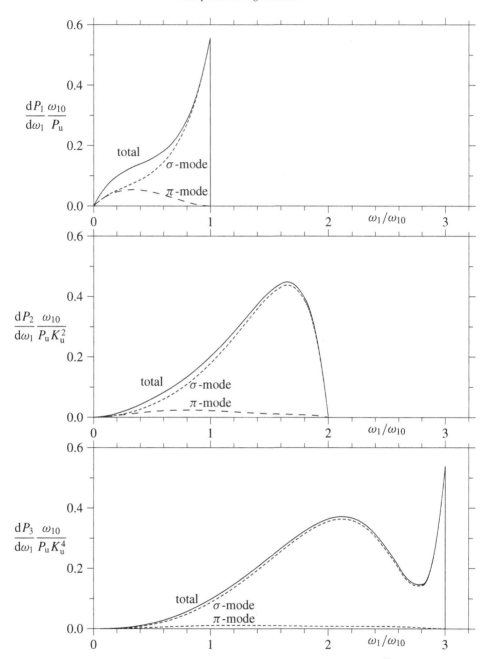

Fig. 8.3. The spectrum of undulator radiation having $K_u = \sqrt{2}$.

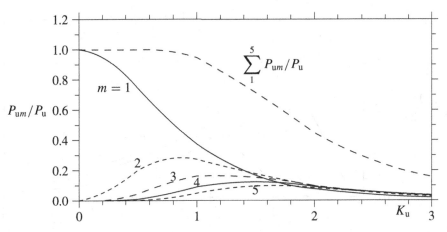

Fig. 8.4. The normalized power in each of the first harmonics m versus K_u.

8.3.5 The properties of the radiation on the axis

The properties of the radiation along the axis, $\theta = 0$, are important since this is the main direction of observation. We will treat it here in more detail and also give the photon distribution, which we have not done for the general case.

We obtain the angular power distribution on the axis by setting $\theta = 0$ in the expression (8.44) and have only odd harmonics with horizontal polarization. The angular power density at $\theta = 0$ is

$$
\frac{\mathrm{d}P_m}{\mathrm{d}\Omega} = \frac{P_u \gamma^2}{K_u^2} \frac{3m^2 K_u^2}{\pi \left(1 + K_u^2/2\right)^3} \left[J_{(m-1)/2}\left(m\frac{K_u^{*2}}{4}\right) - J_{(m+1)/2}\left(m\frac{K_u^{*2}}{4}\right) \right]^2
$$
$$
= \frac{r_0 c m_0 c^2 \gamma^4 k_u^2}{3} \frac{3m^2 K_u^2}{\pi \left(1 + K_u^2/2\right)^3} \left[J_{(m-1)/2}\left(m\frac{K_u^{*2}}{4}\right) - J_{(m+1)/2}\left(m\frac{K_u^{*2}}{4}\right) \right]^2 .
$$

$$(8.46)$$

Since the total power P_u is proportional to K_u^2, we separate out the factor $P_u \gamma^2 / K_u^2$, which is independent of K_u, and obtain the dependence on the undulator parameter and the harmonic number in the rest of the equation. This quantity is plotted in Fig. 8.5 for the first three odd harmonics as a function of K_u. The angular power density of each harmonic increases first with the undulator parameter, then goes through a maximum, and later decays. The higher harmonics dominate at large values of K_u. The fundamental $m = 1$ has its maximum angular power density around $K_u \approx 0.906$ with a value of

$$
\left(\frac{\mathrm{d}P_1}{\mathrm{d}\Omega}\right)_{\max} = \frac{P_u \gamma^2}{K_u^2} 0.238.
$$

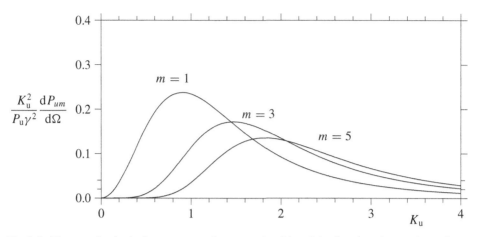

Fig. 8.5. The on-axis, single-frequency angular power densities of the first three harmonics as functions of K_u.

However, this optimum can be misleading since, at $K_u = 0.906$, the fundamental on-axis frequency has decreased to 71% compared with the weak-undulator value. A complete optimization should keep this frequency constant and is more complicated since it involves also a variation of the period length λ_u of the undulator.

The photon flux per unit solid angle is obtained from the above power distribution with the relations (7.51) and (8.42):

$$\frac{d\dot{n}_m}{d\Omega} = \frac{1}{\hbar m \omega_{10}} \frac{dP}{d\Omega} = \frac{\pi \left(1 + K_u^2/2\right)}{mchk_u\gamma^2} \frac{dP}{d\Omega}.$$

The factor $1 + K_u^2/2$ reflects the fact that the frequency decreases with increasing K_u and more photons are radiated for the same power increment. We obtain for the angular photon flux density

$$\frac{d\dot{n}_m}{d\Omega} = \frac{m\alpha_f c \gamma^2 k_u K_u^2}{2\pi \left(1 + K_u^2/2\right)^2} \left[J_{(m-1)/2}\left(m\frac{K_u^{*2}}{4}\right) - J_{(m+1)/2}\left(m\frac{K_u^{*2}}{4}\right) \right]^2. \tag{8.47}$$

Its lowest mode $m = 1$ has a maximum around $K_u \approx 1.2$, which is again only part of a real optimization since the fundamental frequency changes with K_u.

We can convert the above angular distribution into a spectral density using the relation (8.45). Since there is no ϕ-dependence on the axis the integration over this angle gives just a factor of 2π:

$$\frac{dP_m}{d\omega_m} = \frac{\pi}{m\omega_{10}\gamma^{*2}} \frac{dP_m}{d\Omega} = \frac{\pi \left(1 + K_u^2/2\right)}{m\omega_{10}\gamma^2} \frac{dP_m}{d\Omega}.$$

We obtain the spectral power and photon flux density:

$$\frac{dP_m}{d\omega_m} = \frac{P_u}{\omega_{10}} \frac{3m}{\left(1 + K_u^2/2\right)^2} \left[J_{(m-1)/2}\left(m\frac{K_u^{*2}}{4}\right) - J_{(m+1)/2}\left(m\frac{K_u^{*2}}{4}\right) \right]^2$$

$$\frac{d\dot{n}_m}{d\omega_m/\omega_m} = \frac{m\alpha_f c k_u K_u^2}{2\left(1 + K_u^2/2\right)} \left[J_{(m-1)/2}\left(m\frac{K_u^{*2}}{4}\right) - J_{(m+1)/2}\left(m\frac{K_u^{*2}}{4}\right) \right]^2 .$$

The angular spectral power density is obtained by multiplying the angular distribution (8.46) by the spectral function (8.42) adapted to $\theta = 0$,

$$f_N(\Delta\omega_m) = \frac{N_u}{\omega_{10}} \left(\frac{\sin\left(\frac{\Delta\omega_m}{\omega_1}\pi N_u\right)}{\frac{\Delta\omega_m}{\omega_1}\pi N_u} \right)^2 ,$$

giving

$$\frac{d^2 P_m}{d\Omega\, d\omega} = \frac{P_u \gamma^2 3m^2}{\pi\left(1 + K_u^2/2\right)^3} \left[J_{m-1/2}\left(m\frac{K_u^{*2}}{4}\right) - J_{m+1/2}\left(m\frac{K_u^{*2}}{4}\right) \right]^2 f_N(\Delta\omega_m). \quad (8.48)$$

We choose now also the central frequency $\Delta\omega = \omega - m\omega_{10} = 0$, at which the spectral function takes the value

$$f_N(0) = \frac{N_u}{\omega_{10}} = \frac{N_u\left(1 + K_u^2/2\right)}{2k_u c \gamma^2},$$

and obtain from (8.47) the angular spectral photon flux distribution at $\theta = 0$, $\omega = m\omega_{10}$:

$$\frac{d^2\dot{n}_m}{d\Omega\, d\omega/\omega_m} = \frac{m\alpha_f c \gamma^2 k_u K_u^2 N_u}{2\pi\left(1 + K_u^2/2\right)^2} \left[J_{(m-1)/2}\left(m\frac{K_u^{*2}}{4}\right) - J_{(m+1)/2}\left(m\frac{K_u^{*2}}{4}\right) \right]^2 . \quad (8.49)$$

The related angular spectral photon density is plotted in Fig. 8.6 as a function of K_u for the first three odd harmonics. These curves exhibit a similar behavior to that of the angular power densities plotted in Fig. 8.5, but decrease slower, for large values of K_u. This is caused by the fact that, for a fixed absolute frequency increment $d\omega$, the relative value $d\omega/\omega = d\omega/\omega_{10}$ increases with larger undulator parameter since ω_{10} becomes smaller. As a consequence the angular spectral photon density (8.49) is proportional to $(1 + K_u^2/2)^{-2}$, whereas the angular power density (8.46) has a cubic dependence. The on-axis photon spectrum consists of lines at frequencies $m\omega_{10} = 2ck_u\gamma^{*2}$, which are shown in Fig. 8.7 for three different values of the undulator parameter. The number of harmonics increases with K_u and each one goes through a maximum.

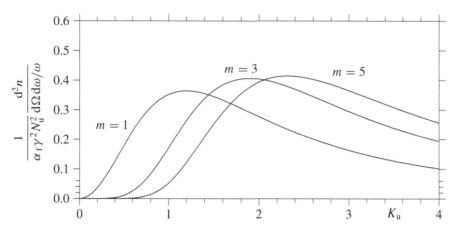

Fig. 8.6. The on-axis, central-frequency angular spectral power densities of the first three harmonics as functions of K_u.

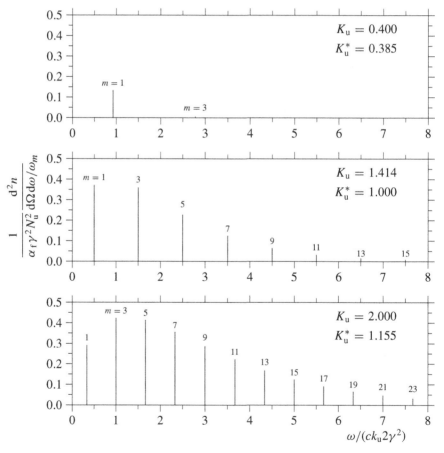

Fig. 8.7. On-axis and central-frequency spectral power densities of strong-undulator-radiation harmonics m for various undulator parameters K_u.

Of practical interest is the photon flux distribution produced by a beam current I, representing I/e electrons traversing the undulator per second, giving, for $\theta = 0$, $\omega = \omega_{10}$,

$$\frac{d^2\dot{n}_{mI}}{d\Omega\,d\omega/\omega_m} = \frac{\alpha_f m^2 \gamma^2 I K_u^2 N_u^2}{e\left(1 + K_u^2/2\right)^2}\left[J_{(m-1)/2}\left(m\frac{K_u^{*2}}{4}\right) - J_{(m+1)/2}\left(m\frac{K_u^{*2}}{4}\right)\right]^2 \tag{8.50}$$

8.3.6 The development with respect to K_u^*

For some applications one likes to operate an undulator at a value of the parameter K_u slightly above unity. For this condition the intensity at the fundamental $m = 1$ is still strong and the higher harmonics not too disturbing. The dependence of the on-axis frequency ω_{m0} on the undulator parameter K_u allows one to adjust the spectrum:

$$\omega_{m0} = m\Omega_u \frac{2\gamma^2}{1 + K_u^2/2} = k_u c 2\gamma^{*2}. \tag{8.51}$$

For this reason it is useful to give an approximation of the radiation, which is valid for $K_u \approx 1$, although the evaluation of the Bessel functions involved is not a major problem. In some cases the approximate expressions give some insight into the general behavior of the radiation.

We apply the power series expansion (B.4) of the Bessel functions and develop some of the properties of the radiation into a power series in K_u^*. However, we separate out the factor

$$\frac{1}{\left(1 + K_u^2/2\right)^2} = \left(1 - K_u^{*2}/2\right)^2,$$

which makes the expressions slightly more compact.

We start with the angular power density (8.43) and develop its first three harmonics each up to its lowest relevant order in K_u^* and give the result (C.1) in Appendix C. A more complete presentation can be found in [19].

To illustrate this pattern of the radiation we present the angular distributions of the first three harmonics of undulator radiation in Fig. 8.8, each to its lowest relevant power in K_u^*, as given by (C.1). As expected, the angular distribution of the horizontal polarization has a maximum on the axis, $\theta = 0$, for the odd harmonics $m = 1$ and $m = 3$ where the vertical polarization vanishes. The even harmonic $m = 2$ has no intensity on the axis, $\theta = 0$, for both modes of polarization. The pattern of the distribution becomes more complicated for the higher harmonics. It should be noted that each harmonic is plotted only to the lowest order in K_u^*, and we expect the distributions to become even more complicated with increasing undulator strength. On comparing this with the exact calculation, shown in Fig. 8.2 for $K_u = \sqrt{2}$, we find that most characteristics of the distribution are contained in the approximation of Fig. 8.8.

For an undulator with many periods $N_u \gg 1$, we obtain the corresponding development of the spectral density by converting the series expression for the angular density (C.1)

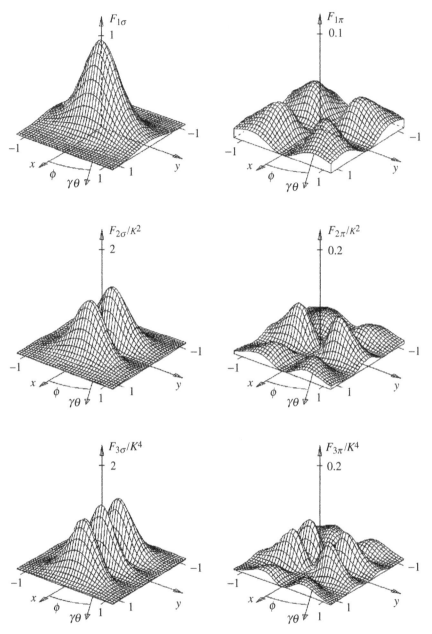

Fig. 8.8. Normalized angular power densities of the horizontal (left) and vertical (right) modes of polarization for the first three harmonics, each approximated to the lowest order in K_u^*.

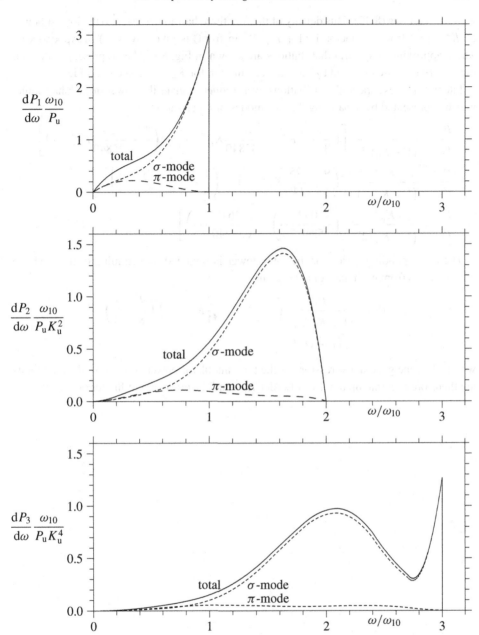

Fig. 8.9. Spectral power distributions of the first three harmonics of undulator radiation, each approximated to the lowest order in K_u^*.

with the relations (8.45). This density of the first three harmonics, each to the lowest power of K_u^* (apart from the factor $1/(1 + K_u^2/2)^2$ in front) is given by (C.2) in Appendix C. These approximate spectral distributions are shown in Fig. 8.9. They represent quite well the pattern of those obtained by an exact calculation for $K_u = \sqrt{2}$ shown in Fig. 8.3.

Integrating these spectral distributions over frequency gives the power in each harmonic, which is presented by separating the two modes of polarization:

$$\frac{P_1}{P_u} = \frac{1}{\left(1 + K_u^2/2\right)^2} \left[\left(\frac{7}{8} - \frac{1}{4}K_u^{*2} + \frac{681}{35840}K_u^{*4} \cdots \right) + \left(\frac{1}{8} - \frac{29}{35840}K_u^{*4} \cdots \right) \right]$$

$$\frac{P_2}{P_u} = \frac{K_u^{*2}}{\left(1 + K_u^2/2\right)^2} \left[\left(\frac{9}{8} - \frac{75}{140}K_u^{*2} \cdots \right) + \left(\frac{1}{8} - \frac{1}{140}K_u^{*2} \cdots \right) \right]$$

$$\frac{P_3}{P_u} = \frac{K_u^{*4}}{\left(1 + K_u^2/2\right)^2} \left[\left(\frac{42039}{35840} \cdots \right) + \left(\frac{3645}{35840} \cdots \right) \right].$$

The corresponding series for the total power is obtained by summing over the above contributions from each harmonic m, giving

$$P_u = \frac{P_u}{\left(1 + K_u^2/2\right)^2} \left(1 + K_u^{*2} + \frac{3}{4}K_u^{*4} \cdots \right) \left(\frac{7}{8} + \frac{1}{8} \right)$$
$$= P_u\left(1 + O\left(K_u^{*6}\right) \right),$$

where P_u is the general expression for the total undulator power. The above developments are therefore accurate up to terms of order K_u^{*4} and the errors are of higher order.

9

The helical undulator

9.1 The trajectory

A helical undulator [50, 59] has close to the axis a spatially periodic transverse magnetic field of constant absolute value and a direction that rotates as a function of the longitudinal coordinate z. In Cartesian coordinates it is given by

$$\mathbf{B} = [B_x, \ B_y, \ B_z] = B_0[-\sin(k_u z), \cos(k_u z), 0]$$

with period length λ_u and wave number $k_u = 2\pi/\lambda_u$ as illustrated in Fig. 9.1.

The motion of a charged particle along the z-axis is determined by the Lorentz force $\mathbf{F} = e[\mathbf{v} \times \mathbf{B}]$, giving

$$\dot{\mathbf{v}} = [\ddot{x}, \ \ddot{y}, \ \ddot{z}] = \frac{eB_0}{m_0\gamma}[-\cos(k_u z)\dot{z}, \ -\sin(k_u z)\dot{z}, (\cos(k_u z)\dot{x} + \sin(k_u z)\dot{y})].$$

The first two components can be integrated,

$$\dot{x}(z) = -\frac{cK_u}{\gamma}\sin(k_u z) + \dot{x}(0), \qquad \dot{y}(z) = -\frac{cK_u}{\gamma}(1 - \cos(k_u z)) + \dot{y}(0),$$

where we used again the undulator parameter

$$K_u = \frac{eB_0}{m_0 c k_u}.$$

We are seeking a trajectory that is symmetric around the z-axis and select the initial conditions

$$\dot{x}(0) = 0, \qquad \dot{y}(0) = cK_u/\gamma,$$

giving the transverse velocity components as functions of the longitudinal coordinate z:

$$\dot{x} = -\frac{cK_u}{\gamma}\sin(k_u z), \qquad \dot{y} = \frac{cK_u}{\gamma}\cos(k_u z).$$

181

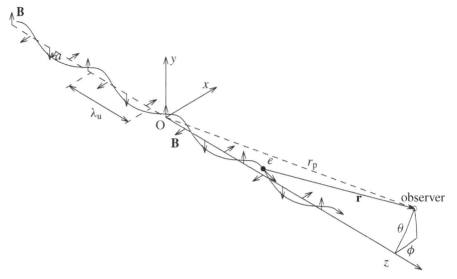

Fig. 9.1. The geometry of the trajectory in a helical undulator.

The longitudinal component can be obtained from the fact that the absolute value of the particle velocity in the magnetic field is constant,

$$v = \beta c = \sqrt{\dot{x}^2 + \dot{y}^2 + \dot{z}^2} = \sqrt{\frac{c^2 K_u^2}{\gamma^2} + \dot{z}^2},$$

which gives a constant velocity for the z-component:

$$\dot{z} = \beta c \sqrt{1 - \frac{K_u^2}{\beta^2 \gamma^2}} = \beta_h^* c \quad \text{and} \quad z(t') = \beta_h^* c t'. \tag{9.1}$$

The absolute value of the angle between the particle velocity and the z-axis is constant, given by

$$\tan \psi_0 = \frac{\sqrt{\dot{x}^2 + \dot{y}^2}}{\dot{z}} = \frac{K_u}{\beta \gamma \sqrt{1 - (K_u/\beta \gamma)^2}} = \frac{K_u}{\beta_h^* \gamma}, \qquad \psi_0 \approx \frac{K_u}{\beta_h^* \gamma}.$$

As in the case of the plane undulator, the character of the radiation is determined by the ratio between the trajectory angle ψ_0 and the natural opening angle $1/\gamma$ of the radiation, which is again determined by the undulator parameter K_u. We distinguish between weak and strong helical-undulator radiation, depending on K_u being smaller or larger than unity.

Using the relation $dz = \beta_h^* c \, dt'$, we can integrate the transverse velocity components and obtain the trajectory of the particle as a function of the longitudinal coordinate z up to second order in K_u/γ,

$$x(z) = \frac{K_u}{\beta_h^* \gamma k_u} \cos(k_u z), \qquad y(z) = \frac{K_u}{\beta_h^* \gamma k_u} \sin(k_u z),$$

or, as a function of the time t',

$$x(z) = \frac{K_\mathrm{u}}{\beta_\mathrm{h}^* \gamma k_\mathrm{u}} \cos(k_\mathrm{u}\beta_\mathrm{h}^* ct'), \qquad y(z) = \frac{K_\mathrm{u}}{\beta_\mathrm{h}^* \gamma k_\mathrm{u}} \sin(k_\mathrm{u}\beta_\mathrm{h}^* ct').$$

The particle moves on a helical trajectory with spatial period λ_u, frequency Ω_u, and radius a:

$$\Omega_\mathrm{u} = k_\mathrm{u}\beta_\mathrm{h}^* c = k_\mathrm{u}\beta c \sqrt{1 - \frac{K_\mathrm{u}}{\beta^2\gamma^2}}, \qquad a = \frac{K_\mathrm{u}}{\beta_\mathrm{h}^* \gamma k_\mathrm{u}}. \tag{9.2}$$

In summary, its position, normalized velocity, and acceleration are

$$\mathbf{R}(t') = [a\cos(\Omega_\mathrm{u}t'),\ a\sin(\Omega_\mathrm{u}t'),\ \beta_\mathrm{h}^* ct']$$

$$\boldsymbol{\beta}(t') = \left[-\frac{K_\mathrm{u}}{\gamma}\sin(\Omega_\mathrm{u}t'),\ \frac{K_\mathrm{u}}{\gamma}\cos(\Omega_\mathrm{u}t'),\ \beta_\mathrm{h}^* \right]$$

$$\dot{\boldsymbol{\beta}}(t') = \frac{K_\mathrm{u}\Omega_\mathrm{u}}{\gamma}[-\cos(\Omega_\mathrm{u}t'),\ -\sin(\Omega_\mathrm{u}t'),\ 0].$$

Like in the case of the strong plane undulator, we introduce the reduced values for the Lorentz factor γ_h^* of the drift velocity and for the undulator parameter K_uh^*, giving, in the ultra-relativistic approximation, the ratio between the trajectory angle ψ_0 and the natural opening angle $1/\gamma_\mathrm{h}^*$ of the radiation:

$$\gamma_\mathrm{h}^* = \frac{1}{\sqrt{1 - \beta_\mathrm{h}^{*2}}} \approx \frac{\gamma}{\sqrt{1 + K_\mathrm{u}^2}}, \qquad K_\mathrm{uh}^* = \frac{K_\mathrm{u}}{\sqrt{1 + K_\mathrm{u}^2}} \approx \gamma_\mathrm{h}^* \psi_0. \tag{9.3}$$

It should be noted that these two quantities have different denominators from those of the corresponding expressions (8.15) for the plane undulator. For this reason we use the notations γ_h^* and K_uh^* for the helical undulator.

The vector $\mathbf{r}(t')$ pointing from the charge to the observer is

$$\mathbf{r}(t') = \mathbf{r}_\mathrm{p} - \mathbf{R}(t')$$

$$= r_\mathrm{p}\left[\sin\theta\cos\phi - \frac{a}{r_\mathrm{p}}\cos(\Omega_\mathrm{u}t'),\ \sin\theta\sin\phi - \frac{a}{r_\mathrm{p}}\sin(\Omega_\mathrm{u}t'),\ \cos\theta - \frac{\beta_\mathrm{h}^* ct'}{r_\mathrm{p}} \right] \tag{9.4}$$

and its absolute value, in the far-distance approximation, taking terms of up to first order in a/r_p and $\beta ct'/r_\mathrm{p}$, is

$$r \approx r_\mathrm{p}\left(1 - \frac{\beta_\mathrm{h}^* ct'\cos\theta}{r_\mathrm{p}} - \frac{a}{r_\mathrm{p}}\sin\theta\cos(\Omega_\mathrm{u}t' - \phi)\right). \tag{9.5}$$

This is used to calculate the relation between the emission and observation times t' and t,

$$t_\mathrm{p} = t' + \frac{r - r_\mathrm{p}}{c} = t'(1 - \beta_\mathrm{h}^*\cos\theta) - \frac{K_\mathrm{uh}^* \sin\theta}{\beta_\mathrm{h}^* \gamma_\mathrm{h}^* k_\mathrm{u}c}\cos(\Omega_\mathrm{u}t'),$$

or, in the ultra-relativistic approximation,

$$t_{\mathrm{p}} = t' \frac{1 + \gamma_{\mathrm{h}}^{*2}\theta^2}{2\gamma_{\mathrm{h}}^{*2}} - \frac{K_{\mathrm{uh}}^{*}\gamma_{\mathrm{h}}^{*}\theta}{\gamma_{\mathrm{h}}^{*2}k_{\mathrm{u}}c}\cos(\Omega_{\mathrm{u}}t'). \tag{9.6}$$

The time period of the particle motion in the undulator $T_0 = 2\pi/\Omega_{\mathrm{u}}$ leads to a radiation period in the observation time of $T_{\mathrm{p}} = 2\pi/\omega_1$ and a relation between the frequencies

$$\omega_1 = \frac{2\gamma_{\mathrm{h}}^{*2}}{1 + \gamma_{\mathrm{h}}^{*2}\theta^2}\Omega_{\mathrm{u}}. \tag{9.7}$$

Like in the case of a plane undulator (7.12), the relation between the two time scales has a linear term and a harmonic modulation. To estimate the latter we multiply the time t_{p} by the frequency ω_1,

$$\omega_1 t_{\mathrm{p}} = \Omega_{\mathrm{u}}t' - \frac{2K_{\mathrm{uh}}^{*}\gamma_{\mathrm{h}}^{*}\theta}{1 + \gamma_{\mathrm{h}}^{*2}\theta^2}\cos(\Omega_{\mathrm{u}}t') = \Omega_{\mathrm{u}}t' - c_{\mathrm{u}}\cos(\Omega_{\mathrm{u}}t'), \tag{9.8}$$

with

$$c_{\mathrm{u}} = \frac{2K_{\mathrm{uh}}^{*}\gamma_{\mathrm{h}}^{*}\theta}{1 + \gamma_{\mathrm{h}}^{*2}\theta^2}. \tag{9.9}$$

For the typical opening angle $\sin\theta \approx 1/\gamma_{\mathrm{h}}^{*}$ or $\cos\theta \approx \beta_{\mathrm{h}}^{*}$, we find for the amplitude of the phase modulation

$$\delta(\omega_1 t_{\mathrm{p}}) \approx \frac{2K_{\mathrm{uh}}^{*}\gamma_{\mathrm{h}}^{*}\theta}{1 + \gamma_{\mathrm{h}}^{*2}\theta^2} \le K_{\mathrm{uh}}^{*}.$$

For a weak helical undulator, $K < 1$, we can neglect this modulation term, but not for the strong case.

The unit vector **n** pointing from the charge to the observer is needed only to lowest order:

$$\mathbf{n} \approx \mathbf{n}_{\mathrm{p}} = [\sin\theta\cos\phi, \ \sin\theta\sin\phi, \ \cos\theta] \approx [\theta\cos\phi, \ \theta\sin\phi, \ 1 - \theta^2/2].$$

For the weak helical undulator we will use the Liénard–Wiechert equation to obtain the radiation in time domain. The corresponding triple vector product in the approximation $K_{\mathrm{u}} < 1$ is

$$[\mathbf{n} \times [(\mathbf{n} - \boldsymbol{\beta}) \times \dot{\boldsymbol{\beta}}]] = \frac{ck_{\mathrm{u}}K_{\mathrm{u}}}{2\gamma^3}$$
$$\times [\cos(\Omega_{\mathrm{u}}t') - \gamma^2\theta^2\cos(2\phi - \Omega_{\mathrm{u}}t'), \ \sin(\Omega_{\mathrm{u}}t') - \gamma^2\theta^2\sin(2\phi - \Omega_{\mathrm{u}}t'), \ 0].$$
$$\tag{9.10}$$

The strong undulator will be treated directly in the frequency domain using equation (2.44) containing the triple vector product

$$[\mathbf{n} \times [\mathbf{n} \times \boldsymbol{\beta}]] = \frac{1}{\gamma_{\mathrm{h}}^{*}}[\gamma_{\mathrm{h}}^{*}\theta\cos\phi + K_{\mathrm{uh}}^{*}\sin(\Omega_{\mathrm{u}}t'), \ \gamma_{\mathrm{h}}^{*}\theta\sin\phi - K_{\mathrm{uh}}^{*}\cos(\Omega_{\mathrm{u}}t'), \ 0].$$

Considering the special symmetry of the helical motion, spherical coordinates are more suitable:

$$[\mathbf{n} \times [(\mathbf{n} - \boldsymbol{\beta}) \times \dot{\boldsymbol{\beta}}]]_{[\theta, \phi, r]} = \frac{ck_u K_u}{2\gamma^3}$$
$$\times [(1 - \gamma^2\theta^2)\cos(\Omega_u t' - \phi), \, (1 + \gamma^2\theta^2)\sin(\Omega_u t' - \phi), \, 0] \quad (9.11)$$

for $K_u < 1$ and

$$[\mathbf{n} \times [\mathbf{n} \times \boldsymbol{\beta}]]_{[\theta, \phi, r]} = \frac{1}{\gamma_h^*}[\gamma_h^*\theta + K_{uh}^* \sin(\Omega_u t' - \phi), \, -K_{uh}^* \cos(\Omega_u t' - \phi), \, 0] \quad (9.12)$$

for an arbitrary K_u but with $K_u/\gamma \ll 1$. The expression in the denominator of the Liénard–Wiechert expression is obtained from (9.8) but is needed only in the weak-undulator approximation:

$$(1 - \mathbf{n} \cdot \boldsymbol{\beta}) = \frac{dt_p}{dt'} = \frac{\Omega_u}{\omega_1}(1 + c_u \sin(\Omega_u t')) \approx \frac{1 + \gamma^2\theta^2}{2\gamma^2}. \quad (9.13)$$

9.2 The radiation emitted in a helical weak undulator

9.2.1 The radiation obtained with the Liénard–Wiechert formula

The radiation field in the time domain is given by the second term of the Liénard–Wiechert equation (2.17) given in Chapter 2:

$$\mathbf{E}(t) = \frac{e}{4\pi\epsilon_0 c}\left\{\frac{[\mathbf{n} \times [(\mathbf{n} - \boldsymbol{\beta}) \times \dot{\boldsymbol{\beta}}]]}{r(1 - \mathbf{n} \cdot \boldsymbol{\beta})^3}\right\}_{\text{ret}}.$$

Using (9.11) for the triple vector product and (9.13) for the denominator, and approximating $r \approx r_p$, we obtain the field directly in spherical coordinates, omitting the vanishing r-component. We give it first as a function of the emission time t',

$$\mathbf{E}_\perp(t') = \frac{ek_u K_u \gamma^3}{\pi\epsilon_0 r_p}$$
$$\times \frac{[(1 - \gamma^2\theta^2)\cos(\Omega_u t' - \phi)\,\boldsymbol{\eta}_\theta + (1 + \gamma^2\theta^2)\sin(\Omega_u t' - \phi)\,\boldsymbol{\eta}_\phi]}{(1 + \gamma^2\theta^2)^3},$$

where we introduce in analogy with (5.25) two orthogonal unit vectors in the θ- and ϕ-directions:

$$\boldsymbol{\eta}_\theta, \quad \boldsymbol{\eta}_\phi, \quad \boldsymbol{\eta}_\theta \cdot \boldsymbol{\eta}_\theta = \boldsymbol{\eta}_\phi \cdot \boldsymbol{\eta}_\phi = 1, \quad \boldsymbol{\eta}_\theta \cdot \boldsymbol{\eta}_\phi = 0. \quad (9.14)$$

Owing to the helical motion of the charge, the field has a phase of the form $\Omega_u t' - \phi$, containing the azimuthal observation angle ϕ.

The oscillating field has its maximum value on the axis,

$$\hat{E}_u = \frac{ek_u K_u \gamma^3}{\pi\epsilon_0 r_p},$$

which is the same as that emitted in a weak plane undulator, (7.31). However, in the plane undulator the field is oscillating in time and has a RMS value of $\hat{E}_u/\sqrt{2}$ on the axis, whereas for the helical undulator this field has a constant value but rotates.

We have the relation (9.6) between the time scales for $K < 1$,

$$t_p = t' \frac{(1 + \gamma^2 \theta^2)}{2\gamma^2},$$

and give this field observed from a large distance r_p as a function of the observation time t or t_p,

$$\Omega_u t' - \phi = -(k_1 r - \omega_1 t + \phi) = \omega_1 t_p - \phi$$

$$\mathbf{E}(r_p, t) = \frac{e k_u K_u \gamma^3}{\pi \epsilon_0 r_p (1 + \gamma^2 \theta^2)^3}$$
$$\times [(1 - \gamma^2 \theta^2) \cos(k_1 r - \omega_1 t + \phi) \boldsymbol{\eta}_\theta - (1 + \gamma^2 \theta^2) \sin(k_1 r - \omega_1 t + \phi) \boldsymbol{\eta}_\phi]$$

$$(9.15)$$

with the wave number $k_1 = \omega_1 c$ of the radiation. Both components of the field are circular functions of the phase $k_1 r + \omega_1 t - \phi$. Their amplitudes are in general different. The field vector $\mathbf{E}(r_p, t)$ propagates in the r-direction and rotates at the same time around $\mathbf{r}$, describing an elliptical helix. The radiation therefore has elliptical polarization. On the axis, $\theta = 0$, there is perfect circular polarization. On a cone $\gamma\theta = 1$ the field component in the θ-direction vanishes with only the ϕ-component left, resulting in linear polarization in the ϕ-direction. At very large angles $\gamma_h^* \theta \gg 1$ a circular polarization of opposite sign is approached. For other angles of observation the polarization is elliptical.

We now describe the field in terms of the reduced observation time t_p,

$$\mathbf{E}(t_p) = \hat{E}_u \frac{[(1 - \gamma^2 \theta^2) \cos(\omega_1 t_p - \phi) \boldsymbol{\eta}_\theta + (1 + \gamma^2 \theta^2) \sin(\omega_1 t_p - \phi) \boldsymbol{\eta}_\phi]}{(1 + \gamma^2 \theta^2)^3}, \qquad (9.16)$$

and arrange this equation differently in such a way as to separate the two circular modes of polarization and bring them into a form equivalent to (5.28):

$$\mathbf{E}(t_p) = \frac{\hat{E}_u \sqrt{2}}{(1 + \gamma^2 \theta^2)^3} \left[\frac{\cos(\omega_1 t_p - \phi) \boldsymbol{\eta}_\theta + \sin(\omega_1 t_p - \phi) \boldsymbol{\eta}_\phi}{\sqrt{2}} \right.$$
$$\left. - \gamma^2 \theta^2 \frac{\cos(\omega_1 t_p - \phi) \boldsymbol{\eta}_\theta - \sin(\omega_1 t_p - \phi) \boldsymbol{\eta}_\phi}{\sqrt{2}} \right]. \qquad (9.17)$$

We assume that there is a large but finite number $N_u \gg 1$ of undulator periods and calculate the radiation field in the frequency domain by using the Fourier transform of (9.17),

$$\tilde{\mathbf{E}}(\omega) = \frac{1}{\sqrt{2\pi}} \int_{-\pi N_u/\omega_1}^{\pi N_u/\omega_1} \mathbf{E}(t_p) e^{-i\omega t_p} \, dt_p,$$

giving from (9.16) the presentation in terms of θ- and ϕ-components:

$$\tilde{\mathbf{E}}(\omega) = \frac{\hat{E}_{\mathrm{u}}}{\sqrt{2\pi}} \frac{\pi N_{\mathrm{u}}}{\omega_1} \frac{\sin(\pi N_{\mathrm{u}} \Delta\omega/\omega_1)}{\pi N_{\mathrm{u}} \Delta\omega/\omega_1} \frac{[(1 - \gamma^2\theta^2)\boldsymbol{\eta}_\theta + \mathrm{i}(1 + \gamma^2\theta^2)\boldsymbol{\eta}_\phi]}{(1 + \gamma^2\theta^2)^3}. \tag{9.18}$$

With (5.32) we separate positive and negative helicities, which are defined for positive frequencies,

$$\tilde{\mathbf{E}}_\perp(\omega) = [\tilde{E}_+, \ \tilde{E}_-] = \frac{\hat{E}_{\mathrm{u}}\sqrt{2}}{\sqrt{2\pi}} \frac{\pi N_{\mathrm{u}}}{\omega_1} \frac{\sin(\pi N_{\mathrm{u}} \Delta\omega/\omega_1)}{\pi N_{\mathrm{u}} \Delta\omega/\omega_1} \frac{[1, \ -\gamma^2\theta^2]}{(1 + \gamma^2\theta^2)^3}. \tag{9.19}$$

with $\Delta\omega = \omega - \omega_1$ and using the approximation valid for $N_{\mathrm{u}} \gg 1$ as for the case of a plane undulator in an earlier section.

9.3 Properties of weak-helical-undulator radiation

9.3.1 The total power

The helical-undulator field $\mathbf{B}$ has a constant absolute value and only its direction changes. For this reason the total power emitted by one electron in the helical undulator is the same as the power P_{s} of synchrotron radiation in long magnets (5.3). However, since we express it here as a function of undulator parameters, we write the power $P_{\mathrm{h}} = P_{\mathrm{s}}$ and the energy $U_{\mathrm{h}} = U_{\mathrm{s}} = P_{\mathrm{h}} L_{\mathrm{u}}/c$, radiated by one electron:

$$
\begin{aligned}
P_{\mathrm{h}} &= \frac{2 r_0 c^3 e^2 E_{\mathrm{e}}^2 B_0^2}{3(m_0 c^2)^3} = \frac{2 r_0 c m_0 c^2 \gamma^2 k_{\mathrm{u}}^2 K_{\mathrm{u}}^2}{3} = \frac{e^2 c \gamma^2 k_{\mathrm{u}}^2 K_{\mathrm{u}}^2}{6\pi \epsilon_0} \\
U_{\mathrm{h}} &= \frac{2 r_0 c^2 e^2 E_{\mathrm{e}}^2 B_0^2 L_{\mathrm{u}}}{3(m_0 c^2)^3} = \frac{2 r_0 m_0 c^2 \gamma^2 k_{\mathrm{u}}^2 K_{\mathrm{u}}^2 L_{\mathrm{u}}}{3} = \frac{e^2 \gamma^2 k_{\mathrm{u}} K_{\mathrm{u}}^2 N_{\mathrm{u}}}{3\epsilon_0}.
\end{aligned}
\tag{9.20}
$$

These expressions have the same form but, due to the constant value of the field, they are a factor of 2 larger than those for a plane undulator, (7.32). The instantaneous power emitted by the particle is constant, but the direction of polarization rotates.

9.3.2 The angular spectral power distribution

We obtain the angular spectral distribution of the emitted power from (7.34),

$$\frac{d^2 P}{d\Omega \, d\omega} = \frac{2 r_{\mathrm{p}}^2 |\tilde{\mathbf{E}}(\omega)|^2}{\mu_0 L_{\mathrm{u}}},$$

using (9.18) to split it into the linear modes of polarization,

$$
\begin{aligned}
\frac{d^2 P}{d\Omega \, d\omega} &= \frac{r_0 c m_0 c^2 k_{\mathrm{u}}^2 K_{\mathrm{u}}^2 \gamma^4}{\pi} \frac{[(1 - \gamma^2\theta^2)^2 + (1 + \gamma^2\theta^2)^2]}{(1 + \gamma^2\theta^2)^5} f_{\mathrm{N}}(\Delta\omega) \\
&= P_{\mathrm{h}} \gamma^2 [F_{\mathrm{h}\theta} + F_{\mathrm{h}\phi}],
\end{aligned}
\tag{9.21}
$$

or, by splitting the terms in the square brackets, into circular modes of polarization. We introduce the corresponding normalized distribution functions to present the angular spectral power distribution in the form

$$
\frac{d^2 P}{d\Omega \, d\omega} = P_h \gamma^2 [F_{h\theta} + F_{h\phi}] f_N(\Delta\omega)
$$

$$
\frac{d^2 P}{d\Omega \, d\omega} = P_h \gamma^2 [F_{h+} + F_{h-}] f_N(\Delta\omega)
$$

(9.22)

with

$$
F_{h\theta}(\theta) = \frac{3}{2\pi} \frac{(1 - \gamma^2\theta^2)^2}{(1 + \gamma^2\theta^2)^5}, \qquad F_{h\phi}(\theta) = \frac{3}{2\pi} \frac{(1 + \gamma^2\theta^2)^2}{(1 + \gamma^2\theta^2)^5}
$$

$$
F_{h+}(\theta) = \frac{3}{\pi} \frac{1}{(1 + \gamma^2\theta^2)^5}, \qquad F_{h-}(\theta) = \frac{3}{\pi} \frac{\gamma^4\theta^4}{(1 + \gamma^2\theta^2)^5},
$$

which satisfy the sum rules

$$
F_{h\theta} + F_{h\phi} = F_{h+} + F_{h-} = F_u
$$

and the same spectral function (7.38) as that used for the plane undulator:

$$
f_N(\Delta\omega) = \frac{N_u}{\omega_1} \left(\frac{\sin(\pi N_u \, \Delta\omega/\omega_1)}{\pi N_u \, \Delta\omega/\omega_1} \right)^2, \qquad \int_{-\infty}^{\infty} f_N(\Delta\omega) \, d\omega = 1.
$$

(9.23)

9.3.3 The angular power distribution

We assume that there is a large number N_u of periods, giving a very narrow frequency band at a given angle θ. Integrating the expression (9.22) over the frequency ω gives the angular power distributions expressed in terms of the linear or circular modes of polarization,

$$
\frac{dP}{d\Omega} = P_h \gamma^2 [F_{u\theta} + F_{u\phi}] = P_h \frac{3\gamma^2}{2\pi} \frac{[(1 - \gamma^2\theta^2)^2 + (1 + \gamma^2\theta^2)^2]}{(1 + \gamma^2\theta^2)^5}
$$

$$
= P_h \gamma^2 [F_{u+} + F_{u-}] = P_h \frac{3\gamma^2}{\pi} \frac{[1 + \gamma^4\theta^4]}{(1 + \gamma^2\theta^2)^5},
$$

(9.24)

which are shown in Fig. 9.2.

Since the angular power distribution is independent of the azimuthal angle, it makes sense to integrate over ϕ, which is just a multiplication by 2π, and we obtain the power as a function of θ,

$$
\frac{dP}{\theta \, d\theta} = 2\pi \frac{dP}{d\Omega},
$$

which is shown later.

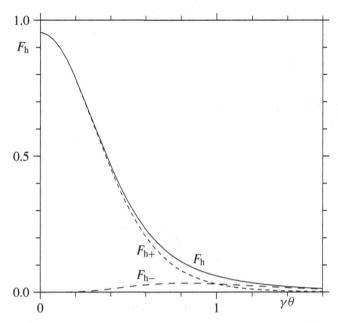

Fig. 9.2. Angular distributions of the total and the two circular modes of polarization of the radiation from a helical undulator after integration over frequency.

9.3.4 The spectral power distribution

For the case of a very large number N_u of periods we observe, at a given angle θ, a very narrow frequency band around ω_1, described by $f_N(\Delta\omega) \approx \delta(\omega - \omega_1)$. On integrating over ω and using the same relation, (7.45), as for the plane undulator to convert angle to frequency, we find

$$\omega \approx \omega_1 = \frac{\omega_{10}}{1 + \gamma^2\theta^2} = \frac{2k_u c \gamma^2}{1 + \gamma^2\theta^2}$$

$$d\omega_1 = -2\omega_{10}\gamma^2 \left(\frac{\omega_1}{\omega_{10}}\right)^2 \theta \, d\theta$$

$$\frac{dP}{d\omega} = \frac{1}{2\omega_{10}\gamma^2}\left(\frac{\omega_{10}}{\omega_1}\right)^2 \int_0^{2\pi} \frac{dP}{d\Omega} \, d\phi = \frac{\pi}{\omega_{10}\gamma^2}\left(\frac{\omega_{10}}{\omega_1}\right)^2 \frac{dP}{d\Omega}.$$

We obtain the spectral power density with respect to the proper frequency, split into the two linear or circular modes of polarization:

$$\frac{dP}{d\omega_1} = \frac{dP_{h\theta}}{d\omega_1} + \frac{dP_{h\phi}}{d\omega_1} = P_h \frac{3}{\omega_{10}}\frac{\omega_1}{\omega_{10}}\left[\left(\frac{1}{2} - 2\frac{\omega_1}{\omega_{10}} + 2\left(\frac{\omega_1}{\omega_{10}}\right)^2\right) + \frac{1}{2}\right]$$

$$\frac{dP}{d\omega_1} = \frac{dP_{h+}}{d\omega_1} + \frac{dP_{h-}}{d\omega_1} = P_h \frac{3}{\omega_{10}}\frac{\omega_1}{\omega_{10}}\left[\left(\frac{\omega_1}{\omega_{10}}\right)^2 + \left(1 - \frac{\omega_1}{\omega_{10}}\right)^2\right].$$

$$(9.25)$$

The latter is shown in Fig. 9.3.

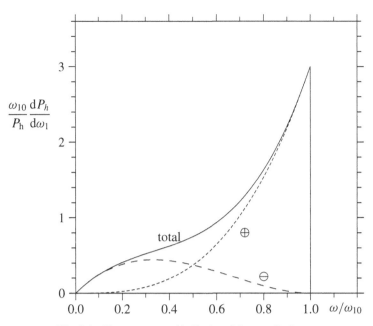

Fig. 9.3. The spectrum of helical-undulator radiation.

9.3.5 The total radiation

By integrating the spectral distribution (9.25) over frequency or the angular distribution (9.24) over the solid angle we recover the total power (9.20) and its distribution between the two linear or circular modes of polarization:

$$P_{\rm h} = P_{{\rm h}\theta} + P_{{\rm h}\phi} = P_{\rm h}\left(\frac{1}{4} + \frac{3}{4}\right), \qquad P_{\rm h} = P_{{\rm h}+} + P_{{\rm h}-} = P_{\rm h}\left(\frac{3}{4} + \frac{1}{4}\right). \tag{9.26}$$

9.3.6 The degree of circular polarization

The degree of circular polarization of the angular spectral distribution (9.22) is given as a function of the angle θ by

$$\frac{F_{{\rm h}+} - F_{{\rm h}-}}{F_{{\rm h}+} + F_{{\rm h}-}} = \frac{1 - \gamma^4\theta^4}{1 + \gamma^4\theta^4}.$$

It is plotted in Fig. 9.4 as a function of $\gamma\theta$. Since in some applications this might have to be weighted with the angular power distribution, also $(\mathrm{d}P/P_{\rm h})/\mathrm{d}(\gamma\theta)$ is shown in Fig. 9.4.

The degree of polarization of the spectral power distribution is obtained from (9.25) and shown in Fig. 9.5:

$$\frac{\mathrm{d}P_{{\rm h}+}/\mathrm{d}\omega_1 - \mathrm{d}P_{{\rm h}-}/\mathrm{d}\omega_1}{\mathrm{d}P_{{\rm h}+}/\mathrm{d}\omega_1 + \mathrm{d}P_{{\rm h}-}/\mathrm{d}\omega_1} = \frac{2\omega/\omega_{10} - 1}{2(\omega/\omega_{10})^2 - 2\omega/\omega_{10} + 1}.$$

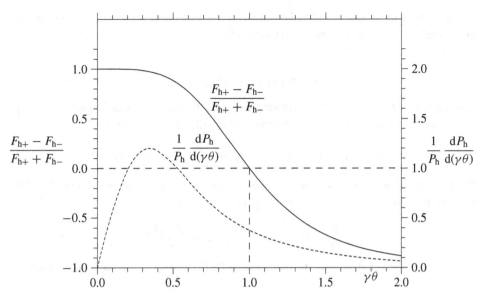

Fig. 9.4. The degree of circular polarization and $dP/d(\gamma\theta)$ as functions of θ.

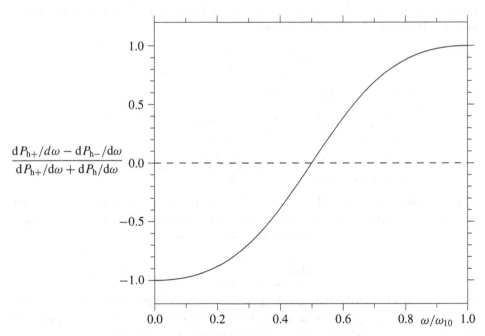

Fig. 9.5. The degree of circular polarization of the spectral distribution as a function of frequency.

It is positive for $\omega/\omega_{10} > 1/2$ and negative below. The frequency $\omega = \omega_{10}/2$ is observed at the angle $\theta = 1/\gamma$ where the polarization is linear.

9.3.7 The on-axis radiation

Like for the plane undulator, the radiation emitted on the axis, $\theta = 0$, at frequency $\omega = \omega_{10}$ is most important for users of the radiation. The distribution functions are

$$F_{\mathrm{h}\theta}(0) = F_{\mathrm{h}\phi}(0) = \frac{3}{2\pi}, \qquad F_{\mathrm{h}+}(0) = \frac{3}{\pi}, \qquad F_{\mathrm{h}-}(0) = 0.$$

On the axis the radiation has perfect circular polarization. According to (9.24) and (7.51), we have the following angular power and photon flux distributions:

$$\frac{\mathrm{d}P}{\mathrm{d}\Omega} = P_{\mathrm{h}}\gamma^2\frac{3}{\pi}, \qquad \frac{\mathrm{d}\dot{n}}{\mathrm{d}\Omega} = \frac{\alpha_{\mathrm{f}}c\gamma^2 k_{\mathrm{u}}K_{\mathrm{u}}^2}{\pi}.$$

These can be converted into spectral distributions (9.22) with respect to the proper frequency ω_1:

$$\frac{\mathrm{d}P}{\mathrm{d}\omega_1} = P_{\mathrm{h}}\frac{3}{\omega_{10}}, \qquad \frac{\mathrm{d}\dot{n}}{\mathrm{d}\omega_1/\omega_{10}} = \alpha_{\mathrm{f}}ck_{\mathrm{u}}K_{\mathrm{u}}^2.$$

For the angular spectral distribution we select again the fundamental frequency ω_{10}, which gives for the spectral function and the power

$$f_{\mathrm{N}}(0) = \frac{N_{\mathrm{u}}}{\omega_{10}}, \qquad \frac{\mathrm{d}^2P}{\mathrm{d}\Omega\,\mathrm{d}\omega} = P_{\mathrm{h}}\frac{3\gamma^2 N_{\mathrm{u}}}{\pi\omega_{10}}$$

and for the photon flux and number of photons emitted by one electron

$$\frac{\mathrm{d}^2\dot{n}}{\mathrm{d}\Omega\,\mathrm{d}\omega/\omega_{10}} = \frac{P_{\mathrm{h}}}{\hbar\omega_{10}}\frac{3\gamma^2 N_{\mathrm{u}}}{\pi} = \frac{e^2\gamma^2 k_{\mathrm{u}}K_{\mathrm{u}}^3 N_{\mathrm{u}}}{2\pi\epsilon_0 h} = \frac{2\alpha_{\mathrm{f}}\gamma^2 ck_{\mathrm{u}}K_{\mathrm{u}}^2 N_{\mathrm{u}}}{2\pi}$$

$$\frac{\mathrm{d}^2 n}{\mathrm{d}\Omega\,\mathrm{d}\omega/\omega_{10}} = \frac{U_{\mathrm{h}}}{\hbar\omega_{10}}\frac{3\gamma^2 N_{\mathrm{u}}}{\pi} = \frac{e^2\gamma^2 K_{\mathrm{u}}^3 N_{\mathrm{u}}^2}{2\pi\epsilon_0 ch} = 2\alpha_{\mathrm{f}}\gamma^2 K_{\mathrm{u}}^2 N_{\mathrm{u}}^2,$$

where P_{h} and U_{h} are expressed by using (9.20). Finally, we consider a beam current I and give the photon flux per unit solid angle and relative frequency band at $\theta = 0$, $\omega = \omega_{10}$:

$$\boxed{\frac{\mathrm{d}^2\dot{n}_I}{\mathrm{d}\Omega\,\mathrm{d}\omega/\omega} = \frac{2\alpha_{\mathrm{f}}\gamma^2 I K_{\mathrm{u}}^2 N_{\mathrm{u}}^2}{e}.} \qquad (9.27)$$

On comparing this with the corresponding expression for the plane weak undulator (7.59), we find that the two expressions have the same form but the helical undulator has twice the flux. This is not astonishing. A helical weak undulator can be regarded as a combination of two plane weak undulators rotated by $90°$ and phase shifted with respect to each other. Apart from the polarization, each of the two gives the same spectrum, which results in twice the power.

9.4 The radiation field from a strong helical undulator

For the strong helical undulator we follow closely [50] and include higher terms in the relation between the emission and observation times t' and t_p given in the ultra-relativistic approximation by (9.8):

$$\omega_1 t_p = \Omega_u t' - c_u \cos(\Omega_u t'), \qquad \omega_1 = \frac{2\gamma_h^{*2}}{1 + \gamma_h^{*2}\theta^2}\Omega_u, \qquad c_u = \frac{2K_{uh}^*\gamma_h^*\theta}{1 + \gamma_h^{*2}\theta^2}.$$

The particle motion in this undulator has a periodic velocity. The emitted radiation is also periodic with frequency ω_1, but now contains higher harmonics. Following the treatment of plane strong undulators, we develop its field $\mathbf{E}_m(t)$ into a Fourier series with components $\mathbf{E}_m$ given in (8.25) and (8.26) :

$$\mathbf{E}(t) = \sum_{m=-\infty}^{\infty} \mathbf{E}_m e^{im\omega_1 t}, \qquad \mathbf{E}_m = \frac{ime\omega_1^2}{8\pi^2\epsilon_0 cr_p}\int_0^{T_0} [\mathbf{n} \times [\mathbf{n} \times \boldsymbol{\beta}]]e^{-im\omega_1 t_p(t')}\,dt'.$$

With the expression (9.12) for the second triple vector product in spherical coordinates and (9.6) for the exponential, we obtain

$$\mathbf{E}_{\perp m(\theta,\phi)} = \frac{iem\omega_1^2}{8\pi^2\epsilon_0 cr_p}\int_0^{T_0} \frac{[\gamma_h^*\theta + K_{uh}^*\sin(\Omega_u t' - \phi), -K_{uh}^*\cos(\Omega_u t' - \phi)]}{\gamma_h^*}$$
$$\times\, e^{-im(\Omega_u t' - c_u\cos(\Omega_u t' - \phi))}\,dt',$$

where the vanishing r-component of the field has been omitted.

As for the strong plane undulator, we express the exponential of a trigonometric function in terms of a series of Bessel functions, (B.21),

$$e^{iz\sin\xi} = \sum_{n=-\infty}^{\infty} J_n(z)e^{in\xi},$$

and obtain

$$e^{-i\omega_1 t_p} = e^{-im(\Omega_u t' - c_u\cos(\Omega_u t' - \phi))} = e^{-im\Omega_u t'}e^{imc_u\sin(\Omega_u t' - \phi + \pi/2)}$$
$$= e^{-im\Omega_u t'}\sum_{n=-\infty}^{\infty} J_n(mc_u)e^{in(\Omega_u t' - \phi)}e^{in\pi/2}$$
$$= e^{-im\phi}\sum_{n=-\infty}^{\infty} i^n J_n(mc_u)e^{i(n-m)(\Omega_u t' - \phi)}.$$

The Fourier component of the field in spherical coordinates becomes

$$\mathbf{E}_{\perp m}(\theta, \phi) = \frac{ime\omega_1^2}{8\pi^2\epsilon_0 cr_p}\int_0^{T_0} \frac{[\gamma_h^*\theta + K_{uh}^*\sin(\Omega_u t' - \phi), -K_{uh}^*\cos(\Omega_u t' - \phi)]}{\gamma_h^*}$$
$$\times\, e^{-im\phi}\sum_{-\infty}^{\infty} i^n J_n(mc_u)e^{i(n-m)(\Omega_u t' - \phi)}\,dt'.$$

This expression contains the following three integrals,

$$I_1 = \int_0^{T_0} e^{i(n-m)(\Omega_u t' - \phi)} dt' = \begin{cases} 0 & \text{if } -m+n \neq 0 \\ T_0 & \text{if } -m+n = 0 \end{cases}$$

$$I_2 = \int_0^{T_0} \sin(\Omega_u t' - \phi) \sin((n-m)(\Omega_u t' - \phi)) dt'$$

$$I_3 = \int_0^{T_0} \cos(\Omega_u t' - \phi) \cos((n-m)(\Omega_u t' - \phi)) dt'$$

giving for the last two

$$I_2 = \begin{cases} T_0/2 & \text{if } n-m=1 \\ -T_0/2 & \text{if } n-m=-1 \\ 0 & \text{otherwise} \end{cases} \qquad I_3 = \begin{cases} T_0/2 & \text{if } n-m=1 \\ T_0/2 & \text{if } n-m=-1 \\ 0 & \text{otherwise.} \end{cases}$$

We obtain for the Fourier components in the two directions of the field

$$E_{m\theta} = -\frac{ek_u\gamma_h^{*3}}{\pi\epsilon_0 r_p(1+\gamma_h^{*2}\theta^2)^2} i^{m+1} m e^{-im\phi}$$
$$\times \left(\frac{K_{uh}^*}{2}(J_{m-1}(mc_u) + J_{m+1}(mc_u)) - \gamma_h^*\theta J_m(mc_u) \right),$$

$$E_{m\phi} = -\frac{iek_u\gamma_h^{*3}}{\pi\epsilon_0 r_p(1+\gamma_h^{*2}\theta^2)^2} i^{m+1} m e^{-im\phi} \frac{K_{uh}^*}{2}(J_{m-1}(mc_u) - J_{m+1}(mc_u)),$$

(9.28)

where we used the relation $\omega_1 = \Omega_u 2\gamma_h^{*2}/(1+\gamma_h^{*2}\theta^2)$, $\Omega_u = \beta_h^* ck_u \approx ck_u$.

The field as a function of the reduced observation time t_p is given by the complex Fourier series obtained from (9.28), containing positive and negative frequencies,

$$\mathbf{E}(t_p) = \sum_{m=-\infty}^{\infty} \mathbf{E}_m e^{im\omega_1 t_p}$$

$$= -\frac{ek_u\gamma_h^{*3}}{\pi\epsilon_0 r_p(1+\gamma_h^{*2}\theta^2)^2} \sum_{m=-\infty}^{\infty} i^{m+1} m e^{im(\omega_1 t_p - \phi)}$$

$$\times \left[\left(\frac{K_{uh}^*}{2}(J_{m-1}(mc_u) + J_{m+1}(mc_u)) - \gamma_h^*\theta J_m(mc_u) \right) \boldsymbol{\eta}_\theta \right.$$

$$\left. + i\frac{K_{uh}^*}{2}(J_{m-1}(mc_u) - J_{m+1}(mc_u))\boldsymbol{\eta}_\phi \right]$$

(9.29)

where we used the unit vectors $\boldsymbol{\eta}_\theta$ and $\boldsymbol{\eta}_\phi$ introduced before in (9.14).

Using the symmetry relations (B.5) and (B.6) of Bessel functions,

$$J_{-n}(x) = (-1)^n J_n(x), \qquad J_n(-x) = (-1)^n J_n(x),$$

we combine terms with $\pm m$ and express the Fourier series in terms of trigonometric functions of positive harmonics $m > 0$ only,

$$
\begin{aligned}
\mathbf{E}(t_{\mathrm{p}}) = {} & \frac{ek_{\mathrm{u}}\gamma_{\mathrm{h}}^{*3}}{\pi\epsilon_0 r_{\mathrm{p}}\left(1 + \gamma_{\mathrm{h}}^{*2}\theta^2\right)^2} \sum_{m=1}^{\infty} m \\
& \times [(K_{\mathrm{uh}}^*(J_{m-1}(mc_{\mathrm{u}}) + J_{m+1}(mc_{\mathrm{u}})) - 2\gamma_{\mathrm{h}}^*\theta J_m(mc_{\mathrm{u}})) \\
& \times \sin(m(\omega_1 t_{\mathrm{p}} - \phi + \pi/2))\,\boldsymbol{\eta}_\theta \\
& + K_{\mathrm{uh}}^*(J_{m-1}(mc_{\mathrm{u}}) - J_{m+1}(mc_{\mathrm{u}})) \\
& \times \cos(m(\omega_1 t_{\mathrm{p}} - \phi + \pi/2))\,\boldsymbol{\eta}_\phi],
\end{aligned} \tag{9.30}
$$

and, by a different arrangement, in circular polarization components:

$$
\begin{aligned}
\mathbf{E}(t_{\mathrm{p}}) = {} & \frac{\sqrt{2}ek_{\mathrm{u}}\gamma_{\mathrm{h}}^{*3}}{\pi\epsilon_0 r_{\mathrm{p}}\left(1 + \gamma_{\mathrm{h}}^{*2}\theta^2\right)^2} \\
& \times \sum_{m=1}^{\infty} m \Bigg[(K_{\mathrm{uh}}^* J_{m-1}(mc_{\mathrm{u}}) - \gamma_{\mathrm{h}}^*\theta J_m(mc_{\mathrm{u}})) \\
& \times \frac{\cos(m(\omega_1 t_{\mathrm{p}} - \phi + \pi/2) - \pi/2)\,\boldsymbol{\eta}_\theta + \sin(m(\omega_1 t_{\mathrm{p}} - \phi + \pi/2) - \pi/2)\,\boldsymbol{\eta}_\phi}{\sqrt{2}} \\
& + (K_{\mathrm{uh}}^* J_{m+1}(mc_{\mathrm{u}}) - \gamma_{\mathrm{h}}^*\theta J_m(mc_{\mathrm{u}})) \\
& \times \frac{\cos(m(\omega_1 t_{\mathrm{p}} - \phi + \pi/2) - \pi/2)\,\boldsymbol{\eta}_\theta - \sin(m(\omega_1 t_{\mathrm{p}} - \phi + \pi/2) - \pi/2)\,\boldsymbol{\eta}_\phi}{\sqrt{2}} \Bigg].
\end{aligned} \tag{9.31}
$$

For a finite length L_{u} of an undulator containing many periods $N_{\mathrm{u}} \gg 1$ of length λ_{u}, the radiation at a given angle θ is no longer monochromatic but has the spectrum

$$
\tilde{\mathbf{E}}(\omega) = \frac{1}{\sqrt{2\pi}} \int_{-\pi N_{\mathrm{u}}/\omega_1}^{\pi N_{\mathrm{u}}\omega_1} \mathbf{E}(t_{\mathrm{p}}) e^{-i\omega t_{\mathrm{p}}}\, dt_{\mathrm{p}}.
$$

From (9.30) we obtain the Fourier transform of the mth harmonic expressed in terms of the two linear modes of polarization,

$$
\begin{aligned}
\tilde{\mathbf{E}}_\perp(\omega) = {} & [\tilde{E}_\theta(\omega),\ \tilde{E}_\phi(\omega)] \\
= {} & \frac{-ek_{\mathrm{u}}\gamma_{\mathrm{h}}^{*3}}{\pi\sqrt{2\pi}\,\epsilon_0 r_{\mathrm{p}}\left(1 + \gamma_{\mathrm{h}}^{*2}\theta^2\right)^2} \\
& \times \sum_{m=1}^{\infty} i^{m+1} m e^{-im\phi}[(K_{\mathrm{uh}}^*(J_{m-1}(mc_{\mathrm{u}}) + J_{m+1}(mc_{\mathrm{u}})) - 2\gamma_{\mathrm{h}}^*\theta J_m(mc_{\mathrm{u}})), \\
& + iK_{\mathrm{uh}}^*(J_{m-1}(mc_{\mathrm{u}}) - J_{m+1}(mc_{\mathrm{u}}))]\frac{\pi N_{\mathrm{u}}}{\omega_1}\frac{\sin(\pi N_{\mathrm{u}}\,\Delta\omega/\omega_1)}{\pi N_{\mathrm{u}}\,\Delta\omega/\omega_1},
\end{aligned} \tag{9.32}
$$

and, with (9.16), in terms of the two circular modes of polarization,

$$
\tilde{\mathbf{E}}_{\perp}(\omega) = [\tilde{E}_{+}(\omega),\ \tilde{E}_{-}(\omega)] = \frac{-ek_u\gamma_h^{*3}}{\pi\sqrt{2\pi}\epsilon_0 r_p(1+\gamma_h^{*2}\theta^2)^2}
$$

$$
\times \sum_{m=1}^{\infty} i^{m+1}m e^{-im\phi}[(K_{uh}^* J_{m-1}(mc_u) - \gamma_h^*\theta J_m(mc_u))(\boldsymbol{\eta}_\theta + i\boldsymbol{\eta}_\phi)
$$

$$
+ (K_{uh}^* J_{m+1}(mc_u) - \gamma_h^*\theta J_m(mc_u))(\boldsymbol{\eta}_\theta - i\boldsymbol{\eta}_\phi)]
$$

$$
\times \frac{\pi N_u}{\omega_1}\frac{\sin(\pi N_u\,\Delta\omega/\omega_1)}{\pi N_u\,\Delta\omega/\omega_1}
\tag{9.33}
$$

with

$$
\Delta\omega = \omega - m\omega_1.
$$

It should be noted that these are complex presentations of the above Fourier-transformed field containing positive and negative frequencies ω but only positive harmonics $m > 0$.

For later discussions of the strong-helical-undulator radiation we need some properties of the field which we derive from the above real Fourier series.

- $m = 1$. We write the fundamental component explicitly:

$$
E_{1\theta}(t_p) = \frac{ek_u\gamma_h^{*3}}{\pi\epsilon_0 r_p(1+\gamma_h^{*2}\theta^2)^2}\cos(\omega_1 t_p - \phi)
$$

$$
\times \left(K_{uh}^*(J_0(c_u)+J_2(c_u)) - 2\gamma_h^*\theta J_1(c_u)\right)
$$

$$
E_{1\phi}(t_p) = \frac{ek_u\gamma_h^{*3}}{\pi\epsilon_0 r_p\left(1+\gamma_h^{*2}\theta^2\right)^2}\sin(\omega_1 t_p - \phi)
\tag{9.34}
$$

$$
\times K_{uh}^*\left(J_0(c_u) - J_2(c_u)\right).
$$

- $\theta = 0$. On the axis all Fourier components vanish except the fundamental $m = 1$, which becomes

$$
\mathbf{E}_{1\theta=0}(t_p) = \frac{ek_u K_{uh}^*\gamma_h^{*3}}{\pi\epsilon_0 r_p}[\cos(\omega_1 t_p - \phi),\ \sin(\omega_1 t - \phi),\ 0],
$$

giving circularly polarized radiation.

- $\gamma_h^*\theta = 1$. We investigate the field on this cone $\gamma_h^*\theta = 1$. Using the expressions for the sum and difference between two Bessel functions two orders apart,

$$
J_{n-1}(z) + J_{n+1}(z) = \frac{2n}{z}J_n(z), \qquad J_{n-1}(z) - J_{n+1}(z) = 2\frac{\mathrm{d}J_n(z)}{\mathrm{d}z},
$$

we find $c_u = K_{uh}^*$ and obtain for the radiation field (9.30)

$$
E_\theta(t_p) = 0
\tag{9.35}
$$

$$
E_\phi(t_p) = \frac{ek_u\gamma_h^{*3}}{2\pi\epsilon_0 r_p}\sum_{m=1}^{\infty}\cos(m(\omega_1 t_p - \phi + \pi/2))\,2K_{uh}^*\frac{\mathrm{d}J_m(mc_u)}{\mathrm{d}(mc_u)}(mc_u).
$$

For $\gamma_h^*\theta = 1$ the θ-component vanishes and we are left with linear polarization in the ϕ-direction.

9.5 Properties of strong-helical-undulator radiation

9.5.1 The total power

The total radiated power and energy are the same as those for the weak helical undulator (9.20).

9.5.2 The angular spectral power distribution

We calculate the angular spectral power distribution of the radiation from a strong undulator with a large number of periods $N_u \gg 1$:

$$\frac{d^2 P}{d\Omega \, d\omega} = \frac{2 r_p^2 |\tilde{E}(\omega)^2|}{\mu_0 L_u}.$$

With the Fourier-transformed field given by (9.32) and the general expression (9.20) for the total P_h, we obtain the angular spectral power distribution in terms of linear and circular components of the polarization,

$$\frac{d^2 P_h}{d\Omega \, d\omega} = \frac{d^2(P_{u\theta} + P_{u\phi})}{d\Omega \, d\omega} = P_h \gamma_h^{*2} \sum_{m=1}^{\infty} (F_{hm\theta} + F_{hm\phi}) f_N(\Delta\omega_m)$$

$$\frac{d^2 P_h}{d\Omega \, d\omega} = \frac{d^2(P_{u+} + P_{u-})}{d\Omega \, d\omega} = P_h \gamma_h^{*2} \sum_{m=1}^{\infty} (F_{hm+} + F_{hm-}) f_N(\Delta\omega_m),$$

(9.36)

with the same spectral function as that found before, (8.42), for the strong plane undulator,

$$f_N(\Delta\omega_m) = \frac{N_u}{\omega_1} \left(\frac{\sin\left(\frac{\Delta\omega}{\omega_1} \pi N_u\right)}{\frac{\Delta\omega}{\omega_1} \pi N_u} \right)^2, \qquad \omega_m = m\omega_1 = m \frac{2\gamma_h^{*2} \Omega_u}{1 + \gamma_h^{*2} \theta^2}$$

$$\frac{\Delta\omega_m}{\omega_1} = \frac{\omega - m\omega_1}{\omega_1}, \qquad \int_{-\infty}^{\infty} f_N(\Delta\omega_m) \, d\omega = 1,$$

and the normalized angular distribution functions separated according to the two linear and the two circular modes of polarization,

$$F_{hm\theta}(\theta) = \frac{3m^2 (K_{uh}^*(J_{m-1}(mc_u) + J_{m+1}(mc_u)) - 2\gamma_h^* \theta J_m(mc_u))^2}{2\pi K_{uh}^{*2} (1 + K_u^2)^2 (1 + \gamma_h^{*2}\theta^2)^3}$$

$$F_{hm\phi}(\theta) = \frac{3m^2 (K_{uh}^*(J_{m-1}(mc_u) - J_{m+1}(mc_u)))^2}{2\pi K_{uh}^{*2} (1 + K_u^2)^2 (1 + \gamma_h^{*2}\theta^2)^3}$$

$$F_{hm+}(\theta) = \frac{6m^2 (K_{uh}^* J_{m-1}(mc_u) - \gamma_h^* \theta J_m(mc_u))^2}{2\pi K_{uh}^{*2} (1 + K_u^2)^2 (1 + \gamma_h^{*2}\theta^2)^3}$$

$$F_{hm-}(\theta) = \frac{6m^2 (K_{uh}^* J_{m+1}(mc_u) - \gamma_h^* \theta J_m(mc_u))^2}{2\pi K_{uh}^{*2} (1 + K_u^2)^2 (1 + \gamma_h^{*2}\theta^2)^3}$$

(9.37)

with the parameter

$$c_u = \frac{2K_{uh}^* \gamma_h^* \theta}{1 + \gamma_h^{*2}\theta^2}.$$

Since helical undulators are mainly used to produce radiation with circular polarization, we will from now on give only the two components of the helicity and omit the expressions for the two linear modes (Fig. 9.6).

9.5.3 The angular power distribution ˙

For a large number of undulator periods $N_u \gg 1$, we have at a given angle a very narrow band of frequencies. Integrating (9.36) over ω gives the angular distribution of the radiated power:

$$\frac{dP}{d\Omega} = P_h \gamma_h^{*2}[F_{hm+}(\theta) + F_{hm-}(\theta)]. \tag{9.38}$$

Most interesting is the radiation on the axis, which has perfect circular polarization and will be discussed in detail later. We showed that the field E_θ vanishes on the cone $\gamma_h^*\theta = 1$, and we are left with linear polarization in the ϕ-direction.

9.5.4 The spectral density of helical-undulator radiation

Like for the strong plane undulator, we assume that there is a large number of undulator periods, giving at a given angle a very narrow spectral width, and approximate the spectral function $f_N(\Delta\omega_m)$ by the δ-function. This results in a relation between θ and frequency $\omega \approx \omega_m$, as for the plane undulator, (8.45), which we can use to convert angle into frequency,

$$\omega \approx \omega_m = m\omega_1 = \frac{m\Omega_u 2\gamma^{*2}}{1 + \gamma_h^{*2}\theta^2}, \qquad d\omega = -2m\omega_{10}\gamma_h^{*2}\left(\frac{\omega_1}{\omega_{10}}\right)^2 \theta\, d\theta,$$

which is used to convert the angular power distribution (9.38) into a spectral density. Owing to the axial symmetry of the helical undulator, the integration of $dP/d\Omega$ over the angle ϕ gives just a factor of 2π. We obtain for the spectral power density

$$\frac{dP_m}{d\omega} = \frac{(1 + \gamma_h^{*2}\theta^2)^2}{2m\omega_{10}\gamma_h^{*2}} \int_0^{2\pi} \frac{dP_m}{d\Omega} d\phi = \frac{\pi(1 + \gamma_h^{*2}\theta^2)^2}{m\omega_{10}\gamma_h^{*2}} \frac{dP_m}{d\Omega}.$$

With the expression (9.38) for the angular distribution, we obtain the spectral power density

$$\frac{dP_m}{d\omega} = P_h \frac{\pi}{m\omega_{10}}\left(\frac{\omega_{10}}{\omega_1}\right)^2 (F_{hm+} + F_{hm-}). \tag{9.39}$$

With the relation

$$\frac{1}{1 + \gamma_h^{*2}\theta^2} = \frac{\omega_1}{\omega_{10}} = \frac{m\omega_1}{m\omega_{10}} = \frac{\omega_m}{\omega_{m0}}$$

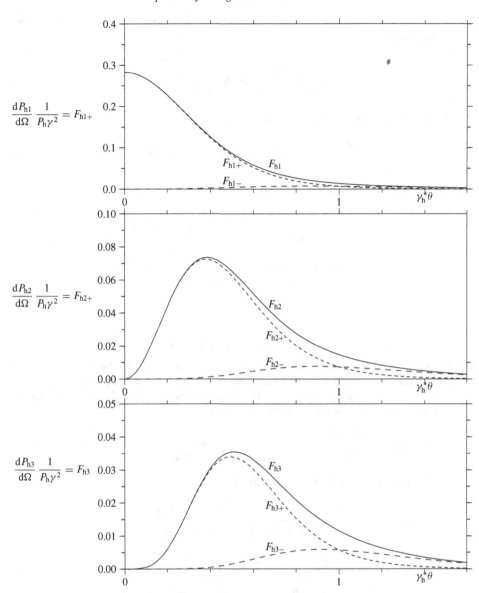

$$\frac{dP_{h1}}{d\Omega}\frac{1}{P_h\gamma^2} = F_{h1+}$$

$$\frac{dP_{h2}}{d\Omega}\frac{1}{P_h\gamma^2} = F_{h2+}$$

$$\frac{dP_{h3}}{d\Omega}\frac{1}{P_h\gamma^2} = F_{h3}$$

Fig. 9.6. Cuts through the angular power distributions of the first three harmonics for $K_u = 1/\sqrt{2}$ ($K_{uh}^* = 1/\sqrt{3}$).

we express the angular distribution functions in terms of frequency:

$$F_{hm+}(\omega_1/\omega_{10}) = \left(\frac{\omega_1}{\omega_{10}}\right)^3 \frac{6m^2(K_{uh}^* J_{m-1}(mc_u) - \gamma_h^*\theta J_m(mc_u))^2}{2\pi K_{uh}^{*2}(1 + K_u^2)^2}$$

$$F_{hm-}(\omega_1/\omega_{10}) = \left(\frac{\omega_1}{\omega_{10}}\right)^3 \frac{6m^2(K_u J_{m+1}(mc_u) - \gamma_h^*\theta J_m(mc_u))^2}{2\pi K_{uh}^{*2}(1 + K_u^2)^2}.$$

The parameter mc_u appearing as argument in the Bessel functions is also expressed in terms of the frequency ratio

$$mc_u = m\frac{2K^*_{uh}\gamma^*_h\theta}{1 + \gamma^{*2}_h\theta^2} = 2mK^*_{uh}\sqrt{\frac{\omega_1}{\omega_{10}}\left(1 - \frac{\omega_1}{\omega_{10}}\right)}.$$

We now express the spectral density in a compact form,

$$\frac{\mathrm{d}P_m}{\mathrm{d}\omega_m} = \frac{P_h}{m\omega_{10}}[S_{hm+} + S_{hm-}] = \frac{P_h}{m\omega_{10}}S_{hm}, \qquad (9.40)$$

with the normalized spectral power functions (Fig. 9.7)

$$S_{hm+} = 3m\frac{\omega_m}{\omega_{10}}\frac{(K^*_{uh}J_{m-1}(mc_u) - \gamma^*_h\theta J_m(mc_u))^2}{K^{*\,2}_{uh}(1 + K^2_u)^2}$$

$$S_{hm-} = 3m\frac{\omega_m}{\omega_{10}}\frac{(K^*_{uh}J_{m+1}(mc_u) - \gamma^*_h\theta J_m(mc_u))^2}{K^{*\,2}_{uh}(1 + K^2_u)^2}.$$

9.5.5 The on-axis radiation

On the axis, $\theta = 0$, the parameter c_u vanishes. As a consequence, all the Bessel functions in the normalized angular distribution functions (9.37) vanish also, except the one of order $(m - 1)/2 = 0$, which gives $J_0(0) = 1$. This also means that only the lowest harmonic $m = 1$ contributes to the on-axis radiation. We have

$$c_u = \frac{2K^*_{uh}\gamma^*_h\theta}{1 + \gamma^{*2}_h\theta^2} = 0, \qquad F_{h1+} = \frac{3}{\pi(1 + K_{uh})^3}, \qquad F_{h1-} = 0.$$

The angular power density (9.38) on the axis becomes

$$\frac{\mathrm{d}P}{\mathrm{d}\Omega} = P_h\frac{3\gamma^2}{\pi(1 + K^2_u)^3} = \frac{2r_0cm_0c^2k^2_uK^2_u\gamma^4}{\pi(1 + K^2_u)^3}.$$

It has a maximum value for $K_u = 1/\sqrt{2}$ or $K^*_{uh} = 1/\sqrt{3}$ of

$$\left(\frac{\mathrm{d}P}{\mathrm{d}\Omega}\right)_{max} = \frac{8r_0cm_0c^2k^2_u\gamma^4}{27\pi}.$$

Since the optimum undulator parameter K_u is relatively small and higher harmonics vanish on the axis, very strong helical undulators are of limited interest.

We assume that the radiation is observed also at the fundamental frequency $\omega = \omega_{10}$, for which the spectral function is

$$f_N(\Delta\omega_m) = \frac{N_u}{\omega_{10}} = \frac{N_u(1 + K^2_u)}{2ck_u\gamma^2},$$

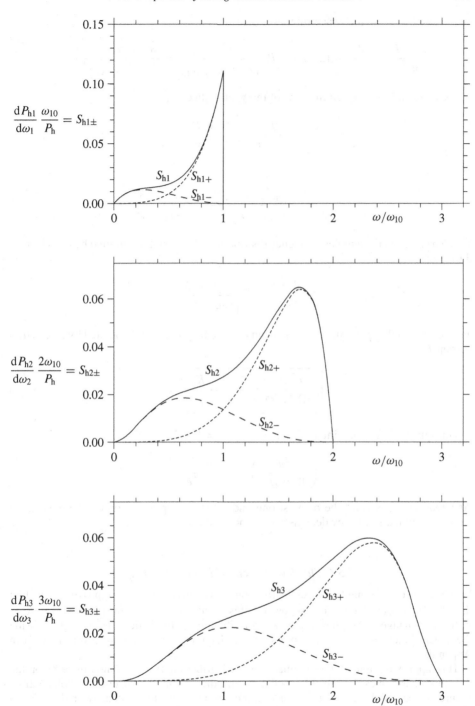

Fig. 9.7. Spectral power distributions of the first three harmonics emitted by a helical undulator with $K_u = 1/\sqrt{2}$ ($K_{uh}^* = 1/\sqrt{3}$).

and obtain for the angular spectral power distribution

$$
\frac{\mathrm{d}^2 P}{\mathrm{d}\Omega\,\mathrm{d}\omega} = \frac{\mathrm{d}P}{\mathrm{d}\Omega}\, f_\mathrm{N}(\Delta\omega_m) = P_\mathrm{h}\frac{3\gamma^2 N_\mathrm{u}}{\pi\left(1 + K_\mathrm{u}^2\right)^3 \omega_{10}} = \frac{r_0 m_0 c^2 \gamma^2 k_\mathrm{u} K_\mathrm{u}^2 N_\mathrm{u}}{\pi\left(1 + K_\mathrm{u}^2\right)^2}.
$$

This can again be expressed in terms of the photon flux,

$$
\frac{\mathrm{d}^2 \dot{n}}{\mathrm{d}\Omega\,\mathrm{d}\omega/\omega} = \frac{1}{\hbar}\frac{\mathrm{d}^2 P}{\mathrm{d}\Omega\,\mathrm{d}\omega},
$$

giving

$$
\frac{\mathrm{d}^2 \dot{n}}{\mathrm{d}\Omega\,\mathrm{d}\omega/\omega} = \frac{r_0 m_0 c^2 \gamma^2 k_\mathrm{u} K_\mathrm{u}^2 N_\mathrm{u}}{\pi\hbar\left(1 + K_\mathrm{u}^2\right)^2} = \frac{\alpha_\mathrm{f} c \gamma^2 k_\mathrm{u} K_\mathrm{u}^2 N_\mathrm{u}}{\pi\left(1 + K_\mathrm{u}^2\right)^2}.
$$

We also calculate the angular frequency distribution of the photons emitted by one electron during one traversal,

$$
\frac{\mathrm{d}^2 n}{\mathrm{d}\Omega\omega/\omega} = \frac{2\alpha_\mathrm{f}\gamma^2 k_\mathrm{u} K_\mathrm{u}^2 N_\mathrm{u}^2}{\left(1 + K_\mathrm{u}^2\right)^2},
$$

and the photon flux per unit solid angle and relative frequency band radiated by an electron current I,

$$
\boxed{\frac{\mathrm{d}^2 \dot{n}_I}{\mathrm{d}\Omega\,\mathrm{d}\omega/\omega} = \frac{2\alpha_\mathrm{f}\gamma^2 I K_\mathrm{u}^2 N_\mathrm{u}^2}{e\left(1 + K_\mathrm{u}^2\right)^2}.}
\tag{9.41}
$$

This quantity has a maximum for $K_\mathrm{u} = 1$, $K_\mathrm{uh}^* = 1/\sqrt{2}$ of

$$
\left(\frac{\mathrm{d}^2 \dot{n}_I}{\mathrm{d}\Omega\,\mathrm{d}\omega/\omega}\right)_\mathrm{max} = \frac{\alpha_\mathrm{f}\gamma^2 I N_\mathrm{u}^2}{2e}.
$$

As already discussed for the plane strong undulator, this optimization can be misleading since the emitted frequency decreases with increasing K_u.

9.5.6 The development with respect to K_uh^*

We develop now the angular power distribution with respect to K_uh^* and approximate the first three harmonics, each to the lowest power of this quantity but leaving the factor $1/(1 + K_\mathrm{u}^2)^2$ in front unchanged. We present the two contributions to the helicity as two terms inside the square brackets. Since the expressions are lengthy, the results are given in (C.3) in Appendix C.

This approximation of the angular power distribution is shown in Fig. 9.8 for the first three harmonics for the total radiation and the two modes of circular polarization. Figure 9.8 clearly shows that the higher harmonics have vanishing intensity on the axis but are distributed around it.

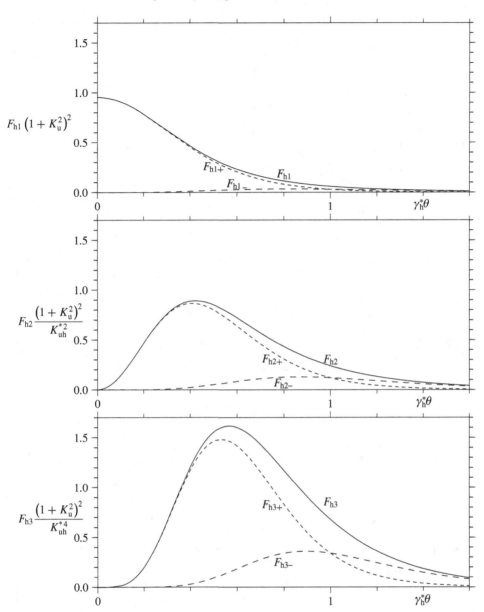

Fig. 9.8. Cuts through the power distributions of the first three harmonics, each to the lowest approximation in K_{uh}^*.

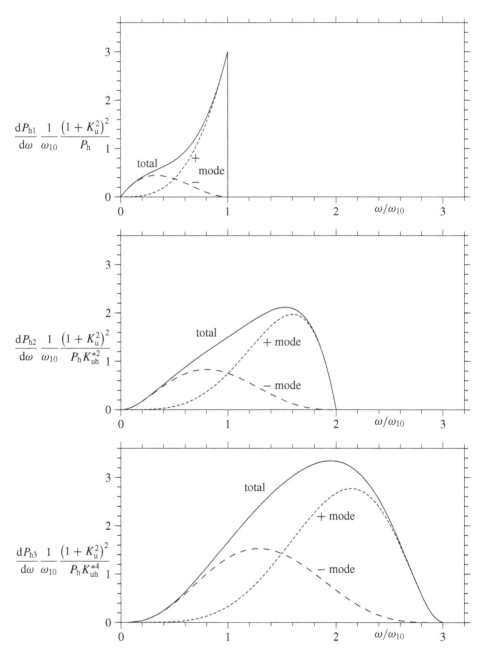

Fig. 9.9. Helical-undulator power spectra of the total radiation and the two modes of circular polarization for the first three harmonics, each approximated to the lowest order in K_{uh}^*.

Integrating the angular distribution (C.3) over ϕ and converting angles into frequencies with the relation (9.39) gives the approximation for the spectral power distribution (C.4) presented in Appendix C.

These approximated spectra are plotted in Fig. 9.9. The basic harmonic $m = 1$ has a sharp edge at the highest frequency, which is due to the fact that the power is large on the axis, $\theta = 0$. The other harmonics vanish on the axis, resulting in a power spectrum that goes to zero at the top frequency.

By integrating the above spectral power densities over frequency ω from 0 to $m\omega_{10}$ or the angular power density over solid angle, we find a development of the power contained in the first three harmonics for the total and the two components of the helicity presented in (C.5) in Appendix C.

We add up all three harmonics and obtain the emitted power, developed up to the fourth power of K_{uh}^*, for the two helicities and their sum:

$$P_+ = P_{\mathrm{h}} \frac{1}{\left(1 + K_{\mathrm{u}}^2\right)^2} \left(\frac{3}{4} + \frac{27}{20} K_{\mathrm{uh}}^{*2} + \frac{1079}{560} K_{\mathrm{uh}}^{*4} + \cdots \right)$$

$$P_- = P_{\mathrm{h}} \frac{1}{\left(1 + K_{\mathrm{u}}^2\right)^2} \left(\frac{1}{4} + \frac{13}{20} K_{\mathrm{uh}}^{*2} + \frac{601}{560} K_{\mathrm{uh}}^{*4} + \cdots \right)$$

$$P = P_{\mathrm{h}} \frac{1}{\left(1 + K_{\mathrm{u}}^2\right)^2} \left(1 + 2 K_{\mathrm{uh}}^{*2} + 3 K_{\mathrm{uh}}^{*4} + \cdots \right) = P_{\mathrm{h}} \left(1 + O(K_{\mathrm{uh}}^{*6}) \right).$$

The errors in the above developments are therefore of order $K_{\mathrm{uh}}^{*\,6}$ and higher.

10

Wiggler magnets

10.1 Introduction

A wiggler is a set of dipole magnets located in a straight section that has a different field strength B_w from that of the bending magnets of the ring. They are arranged and powered in such a way that the overall bending and displacement of the electron orbit vanishes, a condition that is also realized in undulators. For this reason undulators and wigglers are often referred to as *insertion devices* since they are located in a straight section and can be powered at different field levels without disturbing the orbit in the rest of the machine. However, they provide some focusing of the electron beam, which might have to be corrected.

We distinguish basically between two types of wigglers: *wavelength shifters*, having only one period, and *multipole wigglers* with many periods. The wavelength shifter has a short and strong dipole magnet in the center, which is used as the main source of radiation, with longer and weaker magnets on each side to make the overall bending vanish, as illustrated in Fig. 10.1. Varying the field strength changes the critical energy and the radiated power. Usually the central field is much larger than that of the lattice bending magnets and has the purpose of providing very-short-wavelength radiation. For this reason it is called a wavelength shifter. Apart from the different total power and critical frequency, the properties of the emitted radiation are the same as those of that emitted in the bending magnets of the ring.

Multipole wigglers have several periods and resemble strong undulators. They are optimized to provide a large flux with every pole serving as a source. Interference effects, which lead to some quasi-monochromatic peaks in the spectrum, are present but are not an important part of the optimization. Such wigglers are often used also to influence the properties of the beam rather than being sources of radiation. This will be discussed in Chapter 14.

10.2 The wavelength shifter

The wavelength shifter is a wiggler magnet consisting of a center dipole with a high field in between two weaker dipoles, Fig. 10.1. This magnet arrangement should be a true insertion device causing no overall deflection or displacement of the orbit in the rest of the machine.

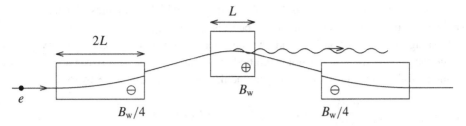

Fig. 10.1. A wavelength shifter consisting of one strong and two weak dipoles.

This can be achieved with

$$\int B_{\rm w}(s)\,{\rm d}s = 0 \quad \text{and} \quad B_{\rm w}(s) = B_{\rm w}(-s),$$

as in the case of undulators. This allows one to vary the field and change the spectrum for one experiment without disturbing others. Only the high-frequency radiation emitted in the center part is used. Owing to its much higher field and the geometrical separation, the radiation from other magnets gives a small contamination to the spectrum. The properties of the emitted radiation are therefore given by the expressions describing normal synchrotron radiation given in Chapter 5. Compared with the ring bending magnets, the high wiggler field $B_{\rm w}$ shifts the spectrum to higher frequencies characterized by the critical frequency $\omega_{\rm cw}$, which is usually higher than $\omega_{\rm c}$ of the lattice:

$$\omega_{\rm cw} = \frac{3c\gamma^3}{2\rho} = \frac{3ec^2 B_{\rm w}\gamma^2}{2m_0 c^2}, \qquad \frac{\omega_{\rm cw}}{\omega_{\rm c}} = \frac{B_{\rm w}}{B}.$$

10.3 The multipole wiggler

Multipole wigglers have several periods with the main purpose of increasing the intensity of the radiation and obtaining a high flux. The interference effect between radiation from different periods, which leads to quasi-monochromatic peaks in undulators, is of secondary importance for wigglers. Being an insertion device with no overall bending of the orbit, the field of a wiggler can be set to optimize the spectrum and the photon flux.

In many cases such wigglers can be treated, within some limitations, as strong undulators. For this the amplitude of the trajectory in the wiggler should not be too large such that the radiation emitted in the field of both polarities will reach the observer. A wiggler is often built up of single-dipole magnets instead of a field with a harmonic dependence in the longitudinal direction. This gives a slightly different particle trajectory and spectrum and is discussed in [50]. For this reason we refer here to the radiated energy $U_{\rm w}$ rather than the power. The number $N_{\rm u}$ of periods in such wigglers is often not very large. Some approximations used for undulators are based on $N_{\rm u} \gg 1$, and their validity for wigglers has to be checked and the spectral function $f_{\rm N}(\Delta\omega)$ adapted to a smaller number of periods if necessary. Finally, we assumed for the undulators that, at a given observation point and frequency, only one harmonic m contributes significantly to the field. For the high harmonics observed and the smaller number of periods this might not be satisfied in wigglers.

Keeping these restrictions in mind, we can calculate the angular spectral energy density of the radiation at the mth harmonic emitted in a plane wiggler with reasonable accuracy using the following expression for strong undulators given in Chapter 8 by (8.38),

$$\frac{d^2 P_m}{d\Omega\, d\omega} = P_u \gamma^{*2} [F_{m\sigma}(\theta, \phi) + F_{m\pi}(\theta, \phi)] f_N(\Delta\omega),$$

with the distribution functions (8.40),

$$F_{m\sigma}(\theta, \phi) = \frac{3m^2}{\pi \left(1 + K_u^2/2\right)^2 K_u^{*2}} \frac{(2\Sigma_{m1}\gamma^*\theta \cos\phi - \Sigma_{m2} K_u^*)^2}{(1 + \gamma^{*2}\theta^2)^3}$$

$$F_{m\pi}(\theta, \phi) = \frac{3m^2}{\pi \left(1 + K_u^2/2\right)^2 K_u^{*2}} \frac{(2\Sigma_{m1}\gamma^*\theta \sin\phi)^2}{(1 + \gamma^{*2}\theta^2)^3},$$

containing the sums Σ_{m1} and Σ_{m2} given by (8.30). The spectral function $f_N(\Delta\omega)$ given by (8.42) was derived for $N_u \gg 1$ but we can give a version that is more accurate for few periods:

$$f_N(\Delta\omega) = \frac{N_u}{\omega_1} \left(\frac{\sin\left(\frac{\Delta\omega}{\omega_1}\pi N_u\right)}{\frac{\Delta\omega}{\omega_1}\pi N_u} \right)^2 \left(1 + \frac{\Delta\omega/\omega_1}{2 + \Delta\omega/\omega_1}\right)^2 .$$

For the total energy U_w radiated by a particle traversing the undulator we had better use here the general expression

$$U_w = \frac{2 r_0 c^2 e^2 E_e^2}{3 (m_0 c^2)^3} \int B_w^2(s)\, ds$$

instead of one specialized to a harmonic trajectory.

11

Weak magnets – a generalized weak undulator

11.1 Properties of weak-magnet radiation

11.1.1 Introduction

We consider a magnet with a symmetry plane z, x in which a charged particle moves mainly in the z-direction. The geometry is the same as the one given in Fig. 6.1 but the undulator is replaced by a magnet with a more general field of the form $B_y(z)$. We make the same approximations as for the weak undulator and assume that we have an ultra-relativistic motion and that the magnetic field is sufficiently weak that the maximum angle of the particle trajectory with respect to some suitable axis is smaller than the natural opening angle of the emitted radiation:

$$\hat{x}' \approx \psi_0 \approx \hat{\beta}_x < 1/\gamma.$$

This justifies a paraxial approximation:

$$\frac{1}{\rho(z)} = \frac{eB_y(z)c}{m_0c^2\gamma} = \frac{d^2x/dz^2}{(1+(dx/dz)^2)^{3/2}} \approx \frac{1}{c^2}\frac{d^2x}{dt'^2} = \frac{\dot{\beta}_x}{c}.$$

Furthermore, we assume the radiation to be observed from a distance r_p much larger than the length L_u of the generalized undulator:

$$r_p \gg L_u.$$

Finally, this magnet is an insertion device and thus should not affect the orbit elsewhere in the ring, giving the condition for the trajectory $x(z)$ and its derivative $x'(z)$ at its entry and exit:

$$x(L_u/2) = x(-L_u/2) = 0, \qquad x'(L_u/2) = x'(-L_u/2) = 0.$$

This is fulfilled for a field with a vanishing integral and symmetry around the center $z = 0$:

$$\int B_z(z)\,dz = 0, \qquad B_z(z) = B_z(-z).$$

The second condition is sufficient but not necessary and it simplifies some other aspects, so we will use it in all examples.

The field of a weak magnet can be presented directly as $B_y(z)$ or in terms of spatial Fourier transform $\tilde{B}_y(k_g)$. They are related by

$$\tilde{B}_y(k_g) = \frac{1}{\sqrt{2\pi}} \int_{-\infty}^{\infty} B_y(z) e^{-ik_g z}\, dz$$

$$B_y(z) = \frac{1}{\sqrt{2\pi}} \int_{-\infty}^{\infty} \tilde{B}_y(k_g) e^{ik_g z}\, dk_g. \tag{11.1}$$

The second expression represents a decomposition of the field $B_y(z)$ into infinitely long weak undulators with wave numbers $k_g = 2\pi/\lambda_g$. We can calculate the radiation field due to each component using the undulator equation. Integrating all these contributions over k_g with the proper phase and weighted with the field strength $\tilde{B}_y(k_g)$ gives the total radiation field emitted in this weak magnet.

11.1.2 The trajectory

In the notation used here the magnet has spatial Fourier components with wave number k_g that replace k_u used for a harmonic undulator. Each component represents a very long undulator, which radiates at a given angle a very sharp line with frequency

$$\omega \approx \omega_1 = \frac{2ck_g\gamma^2}{1 + \gamma^2\theta^2}.$$

The spectrum of the final radiation field is made up of the contributions by each of these components.

We determine the particle trajectory using the same approximation as for weak undulators (7.15):

$$\mathbf{n} \approx \mathbf{n}_p \approx [\theta\cos\phi,\, \theta\sin\phi,\, 1 - \theta^2/2]$$

$$\boldsymbol{\beta} = [\beta_x,\, 0,\, \beta_z] \approx \beta[0,\, 0,\, 1]$$

$$\dot{\boldsymbol{\beta}} = [\dot{\beta},\, 0,\, 0] = -\frac{ec^2 B_y(z)}{m_0 c^2 \gamma}[1,\, 0,\, 0]$$

$$1 - \mathbf{n}\cdot\boldsymbol{\beta} = 1 - \beta\cos\theta = \frac{1 + \gamma^2\theta^2}{2\gamma^2}$$

$$[\mathbf{n} \times [(\mathbf{n} - \boldsymbol{\beta}) \times \dot{\boldsymbol{\beta}}]] = \frac{c^2 e B_y(z)}{2m_0 c^2 \gamma^3}[1 - \gamma^2\theta^2\cos(2\phi),\, -\gamma^2\theta^2\sin(2\phi)].$$

11.1.3 The radiation from weak magnets

We get the radiation field from the Liénard–Wiechert expression,

$$\mathbf{E}(t) = \frac{e}{4\pi\epsilon_0}\left\{\frac{[\mathbf{n} \times [(\mathbf{n} - \boldsymbol{\beta}) \times \dot{\boldsymbol{\beta}}]]}{cr(1 - \mathbf{n}\cdot\boldsymbol{\beta})^3}\right\}_{\text{ret}},$$

taking only the lowest order of the field strength and using again the classical electron radius $r_0 = e^2/(4\pi\epsilon_0 m_0 c^2)$:

$$\mathbf{E}_\perp(t) = \frac{4r_0 c\gamma^3}{r_p}\left\{B(z)\frac{[1-\gamma^2\theta^2\cos(2\phi),\ -\gamma^2\theta^2\sin(2\phi)]}{(1+\gamma^2\theta^2)^3}\right\}_{\text{ret}}.$$

To express this field in terms of the observation time t, or $t_p = t - r_p/c$, and frequency ω we use the relations

$$z = \beta ct' \approx ct' = \frac{2\gamma^2}{1+\gamma^2\theta^2}ct_p, \qquad k_g = \frac{1+\gamma^2\theta^2}{2c\gamma^2}\omega, \qquad k_g z = \omega t_p \qquad (11.2)$$

and obtain the radiation field emitted in this weak-field undulator in the time domain:

$$\mathbf{E}_\perp(t_p) = \frac{4r_0 c\gamma^3}{r_p}\frac{[1-\gamma^2\theta^2\cos(2\phi),\ -\gamma^2\theta^2\sin(2\phi)]}{(1+\gamma^2\theta^2)^3}B\left(\frac{2\gamma^2}{1+\gamma^2\theta^2}ct_p\right). \qquad (11.3)$$

We obtain this field in the frequency domain [60] by taking a Fourier transform:

$$\tilde{\mathbf{E}}_\perp(\omega) = \frac{4r_0 c\gamma^3}{r_p}\frac{[1-\gamma^2\theta^2\cos(2\phi),\ -\gamma^2\theta^2\sin(2\phi)]}{(1+\gamma^2\theta^2)^3}$$

$$\times \frac{1}{\sqrt{2\pi}}\int_{-\infty}^\infty B\left(\frac{2\gamma^2}{1+\gamma^2\theta^2}ct_p\right)e^{-i\omega t_p}\,dt_p.$$

With the relations (11.2) we can express the Fourier integral appearing above:

$$\frac{1}{\sqrt{2\pi}}\int_{-\infty}^\infty B\left(\frac{2\gamma^2}{1+\gamma^2\theta^2}ct_p\right)e^{-i\omega t_p}\,dt_p = \frac{1+\gamma^2\theta^2}{2c\gamma^2}\frac{1}{\sqrt{2\pi}}\int_{-\infty}^\infty B(z)e^{-ik_g z}\,dz$$

$$= \frac{1+\gamma^2\theta^2}{2c\gamma^2}\tilde{B}(k_g).$$

With (11.2) we express the wave number k_g in terms of the frequency ω and obtain the radiation field in the frequency domain directly from $\tilde{B}_y(k_g)$:

$$\tilde{\mathbf{E}}_\perp(\omega) = \frac{2r_0\gamma}{r_p}\frac{[1-\gamma^2\theta^2\cos(2\phi),\ -\gamma^2\theta^2\sin(2\phi)]}{(1+\gamma^2\theta^2)^2}\tilde{B}\left(\frac{1+\gamma^2\theta^2}{2c\gamma^2}\omega\right). \qquad (11.4)$$

However, in many cases it is easier to calculate first the radiation field $\tilde{\mathbf{E}}(t_p)$ and Fourier transform it afterwards. At a fixed angle θ, the time dependence of this field has the same form as the spatial dependence of the magnetic field and, as a consequence, the frequency-domain field has the same form as the spatial Fourier transform of the magnetic field with the scaling ratios

$$ct_p = \frac{1+\gamma^2\theta^2}{2\gamma^2}z, \qquad \omega = \frac{2\gamma^2}{1+\gamma^2\theta^2}ck_g.$$

The total energy radiated by a charge e in this weak magnet and its angular spectral distribution are given by the general expressions

$$U_0 = \frac{2r_0e^2c^2E_e^2}{3(m_0c^2)^3}\int_{-\infty}^{\infty}B^2(z)\,\mathrm{d}z, \qquad \frac{\mathrm{d}^2U}{\mathrm{d}\Omega\,\mathrm{d}\omega} = \frac{2r_p^2|\tilde{E}(\omega)|^2}{\mu_0c}. \qquad (11.5)$$

From the latter and (11.4) we obtain

$$\boxed{\begin{aligned}\frac{\mathrm{d}^2U}{\mathrm{d}\Omega\,\mathrm{d}\omega} &= \frac{4r_0ce^2E_e^2\gamma^2}{3(m_0c^2)^3}(F_{u\sigma}+F_{u\pi})\frac{1+\gamma^2\theta^2}{2\gamma^2}\left|\tilde{B}\left(\frac{1+\gamma^2\theta^2}{2c\gamma^2}\omega\right)\right|^2 \\[2mm] &= U_0\frac{F_{u\sigma}+F_{u\pi}}{c\int_{-\infty}^{\infty}B_y^2(z)\,\mathrm{d}z}(1+\gamma^2\theta^2)\left|\tilde{B}\left(\frac{1+\gamma^2\theta^2}{2c\gamma^2}\omega\right)\right|^2,\end{aligned}} \qquad (11.6)$$

where we use the particle energy $E_e = m_0c^2\gamma$ and the angular distribution functions (7.37) introduced before, $F_{u\sigma}$, $F_{u\pi}$, and $F_u = F_{u\sigma}+F_{u\pi}$ for the horizontal and vertical polarizations and for the total undulator radiation. The above angular spectral energy distribution contains positive frequencies only:

$$F_{u\sigma}(\theta,\phi) = \frac{3}{\pi}\frac{(1-\gamma^2\theta^2\cos(2\phi))^2}{(1+\gamma^2\theta^2)^5}, \qquad F_{u\pi}(\theta,\phi) = \frac{3}{\pi}\frac{\sin^2(2\phi)}{(1+\gamma^2\theta^2)^5}.$$

On the basis of Parseval's equation

$$\int_{-\infty}^{\infty}B^2(z)\,\mathrm{d}z = \int_{-\infty}^{\infty}|\tilde{B}|^2(k_g)\,\mathrm{d}k_g = 2\int_0^{\infty}|\tilde{B}|^2(k_g)\,\mathrm{d}k_g$$

we included the necessary factor of 2. The above expression gives the spectral angular energy distribution of the radiation emitted in a general weak magnet in which the angle of the particle trajectory nowhere exceeds the natural emission angle $1/\gamma$. Since the magnetic field $B_y(z)$ can have strong variations it is more suitable to give the energy rather than the power distribution of the radiation.

The angular energy density is obtained by integrating (11.6) over the frequency ω. With the relation (11.2) and Parseval's equation we obtain

$$\int_{-\infty}^{\infty}\frac{1+\gamma^2\theta^2}{2c\gamma^2}\left|\tilde{B}\left(\frac{1+\gamma^2\theta^2}{2c\gamma^2}\omega\right)\right|^2\mathrm{d}\omega = \int_{-\infty}^{\infty}|\tilde{B}(k_g)|^2\,\mathrm{d}k_g = \int_{-\infty}^{\infty}B^2(z)\,\mathrm{d}z$$

and

$$\frac{\mathrm{d}U}{\mathrm{d}\Omega} = U_0\gamma^2(F_{u\sigma}+F_{u\pi}).$$

This angular energy distribution of the total radiation is the same for all generalized plane weak undulators. However, the spectral distribution depends on the particular shape of the magnetic field. Furthermore, the angular distribution of the radiation selected at a certain frequency ω depends also on the shape of the field. This makes it possible to obtain a desired spectrum or a certain angular distribution at a given frequency by a suitable choice of the undulator field.

In the following we distinguish between weak magnets of relatively limited length, called short magnets, and generalized weak undulators in which the fundamental harmonic field is modulated by a slowly varying function. Both magnets should be weak such that the maximum angle ψ_0 of the particle trajectory is smaller than the typical opening angle of the emitted radiation $\psi_0 < 1/\gamma$.

11.2 Short magnets

11.2.1 Introduction

These magnets are short and do not have a quasi-periodic structure. Their use has been considered in order to obtain a large opening angle or a shorter wavelength [60] under some conditions. We assume that we have plane magnets with a (x, z)-symmetry plane where the magnetic field has only a y-component. The condition for vanishing overall deflection and displacement of the particle trajectory is satisfied with

$$\int_{-\infty}^{\infty} B_y(z)\,dz = 0, \qquad B_y(-z) = B_y(z),$$

as for undulators. The condition $\psi_0 \approx \hat{x}' < 1/\gamma$ limits the strength of the magnetic field for a given length of the magnet.

11.2.2 Qualitative properties of the short-magnet radiation

As in Chapter 6, we use qualitative arguments to estimate some properties of the radiation. We start with the emitted pulse length or frequency and consider a magnet of length L with an observer located about at the extension of the particle trajectory as shown in Fig. 11.1.

By repeating the calculations that we performed in the first chapter, we calculate the length of the radiation pulse received by the observer. The first detected photon was emitted at the entrance of the magnet and the last one originated at its exit. The full length of the pulse is given by the difference in time which the radiation and the electron take to traverse the magnet of length L. We neglect the increase in length of the trajectory due to the small deviation from a straight line of the electron trajectory in this weak magnet and obtain for the pulse length

$$\Delta t_{sm} = t_e - t_\gamma \approx \frac{L}{\beta c} - \frac{L}{c} \approx \frac{L}{\beta c}(1 - \beta) \approx \frac{L}{2c\gamma^2}$$

and for the typical wavelength and frequency

$$\lambda_{sm} \approx \frac{L}{2\gamma^2}, \qquad \omega_{sm} \approx \frac{2\pi}{\Delta t_{sm}} = \frac{4\pi c\gamma^2}{L}. \tag{11.7}$$

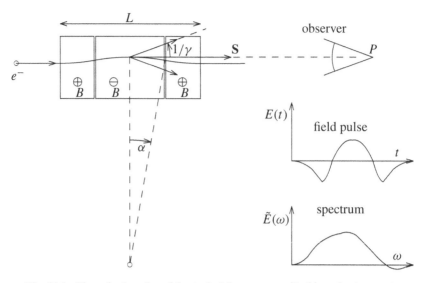

Fig. 11.1. The pulse length and the typical frequency emitted in a short magnet.

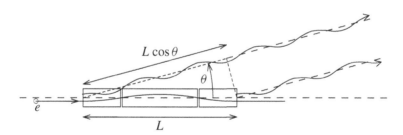

Fig. 11.2. The typical opening angle of short-magnet radiation at a frequency ω.

The result of this calculation is similar to that obtained for weak undulators with L representing the period length. However, in contrast to undulator radiation, the spectrum emitted in a short magnet is very broad and different frequencies can be received at the same angle.

Next we estimate the opening angle θ_{sm} of the radiation emitted at a given frequency ω. We consider a short magnet and ask at what angle the radiation contributions emitted at the entrance and exit of this magnet cancel out for a certain wavelength. From Fig. 11.2 we find the time difference between these two contributions to the wave emitted at an angle θ,

$$\Delta t = \frac{L}{\beta c} - \frac{L\cos\theta}{c} = \frac{1}{\beta c}(1 - \beta\cos\theta) \approx \frac{L}{c}\frac{1 + \gamma^2\theta^2}{2\gamma^2} = \left(n + \frac{1}{2}\right)\frac{\lambda}{c},$$

where we take for the order n the lowest value which gives a positive value for $\gamma^2\theta^2$. We choose a typical wavelength $\lambda = L/(2\gamma^2)$ and find with $n = 1$ an opening angle of

$$\theta_{\mathrm{sm}} \approx \frac{1}{\sqrt{2}\gamma}. \tag{11.8}$$

Since the spectrum is broad, we can choose a longer wavelength and obtain a larger opening angle.

Short magnets are sometimes used to obtain radiation of relatively high frequency having a large opening angle. In this case we might choose $\omega \ll \omega_{\mathrm{sm}}$ and get $\gamma^2\theta^2 \gg 1$. Using the corresponding approximation, we find

$$\theta \approx \frac{1}{\gamma}\sqrt{\frac{\omega_{\mathrm{sm}}}{2\omega}} \approx \sqrt{\frac{2\lambda}{L}}, \tag{11.9}$$

where we used the expression (11.7) for the typical frequency ω_{sm} of short-magnet radiation.

This opening angle is an appropriate description of the distribution of short-magnet radiation and is valid for the horizontal plane of deflection as well as for the perpendicular direction. Its dependence on the length of the magnet is given by the same physical principles as those relevant for a long array of antennas that can confine emitted radio waves to a small opening angle.

11.3 The modulated undulator radiation

11.3.1 Introduction

We consider a weak magnetic field with a basic period of length λ_u and a slow superimposed modulation [61, 62]. Assuming that we have a symmetry condition $B_y(-z) = B_y(z)$, the field is of the form

$$B_y(z) = B_0\, f_{\mathrm{m}}(z) g(z) = B_0\, f_{\mathrm{m}}(z)\cos(k_u z),$$

where the modulation function $f_{\mathrm{m}}(z)$ is assumed to change little over a period length λ_u.

We calculate the Fourier transform of the periodic part, assuming first that we have a finite length $L_u = N_u\lambda_u$:

$$\tilde{g}(k_g) = \frac{1}{\sqrt{2\pi}}\int_{-L_u/2}^{L_u/a}\cos(k_u z)\cos(k_g z)\,dz$$

$$= \frac{1}{\sqrt{2\pi}}\left(\frac{\sin((k_g - k_u)\pi N_u/k_u)}{k_g - k_u} + \frac{\sin((k_g + k_u)\pi N_u/k_u)}{k_g + k_u}\right).$$

For the interesting case of $k_g \approx k_u$ and a large number of periods $N_u \gg 1$, we can neglect the second term and use the form

$$\tilde{g}(k_g) \approx \frac{1}{\sqrt{2\pi}}\frac{\pi N_u}{k_u}\frac{\sin((k_g - k_u)\pi N_u/k_u)}{(k_g - k_u)\pi N_u/k_u} \quad\text{with}\quad \int_{-\infty}^{\infty}\tilde{g}(k_g)\,dk_g = \sqrt{\frac{\pi}{2}}.$$

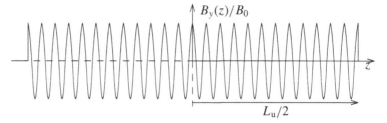

Fig. 11.3. An undulator field $B_y(z)$ with sharp termination at finite length $\pm L_u/2$.

For $N_u \gg 1$ the above expression is very large around $k_g \approx k_u$ but quite small elsewhere. We can approximate the Fourier transform of $g(t)$ by a Dirac δ-function:

$$\tilde{g}(k_g) = \frac{\pi}{\sqrt{2\pi}}\delta(k_g - k_u).$$

To calculate the Fourier transform $\tilde{B}(k_g)$ of the field we use the convolution theorem,

$$\tilde{B}_y(k_g) = \frac{B_0}{\sqrt{2\pi}}\int_{-\infty}^{\infty} f_m(z)g(k_g)e^{-ik_g z}\,dz = \frac{B_0}{\sqrt{2\pi}}\int_{-\infty}^{\infty} \tilde{f}_m(k_d)\tilde{g}(k_g - k_d)\,dk_d,$$

giving

$$\tilde{B}_y(k_g) = \frac{B_0}{2}\int_{-\infty}^{\infty} \tilde{f}_m(k_d)\delta(k_g - k_u - k_d)\,dk_d = \frac{B_0}{2}\tilde{f}_m(k_g - k_u). \tag{11.10}$$

This expression gives directly the Fourier transform of the modulated undulator field. However, this method does not present a significant advantage over calculating the Fourier transformation of the product between the two functions in the standard way.

11.3.2 The undulator of finite length

We have already treated an undulator of finite length. We do it again as an example of the weak-magnet formalism. We take a harmonic undulator with a large number of periods, $N_u \gg 1$, of length λ_u, and total length $L_u = N_u\lambda_u$. The magnet field, shown in Fig. 11.3, has the form

$$B_y(z) = B_0\cos(k_u z) \text{ for } -N_u\lambda_u \le z \le N_u\lambda_u; \qquad \text{otherwise } B_y(z) = 0.$$

The modulating function $f_m(z)$ and its Fourier transform are

$$f_m(z) = \begin{cases} 1 & \text{for } |z| < L_u/2 \\ 0 & \text{for } |z| > L_u/2 \end{cases}$$

$$\tilde{f}(k_g) = \frac{2}{\sqrt{2\pi}}\frac{\pi N_u}{k_u}\frac{\sin((k_g - k_u)\pi N_u/k_u)}{(k_g - k_u)\pi N_u/k_u},$$

where we used $L_u = N_u \lambda_u = 2\pi N_u / k_u$. With (11.10) this gives for the Fourier-transformed field

$$\tilde{B}_y(k_g) = \frac{1}{\sqrt{2\pi}} \int_{-\infty}^{\infty} B_y(z) e^{-ik_g z} \, dz \approx \frac{B_0}{\sqrt{2\pi}} \frac{N_u \pi}{k_u} \frac{\sin((k_g - k_u)N_u \pi / k_u)}{(k_g - k_u)N_u \pi / k_u}.$$

We calculate the total and maximum deflection angles of an ultra-relativistic charge e of energy E_e traversing the undulator and the integral over the square of the magnetic field:

$$\psi_0 = \frac{ec}{E_e} \int_0^{\lambda_u/4} B(z) \, dz = \frac{e B_0}{m_0 c \gamma k_u} = \frac{K_u}{\gamma} < \frac{1}{\gamma} \ll 1$$

$$\phi_0 = \frac{ec}{E_e} \int_{-\infty}^{\infty} B_y(z) \, dz = 0$$

$$\int_{-L_u/2}^{L_u/2} B_y^2(z) \, dz = B_0^2 \frac{L_u}{2}.$$

We use the relations (11.2) and

$$k_u = \omega_1 \frac{1 + \gamma^2 \theta^2}{2c\gamma^2}, \qquad \Delta\omega = \omega - \omega_1, \qquad K_u = \frac{e B_0}{m_0 c k_u} < 1.$$

We obtain the radiation field from (11.3) in time for $|\omega_1 t_p| \leq \pi N_u$ and from (11.4) in the frequency domain,

$$\mathbf{E}_\perp(t_p) = \frac{4 r_0 c \gamma^3 B_0}{r_p} \frac{[1 - \gamma^2 \theta^2 \cos(2\phi), -\gamma^2 \theta^2 \sin(2\phi)]}{(1 + \gamma^2 \theta^2)^3} \cos(\omega_1 t_p)$$

$$\tilde{\mathbf{E}}_\perp(\omega) = \frac{4 r_0 c \gamma^3 B_0}{\sqrt{2\pi} r_p} \frac{[1 - \gamma^2 \theta^2 \cos(2\phi), -\gamma^2 \theta^2 \sin(2\phi)]}{(1 + \gamma^2 \theta^2)^3} \frac{\pi N_u}{\omega_1} \frac{\sin(\Delta\omega \pi N_u / \omega_1)}{\Delta\omega \pi N_u / \omega_1},$$

which agrees with the expression (7.30) obtained before directly.

The total radiated energy is (11.5)

$$U_0 = \frac{2 r_0 e^2 c^2 E_e^2}{3(m_0 c^2)^3} \int_{-\infty}^{\infty} B^2(z) \, dz = \frac{r_0 e^2 c^2 E_e^2 B_0^2 L_u}{3(m_0 c^2)^3} = \frac{2\pi r_0 m_0 c^2 \gamma^2 k_u K_u^2 N_u}{3}$$

and its angular spectral distribution is

$$\frac{d^2 U}{d\Omega \, d\omega} = U_0 \gamma^2 [F_{u\sigma}(\theta, \phi) + F_{u\pi}(\theta, \phi)] f_N(\Delta\omega) \tag{11.11}$$

with $f_N(\Delta\omega)$ being the normalized spectrum function (7.38):

$$\int \frac{N_u}{\omega_1} \left(\frac{\sin(\pi N_u \Delta\omega / \omega_1)}{\pi N_u \Delta\omega / \omega_1} \right)^2 d\Delta\omega = 1.$$

The angular spectral energy density of the radiation at the fundamental frequency is obtained by setting $\omega = \omega_{10}$ in the general expression (11.11) and expressing the relative frequency deviation by

$$\frac{\Delta\omega}{\omega_1} = \frac{\omega_{10} - \omega_1}{\omega_1} = \gamma^2\theta^2.$$

For a large number of periods $N_u \gg 1$ the square of the normalized opening angle of the filtered radiation is small, $\gamma^2\theta^2 \ll 1$, and the angular distribution can be approximated as $F_u \approx F_{u\sigma} \approx 3/\pi$. We obtain for the angular spectral energy distribution at the fundamental frequency

$$\frac{d^2 U}{d\Omega\,d\omega}(\omega_{10}) = U_0\frac{3\gamma^2}{\pi}\frac{N_u}{\omega_{10}}\left(\frac{\sin(\pi N_u\gamma^2\theta^2)}{\pi N_u\gamma^2\theta^2}\right)^2. \qquad (11.12)$$

This distribution was discussed earlier in (7.41) and is plotted here on the left-hand side of Fig. 11.4 against $\sqrt{\pi N_u}\gamma\theta$. Integration over the solid angle gives the spectral energy density at ω_{10},

$$\frac{dU}{d\omega}(\omega_{10}) = U_0\frac{3}{2\omega_{10}},$$

which agrees with the earlier results.

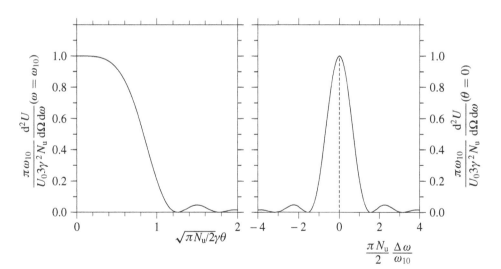

Fig. 11.4. The normalized angular spectral energy density of the radiation from a weak undulator of finite length having N_u periods: left; at the fundamental frequency ω_{10} as a function of angle θ, and right; on the axis, $\theta = 0$, as a function of the frequency deviation $\Delta\omega = \omega - \omega_{10}$.

The variance of the angular distribution at ω_{10},

$$\langle\theta^2\rangle = \frac{\displaystyle\int_0^\infty \theta^2 \frac{\mathrm{d}^2 U}{\mathrm{d}\Omega\,\mathrm{d}\omega}\,\theta\,\mathrm{d}\theta\,\mathrm{d}\phi}{\dfrac{\mathrm{d}U}{\mathrm{d}\omega}} \to \infty,$$

diverges, as we have discussed already in Chapter 7. This is caused by the unrealistic sharp termination of the undulator field at $\pm L_u/2$. This leads to an excessive amount of radiation at high frequencies, which in turn gives a wide angular distribution after filtering at ω_{10}. A realistic undulator has a smooth field termination at the end and does not produce such a high-frequency tail. We have fitted the distribution (11.12) with the function (7.43) having at the center the same value and the same first non-vanishing derivative in order to obtain an approximate value (7.44) for θ_{RMS}.

The spectral energy density on the axis is obtained by setting $\theta = 0$, giving $F_u = F_{u\sigma} = 3/\pi$ and $\omega_1 = \omega_{10}$ in (11.11),

$$\frac{\mathrm{d}^2 U}{\mathrm{d}\Omega\,\mathrm{d}\omega}(\omega_{10}) = U_0 \frac{3\gamma^2}{\pi}\frac{N_u}{\omega_{10}}\left(\frac{\sin(\pi N_u/\Delta\omega/\omega_{10})}{\pi N_u \Delta\omega/\omega_{10}}\right)^2$$

with $\Delta\omega = \omega - \omega_{10}$. This distribution is shown on the right-hand side of Fig. 11.4.

11.3.3 The undulator radiation with amplitude modulation

We consider an undulator field, shown in Fig. 11.5, of the form

$$B_y(z) = B_0\cos(k_0 z)\,(1 + a\cos(k_m z)) \text{ for } |z| \le L_u/2.$$

This represents a harmonic field with the fundamental wave number $k_0 = 2\pi/\lambda_0$, which is amplitude modulated by a relative magnitude a and wave number $k_m = 2\pi/\lambda_m$. We assume that this undulator of length L_u contains a large integer number of both wavelengths:

$$\frac{L_u}{\lambda_0} = N_0 \gg 1, \qquad \frac{L_u}{\lambda_m} = N_m \gg 1, \qquad N_0 - N_m \gg 1.$$

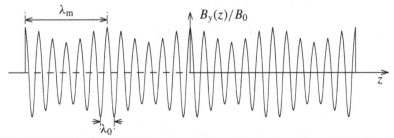

Fig. 11.5. The field of a finite-length undulator with a harmonic field modulation.

The field can be expressed as a function of z of the form

$$
\begin{aligned}
B_y(z) &= B_0 \left(\cos(k_0 z) + \frac{a}{2} (\cos((k_0 - k_m)z) + \cos((k_0 + k_m)z)) \right) \\
&= B_0 \left(\cos(k_0 z) + \frac{a}{2} (\cos(k_- z) + \cos(k_+ z)) \right) \qquad \text{for } |z| \le L_u/2
\end{aligned}
$$

with

$$
k_- = k_0 - k_m, \qquad k_+ = k_0 + k_m, \qquad N_- = N_0 - N_m, \qquad N_+ = N_0 + N_m.
$$

The Fourier-transformed magnetic field as a function of k_u is

$$
\begin{aligned}
\tilde{B}_y(k_u) \approx B_0 \sqrt{\frac{\pi}{2}} \Bigg(&\frac{N_0}{k_0} \frac{\sin((k_u - k_0)\pi N_0/k_0)}{(k_u - k_0)\pi N_0/k_0} \\
&+ \frac{a}{2} \left(\frac{N_-}{k_-} \frac{\sin((k_u - k_-)\pi N_-/k_-)}{(k_u - k_-)\pi N_-/k_-} + \frac{N_+}{k_+} \frac{\sin((k_u - k_+)\pi N_+/k_+)}{(k_u - k_+)\pi N_+/k_+} \right) \Bigg).
\end{aligned}
$$

It consists of a carrier with wave number around k_0 and two side bands spaced by $\pm k_m$ around it, all having a certain width.

Using the relations

$$
z = \frac{2\gamma^2}{1 + \gamma^2\theta^2} ct_p \le \frac{L_u}{2}, \qquad ck_0 = \frac{2\gamma^2}{1 + \gamma^2\theta^2} \omega_1, \qquad ck_m = \frac{2\gamma^2}{1 + \gamma^2\theta^2} \omega_m
$$

and $\omega^- = \omega_1 - \omega_m$, $\omega^+ = \omega_1 + \omega_m$, we obtain the radiation field from (11.3) in the time domain,

$$
\mathbf{E}_\perp(t_p) = \frac{4r_0 c\gamma^3 B_0}{r_p} \frac{[1 - \gamma^2\theta^2 \cos(2\phi), -\gamma^2\theta^2 \sin(2\phi)]}{(1 + \gamma^2\theta^2)^3} \cos(\omega_1 t_p)(1 + a \cos(\omega_{m1} t_p)),
$$

and from (11.4) in the frequency domain,

$$
\begin{aligned}
\tilde{\mathbf{E}}_\perp(\omega) = \frac{4\pi r_0 c\gamma^3 B_0}{\sqrt{2\pi} r_p} &\frac{[1 - \gamma^2\theta^2 \cos(2\phi), -\gamma^2\theta^2 \sin(2\phi)]}{(1 + \gamma^2\theta^2)^3} \\
&\times \Bigg(\frac{N_0}{\omega_1} \frac{\sin\left(\dfrac{\omega - \omega_1}{\omega_1} \pi N_0\right)}{\dfrac{\omega - \omega_1}{\omega_1} \pi N_0} + \frac{a}{2} \Bigg(\frac{N_-}{\omega_-} \frac{\sin\left(\dfrac{\omega - \omega_-}{\omega_-} \pi N_-\right)}{\dfrac{\omega - \omega_-}{\omega_-} \pi N_-} \\
&+ \frac{N_+}{\omega_+} \frac{\sin\left(\dfrac{\omega - \omega_+}{\omega_+} \pi N_+\right)}{\dfrac{\omega - \omega_+}{\omega_+} \pi N_+} \Bigg) \Bigg).
\end{aligned}
$$

From (11.5) we obtain the total radiated energy,

$$
U_0 = \frac{2r_0 e^2 c^2 E_e^2}{3(m_0 c^2)^3} \int_{-\infty}^{\infty} B^2(z) \, dz = \frac{r_0 e^2 c^2 E_e^2 B_0^2 L_u}{3(m_0 c^2)^3} \left(1 + \frac{a^2}{2} \right),
$$

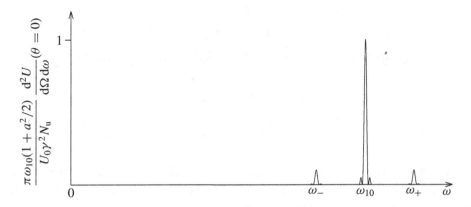

Fig. 11.6. The spectral angular energy distribution on the axis, $\theta = 0$, of the radiation from an undulator with $N_0 = 96$ periods having a harmonic modulation of index $a = 0.2$ and $N_m = 16$ periods.

and its angular spectral distribution,

$$
\frac{\mathrm{d}^2 U}{\mathrm{d}\Omega\,\mathrm{d}\omega} = U_0 \gamma^2 (F_{u\sigma} + F_{u\pi}) \frac{1}{1 + a^2/2}
$$

$$
\times \left(\frac{N_0}{\omega_1} \left(\frac{\sin\left(\frac{\omega - \omega_1}{\omega_1} \pi N_0\right)}{\frac{\omega - \omega_1}{\omega_1} \pi N_0} \right)^2 + \frac{a^2}{4} \left(\frac{N_-}{\omega_-} \left(\frac{\sin\left(\frac{\omega - \omega_-}{\omega_-} \pi N_-\right)}{\frac{\omega - \omega_-}{\omega_-} \pi N_-} \right)^2 \right. \right.
$$

$$
\left. \left. + \frac{N_+}{\omega_+} \left(\frac{\sin\left(\frac{\omega - \omega_+}{\omega_+} \pi N_+\right)}{\frac{\omega - \omega_+}{\omega_+} \pi N_+} \right)^2 \right) \right).
$$

The spectrum of the amplitude-modulated undulator has a carrier at ω_1 and two side bands as illustrated in Fig. 11.6.

11.3.4 The undulator radiation with Lorentzian modulation

As another example of interest we consider a weak harmonic undulator with period length λ_u having a slow modulation with a Lorentzian function [63], which is shown in Fig. 11.7,

$$
B(z) = B_0 \frac{\cos(k_u z)}{1 + (z/z_0)^2}.
$$

The number N_u of periods within the characteristic length $2z_0$ should be large. We obtain from (11.10)

$$
\tilde{B}(k_g) = \sqrt{\frac{\pi}{2}} \frac{B_0 z_0}{2} e^{-|k_g - k_u|\pi N_u/k_u}. \tag{11.13}
$$

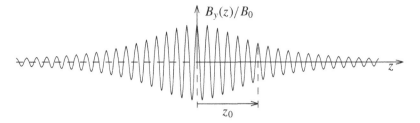

Fig. 11.7. An undulator field $B_y(z)$ having a Lorentzian modulation.

We use the undulator parameter $K_0 = eB_0/(m_0 ck_u) < 1$ in the center, assume that we have a weak undulator, and obtain the maximum and total deflecting angles ψ_0 and ϕ_0, and the integral over the square of the magnetic field:

$$\psi_0 = \frac{ec}{E_e} \int_0^{\lambda_u/4} B_y(z)\,dz \approx \frac{eB_0}{m_0 c\gamma k_u} \approx \frac{K_u}{\gamma} < \frac{1}{\gamma}$$

$$\phi_0 = \frac{ec}{E_e} \int_{-\infty}^{\infty} B_y(z)\,dz = \frac{\pi^2 K_0 N_u}{\gamma} e^{-\pi N_u}$$

$$\int_{-\infty}^{\infty} B_y^2(z)\,dz = \frac{\pi B_0^2 z_0}{4}(1 + (1 + 2k_u z_0)e^{-2k_u z_0}) \approx \frac{\pi B_0^2 z_0}{4}.$$

We use the relations (11.2) and

$$ck_u = \frac{1 + \gamma^2\theta^2}{2\gamma^2}\omega_1, \qquad z_0 = \frac{2\gamma^2}{1 + \gamma^2\theta^2} ct_0$$

to obtain, with (11.13), the radiation field, from (11.3) in time, and from (11.4) in the frequency domain:

$$\mathbf{E}_\perp(t_p) = \frac{e\gamma^3 k_u K_0}{\pi\epsilon_0 r_p} \frac{[1 - \gamma^2\theta^2 \cos(2\phi),\ -\gamma^2\theta^2 \sin(2\phi)]}{(1 + \gamma^2\theta^2)^3} \frac{\cos(\omega_1 t_p)}{1 + t_p^2/t_0^2}$$

$$\tilde{\mathbf{E}}_\perp(\omega) = \frac{e\gamma^3 k_u K_0}{\pi\epsilon_0 r_p} \frac{[1 - \gamma^2\theta^2 \cos(2\phi),\ -\gamma^2\theta^2 \sin(2\phi)]}{(1 + \gamma^2\theta^2)^3} \sqrt{\frac{\pi}{2}} \frac{\pi N_u}{2\omega_1} \exp\left(-\left|\frac{\omega - \omega_1}{\omega_1}\right| \pi N_u\right).$$

The total energy radiated by an electron of energy $E_e = m_0 c^2 \gamma$ traversing this undulator is, according to (11.5),

$$U_0 = \frac{2r_0 e^2 c^2 E_e^2 B_0^2 z_0}{3(m_0 c^2)^3} \frac{\pi}{4} = \frac{\pi^2 r_0 m_0 c^2 \gamma^2 k_u K_0^2 N_u}{6},$$

and its angular spectral energy distribution is, according to (11.5),

$$\frac{d^2 U}{d\Omega\,d\omega} = U_0 \gamma^2 (F_{u\sigma} + F_{u\pi}) \frac{\pi N_u}{\omega_1} e^{-2\pi N_u |\omega - \omega_1|/\omega_1}. \qquad (11.14)$$

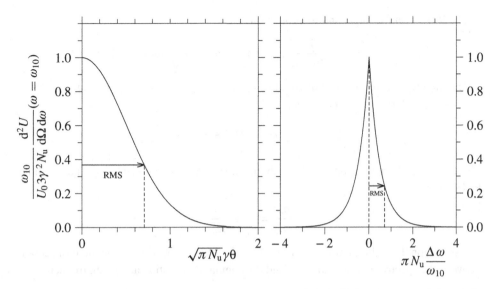

Fig. 11.8. The normalized angular spectral energy density of the radiation from a weak harmonic undulator with a Lorentzian modulation: left; at the fundamental frequency ω_{10} as a function of the angle θ; and right; on the axis, $\theta = 0$, as a function of the frequency deviation $\Delta\omega = \omega - \omega_{10}$.

Integrating over frequency gives the normalization

$$\frac{\pi N_u}{\omega_1} \int e^{-2\pi N_u |\omega-\omega_1|/\omega_1} \, d\omega = \frac{\pi N_u}{\omega_1} \left(\int_0^{\omega_1} e^{2\pi N_u(\omega-\omega_1)/\omega_1} \, d\omega + \int_{\omega_1}^{\infty} e^{-2\pi N_u(\omega-\omega_1)/\omega_1} \, d\omega \right) = 1.$$

At a fixed angle θ the spectrum decays exponentially above and below the proper frequency ω_1 with a full width at $1/e$ of the maximum height of $\delta\omega/\omega_1 = 1/(\pi N_u)$.

We filter the radiation at the fundamental frequency by setting $\omega = \omega_{10}$ in (11.14) and approximate for $N_0 \gg 1$, resulting in $\gamma^2\theta^2 \ll 1$. We are left with the σ-mode only, which has the field and energy distributions

$$\tilde{E}_x(\omega_{10}) = \frac{\sqrt{2\pi}\, e\gamma^3 k_u K_0 N_u}{4\epsilon_0 r_p \omega_{10}} e^{-\pi N_u \gamma^2\theta^2}$$

$$\frac{d^2U}{d\Omega\, d\omega}(\omega_{10}) = U_0 \frac{3\gamma^2 N_u}{\omega_{10}} e^{-2\pi N_u \gamma^2\theta^2}.$$

(11.15)

The radiation emitted by an undulator with a Lorentzian modulation selected at the fundamental frequency has a Gaussian angular distribution, as shown on the left-hand side of Fig. 11.8. Integrating over the solid angle gives the spectral energy density at the fundamental frequency

$$\frac{dU}{d\omega}(\omega_{10}) = U_0 \frac{3}{2\omega_{10}}.$$

(11.16)

The RMS opening angle and its projections are

$$\sigma_\theta = \frac{1}{\sqrt{2\pi N_0}\gamma}, \qquad \sigma_{x'} = \sigma_{y'} = \frac{1}{2\sqrt{\pi N_0}\gamma}.$$

The angular spectral energy distribution on the axis is obtained by setting $\theta = 0$ and $\omega_1 = \omega_{10}$ in (11.14), giving $\omega_1 = \omega_{10}$:

$$\frac{\mathrm{d}^2 U}{\mathrm{d}\Omega\,\mathrm{d}\omega}(\omega_{10}) = U_0 \frac{3\gamma^2 N_\mathrm{u}}{\omega_{10}} \mathrm{e}^{-2\pi N_\mathrm{u}|\omega-\omega_{10}|/\omega_{10}}.$$

This spectrum decays exponentially on both sides of the fundamental frequency with an RMS width of

$$\frac{\sigma_\omega}{\omega_{10}} = \frac{1}{\sqrt{2\pi N_0}}.$$

It is shown on the right-hand side of Fig. 11.8. This example illustrates the interconnection between the spectrum of the radiation and the angular distribution at a given frequency.

11.4 The Compton back scattering and quantum correction

In some experiments a laser beam collides head-on with electrons circulating in a storage ring, Fig. 11.9. The laser is an electromagnetic wave having magnetic and electric fields perpendicular to each other and to the direction of propagation. To make later the connection to undulators, we call the laser wavelength λ_ℓ, its wave number $k_\ell = 2\pi/\lambda_\ell$, and its frequency $\omega_\ell = k_\ell c$. For the electron the field of this wave resembles that of an undulator, with two differences: the laser beam has an E-Field and a B-field both deflecting the electron, and the laser beam moves against electrons, which makes them oscillate at a frequency Ω_u that is about twice the frequency of the light. While the electron advances $\lambda_\ell/2$ the light wave moves by the same amount against the electron and the two have the same relative phase as at the beginning:

$$\Omega_\mathrm{u} = (1 + \beta)k_\ell c \approx 2k_\ell c.$$

The emitted radiation therefore has twice the frequency compared with that for a static undulator of the same period length $\lambda_\mathrm{u} = \lambda_\ell$.

Since the electromagnetic field of the laser represents a deflecting force for the electron, very similar to the static magnetic field of an undulator, we could calculate the resulting radiation the same way in both cases. However, we will consider the laser field as a stream of photons of energy $\hbar\omega_\ell = \hbar c k_\ell$ and investigate the kinematics of the Compton scattering between photons and electrons.

We assume that we have an ultra-relativistic electron beam with $\gamma \gg 1$ moving in the z-direction and colliding head-on with the laser wave moving in the $-z$-direction, Fig. 11.9.

The electron has energy and momentum

$$E_\mathrm{e} = m_0 c^2 \gamma, \qquad \mathbf{p} = p_0[0,\ 0,\ 1]$$

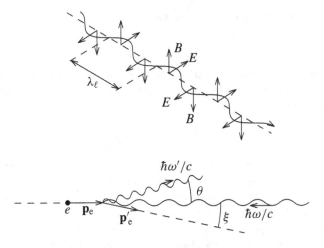

Fig. 11.9. Compton scattering between a laser and an electron beam.

with $p_0 = m_0 c \beta \gamma \approx m_0 c \gamma$. For the photon these parameters are

$$E_\ell = \hbar \omega_\ell, \qquad \mathbf{p}_\ell = \frac{\hbar \omega_\ell}{c}[0, 0, -1].$$

We calculate the radiation as Compton scattering under conservation of energy and momentum with γ, $\gamma' \gg 1$,

$$m_0 c^2 \gamma + \hbar \omega_\ell = m_0 c^2 \gamma' + \hbar \omega_1$$

$$m_0 c \beta \gamma - \frac{\hbar \omega_\ell}{c} = m_0 c \beta' \gamma' \cos \xi + \frac{\hbar \omega_1}{c} \cos \theta$$

$$0 = -m_0 c \beta' \gamma' \sin \xi + \frac{\hbar \omega_1}{c} \sin \theta,$$

where ω_1, γ', and β' are the values after scattering. For given initial parameters ω_ℓ and γ we can select a scattering angle θ of the photon and calculate its frequency ω_1 as well as the angle ξ and Lorentz factor γ' of the electron after scattering. We find for the frequency

$$\omega_1 = \omega_\ell \frac{4\gamma^2}{1 + \gamma^2 \theta^2 + 4 \hbar \omega_\ell \gamma / (m_0 c^2)}, \qquad \omega_{10} = \omega_\ell \frac{4\gamma^2}{1 + 4 \hbar \omega_\ell \gamma / (m_0 c^2)} \qquad (11.17)$$

with ω_{10} being the highest frequency which is observed in the forward direction $\theta = 0$. This equation has a similar form to that giving the frequency of the radiation emitted in an undulator of wave number and wavelength given by $k_u = 2\pi / \lambda_u$:

$$\omega_1 = \frac{2\pi c}{\lambda_u} \frac{2\gamma^2}{1 + K_u^2 / 2 + \gamma^2 \theta^2}. \qquad (11.18)$$

On comparing the two we find three important differences.

- The emitted frequency for Compton scattering is a factor of two higher than that for undulator radiation. This is expected since the photon and ultra-relativistic electron beam move against each other as discussed before.

- Compton scattering has an extra term in the denominator,

$$\frac{4\hbar\omega_\ell}{m_0c^2} = \frac{4\hbar\omega_\ell\gamma^2}{m_0c^2\gamma} = \frac{\hbar\omega_{10}}{m_0c^2\gamma}, \tag{11.19}$$

which gives the ratio between the energy of the photon emitted in the forward direction and the energy of the electron. We express (11.17) for emission in the forward direction, multiply it by $\hbar$ and arrange it differently,

$$\hbar\omega_{10} = m_0c^2\gamma\frac{4\gamma^2\hbar\omega_\ell}{m_0c^2\gamma + 4\gamma^2\hbar\omega_\ell},$$

which now gives the photon energy $\hbar\omega$ emitted in the forward direction as a function of the laser photon energy $\hbar\omega_\ell$. As long as $4\gamma^2\hbar\omega_\ell \ll m_0c^2\gamma$ the energy of the emitted photons increases about linearly with that of the laser photons. However, as the latter becomes larger this increase becomes slower and approaches in the extreme case $\hbar\omega_{10} \to m_0c^2\gamma$. This is of course expected since the electron is not able to emit a photon of higher energy than it has itself.

However, in the expression (11.18) for undulators there is no such limitation and, for an extremely small period length λ_u, the radiated photon could have a larger energy than that of the electron. Quantum effects were neglected in deriving the undulator-radiation spectrum. The relation $E_\gamma = \hbar\omega$ was applied only at the end in order to convert the frequency spectrum into a photon distribution. In this process the recoil momentum acting on the electron was ignored, but it is included in Compton scattering. The extra term (11.19) is often called the quantum correction. Such a modification should be included for undulator and synchrotron radiation. A proper quantum-mechanical treatment is not covered here but given in many publications [36].

In a qualitative approach to understanding undulator radiation we went into a system that moves with the drift velocity of the electron as illustrated in Figs. 6.5 and 6.6. In this frame the undulator represents a periodic field, moving against the electron with nearly the speed of light, which greatly resembles the situation in Compton scattering. However, there is a fundamental difference between the two. For the moving undulator there exists a frame, the laboratory frame, in which it becomes a static magnetic field. Such a frame does not exist for the laser beam used in Compton scattering.

- The undulator-radiation frequency contains a term $K_u^2/2$ in the denominator that corrects for strong-field effects. This term is missing from the Compton effect since it is usually negligible. Laser back scattering is done with visible radiation, which has a small wavelength λ_ℓ. Unless the field is extremely large, the corresponding undulator parameter is very small. Since we obtain deflection from the magnetic and electric fields, we have to include a factor of 2 in the deflecting force but a shorter period seen by the electron:

$$K_\ell = 2\frac{e\hat{B}}{m_0ck_\ell} = \frac{e\hat{E}}{m_0c^2k_\ell} = 2\frac{e\hat{B}\lambda_\ell}{2\pi m_0c}.$$

For green light with a wavelength $\lambda_\ell = 500$ nm we need a field of $\hat{B} = 21.4$ T or $\hat{E} = 6.4 \times 10^{12}$ V/m^{-1} in order to reach $K_\ell = 1$ and observe some of the strong-undulator effects. This field is very large but not impossible to reach in the focus of a strong laser beam. This experiment has been carried out and the higher harmonics were observed in Compton scattering [64].

Part IV
Applications

12

Optics of SR – imaging

12.1 Imaging with SR – a qualitative treatment

12.1.1 The limitation on resolution caused by diffraction and the depth-of-field effect

Synchrotron radiation is often used to form an image of the cross section of the beam with the arrangement shown in Fig. 12.1. For simplicity we take a single lens of focal length $f = r_p/2$ at the distance r_p from the source to form a $1:1$ image at the same distance r_p beyond the lens. This is no restriction since an image of a different magnification can be projected back to the source to obtain the resolution in terms of the size of the beam. With the small opening angle σ'_γ of the radiation, only the central part of the lens is illuminated. The situation is therefore similar to that of optical imaging with a lens of small size, which leads to a resolution that is limited by diffraction [65–68]. For a circular full lens aperture D this resolution is

$$d \approx \frac{1.22\lambda}{D/r_p} \approx \frac{1.22\lambda}{4\sigma'_\gamma}, \qquad (12.1)$$

where d is defined as the distance from the center of the image to the first minimum of the diffraction pattern. In normal light optics this aperture has a sharp edge, whereas for synchrotron radiation the angular distribution is smooth and determined by the nature of the source itself. We use here a typical opening angle $\sigma'_\gamma = D/(4r_p)$ to relate the case of synchrotron radiation to that of diffraction by an aperture-limited lens.

Synchrotron-radiation sources have a finite longitudinal extension ℓ_r, which leads to a limitation of resolution by the depth-of-field effect in forming an image, as illustrated in Fig. 12.2. We consider three point sources A, B, and C along the longitudinal extent of the source. Point B is located at the nominal distance $r_p = 2f$ from the lens and its image point is at the same distance r_p at the other side. The other point sources A and C have their image points at approximate distances $\pm\ell_r/2$ from the first one and form a spot of finite size d_f in the image plane. Using again $\pm 2\sigma'_\gamma$ for the maximum angle, we find the resolution due to the depth-of-field effect:

$$d_f \approx \frac{\sigma'_\gamma \ell_r}{2}. \qquad (12.2)$$

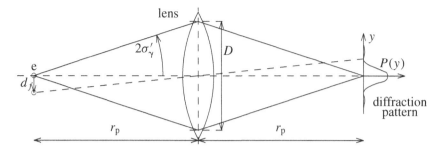

Fig. 12.1. Imaging the cross section of the beam with synchrotron radiation.

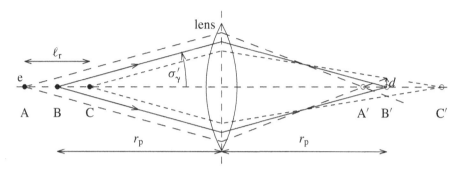

Fig. 12.2. The depth-of-field effect.

We will now investigate these two limitations to the image formed with synchrotron, undulator, and weak-magnet radiation.

12.1.2 Diffraction and the depth-of-field effect for SR from long magnets

Synchrotron radiation emitted in a long magnet of bending radius ρ is often used to form an image of the cross section of the beam. For most radiation sources the main part of the spectrum lies in the ultraviolet or x-ray region while for practical reasons the image is formed with visible light. Therefore we consider here the lower part of the spectrum, which has an opening angle for the horizontal polarization (5.21) and a resulting source length given by

$$\sigma_\gamma' \approx 0.41(\lambda/\rho)^{1/3}, \qquad \ell_r \approx 4\sigma_\gamma'\rho = 1.64(\lambda\rho^2)^{1/3}.$$

The resolution due to diffraction becomes, according to (12.1),

$$d = \frac{1.22\lambda}{4\sigma_\gamma'} = 0.74(\lambda^2\rho)^{1/3}.$$

We find that the resolution improves with shorter wavelength and with smaller radius of curvature. Because of the latter dependence, synchrotron-radiation monitors have poor resolution in large machines. Using as short a wavelength as possible can help improve the resolution. Using special magnets with strong curvature could help too, but the weak dependence of the resolution on ρ makes this approach less attractive.

For the depth-of-field effect we obtain from (12.2) the resolution

$$d_f \approx 2\sigma_\gamma'^2 \rho = 0.34 (\lambda^2 \rho)^{1/3} .$$

It is interesting to note that the depth-of-field effect leads to a resolution that has the same parameter dependence as and a similar magnitude to that for diffraction. This connection will be discussed later.

12.1.3 Diffraction and the depth-of-field effect for undulator radiation

We consider the radiation from a weak undulator of length L_u having N_u periods of length λ_u and filter out the fundamental frequency ω_{10}. This results in the distribution (7.41) derived in Chapter 7 and plotted in Fig. 7.9. Since its RMS angle diverges due to the high frequencies produced by the assumed sharp termination of the undulator field, we made a fit (7.43) through the central part of the angular distribution. Its RMS angle is given by (7.44), which we use now as an approximation:

$$\sigma_\gamma' \approx 0.56 \frac{1}{\gamma \sqrt{N_u}}.$$

We obtain from (12.1) the resolution due to diffraction,

$$d = \frac{1.22\lambda}{4\sigma_\gamma'} = 0.54\lambda_{10}\gamma\sqrt{N_u} = 0.39\sqrt{\lambda_{10}L_u},$$

where we used the relation $\lambda_u = 2\gamma^2\lambda_{10}$.

For the depth-of-field effect we find from (12.2)

$$d_f \approx \frac{\sigma_\gamma' L_u}{2} \approx 0.28 \frac{L_u}{\gamma\sqrt{N_u}} = 0.40\sqrt{\lambda_{10}L_u},$$

which is again of the same form and magnitude as the limit on resolution for diffraction.

Since the resolution is proportional to $\sqrt{L_u}$ we would like to work with a short undulator. However, we obtained the above expression with the assumption $N_u \gg 1$. A more detailed calculation is necessary in order to treat the more general case of an undulator having few periods and to optimize the resolution.

12.1.4 Diffraction and the depth-of-field effect for short-magnet radiation

We consider a short magnet of length L_s that radiates, according to (11.7) and (11.8), a broad spectrum with a typical wavelength and opening angle of

$$\lambda_{sm} \approx \frac{L_s}{2\gamma^2}, \qquad \sigma_\gamma' \approx \theta_{sm} \approx \frac{1}{\sqrt{2}\gamma}.$$

We obtain from (12.1) the resolution due to diffraction:

$$d = \frac{1.22\lambda}{4\sigma'_\gamma} = 0.43\lambda_{sm}\gamma \approx 0.31\sqrt{\lambda_{sm}L_s}.$$

The depth-of-field effect gives, according to (12.2),

$$d_f \approx \frac{\sigma'_\gamma L_s}{2} \approx 0.36\frac{L_s}{\gamma} = 0.5\sqrt{\lambda_{sm}L_s}.$$

These results are similar to the ones obtained for the undulator.

12.1.5 Discussion

In large storage rings the image of the cross section of the beam formed with synchrotron radiation has a limitation on its resolution due to diffraction. Sometimes the question of whether a large angular spread of the particles in the beam helps the resolution since a larger part of the lens is now illuminated arises. However, this is not the case since the diffraction results in a finite-sized image of each particle. In most cases the waves emitted by different particles have no systematic phase relation and do not interfere to form a diffraction pattern. An exception to this can occur if the electrons are grouped together into a length smaller than the emitted wavelength, which will be discussed later.

The depth-of-field effect is due to the longitudinal extent of the radiation source. At first sight it seems astonishing that the resolution due to diffraction and that due to the depth-of-field effect have the same parameter dependence and are of similar magnitude. The reason for this lies in the relation between the length of the source and the opening angle. For synchrotron radiation from long magnets the opening angle $\approx 1/\gamma$ determines the length of the source, as illustrated in Fig. 1.2 of Chapter 1. For undulator and weak-magnet radiation we have similar relations. Since these two effects are so closely related, they should be treated not as two separate effects but rather as a single one. It is therefore sufficient to consider just diffraction and the depth-of-field effect will implicitly be included.

Synchrotron radiation is often used as a diagnostic tool to measure the size and/or angular spread of the electron beam from which it originates. The finite opening angle and image resolution represent a limitation to this measurement. The product of the two contributions is a quantitative measure of this effect and is often called the emittance of the photon beam in analogy with the emittance defined for electron beams. This will be discussed later in more detail.

12.2 Imaging with SR – a quantitative treatment

12.2.1 The Fraunhofer diffraction

We now treat the diffraction in a quantitative way and consider a 1 : 1 image formed with synchrotron radiation by a single lens as illustrated in Fig. 12.3. Since we consider only

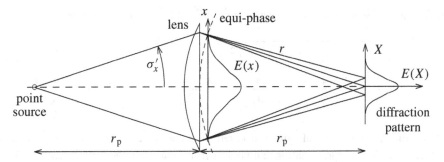

Fig. 12.3. Fraunhofer diffraction in imaging with synchrotron radiation.

very small emission angles with respect to the z-axis, we treat the radiation field in this chapter as a two-component vector. The emitted Fourier-transformed components of the field have horizontal and vertical angular distributions of the forms

$$\tilde{E}_x = \tilde{E}_x(x'_\gamma, y'_\gamma), \qquad \tilde{E}_y = \tilde{E}_y(x'_\gamma, y'_\gamma).$$

At the lens these are transformed into spatial distributions:

$$\tilde{E}_x(x, y) = \tilde{E}_x(r_p x'_\gamma, r_p y'_\gamma), \qquad \tilde{E}_y(x, y) = \tilde{E}_x(r_p x'_\gamma, r_p y'_\gamma).$$

The point source in Fig. 12.3 is projected onto the image point at the distance r_p from the lens. In this case all rays between source and image points have the same optical length, i.e. all photons take the same time to go from the source to the image. The dashed circular arc with radius r_p around the image point represents therefore a cut through an equi-phase surface. The physics would be clearer if a focusing mirror were used to form a $1:1$ image right back at the source. In this case the surface of the mirror itself would represent an equi-phase surface. However, this situation would be difficult to realize and to present in a drawing.

To calculate the image we use Huygens's principle [65], which considers each point on the equi-phase surface at the lens as a source of a radiation field of strength proportional to $[\tilde{E}_x(x, y), \tilde{E}_y(x, y)]$. This is explained in most books on optics and only the results are given here. We restrict ourselves to a scalar field E, which can stand for either the horizontal or the vertical component of the field. The field contribution emitted by each secondary source point $[x, y, z]$ on the equi-phase surface propagates towards the image plane $[X, Y]$ in the form of a wave

$$\delta\tilde{E}(X, Y)\,e^{i(kr-\omega t)} = -\frac{i}{\lambda r}\tilde{E}(x, y)e^{i(kr-\omega t)}\,dx\,dy, \tag{12.3}$$

where $k = 2\pi/\lambda$ is the wave number of the radiation. The factor i indicates a phase change of $\pi/2$, which is of no interest for our application, and r is the distance between the secondary source and the observation point in the image plane:

$$r^2 = (r_p - z)^2 + (x - X)^2 + (y - Y)^2.$$

On the equi-phase surface we have $(r_p - z)^2 = r_p^2 - (x^2 + y^2)$, which gives

$$
\begin{aligned}
r &= \sqrt{r_p^2 - 2(xX + yY) + X^2 + Y^2} \\
&= r_p\left(1 - \frac{xX + yY}{r_p^2} + \frac{X^2 + Y^2}{2r_p^2} - \frac{(xX + yY)^2}{2r_p^4} \cdots\right) \approx r_p - \frac{xX + yY}{r_p}.
\end{aligned}
$$

Since the opening angle of synchrotron radiation and the extent of the image are small, we make the above approximation and neglect from now on higher-order terms in (X/r_p) and (Y/r_p). This corresponds to the Fraunhofer approximation in optics. Furthermore, we approximate r in the denominator of (12.3) by r_p.

We obtain the field in the image plane by integrating the contributions (12.3) from each surface element of the secondary source,

$$
\begin{aligned}
\tilde{E}(X, Y)e^{-i\omega t} &= -\frac{i}{\lambda}e^{-i\omega t}\int_{-\infty}^{\infty}\int_{-\infty}^{\infty}\frac{\tilde{E}(x, y, \omega)}{r_p}\exp\left[i\left(kr_p - \frac{kxX}{r_p} - \frac{kyY}{r_p}\right)\right]dx\,dy \\
&= -\frac{ir_p}{\lambda}e^{-i\omega t}\int_{-\infty}^{\infty}\int_{-\infty}^{\infty}\tilde{E}(x, y)e^{-ik(x'kX + y'kY)}dx'dy'.
\end{aligned} \tag{12.4}
$$

In the second step we omitted the unimportant phase factor $\exp(ikr_p)$ and replaced the coordinates (x, y) at the lens by the emission angles (x', y') at the source:

$$
x \approx x'r_p \approx r_p\theta\cos\phi, \qquad y \approx y'r_p \approx r_p\theta\sin\phi.
$$

This integral represents a Fourier transform. In other words, the field distribution $\tilde{E}(X, Y)$ in the image plane is just proportional to the two-dimensional Fourier transform of the field distribution on the equi-phase surface at the lens, or of the angular distribution of the emitted radiation,

$$
\tilde{E}(X, Y) \propto \mathcal{F}\tilde{E}(x', y').
$$

Sometimes it is convenient to give the emission angles in spherical coordinates θ and ϕ and to replace the image coordinates (X, Y) by polar ones $(\mathcal{R}, \Phi)$ with the relations

$$
X = \mathcal{R}\cos\Phi, \qquad Y = \mathcal{R}\sin\Phi, \tag{12.5}
$$

which yields

$$
\tilde{E}(\mathcal{R}, \Phi) = -\frac{ir_p}{\lambda}\int_0^{\infty}\int_0^{2\pi}\tilde{E}(\theta, \phi)e^{-ik\theta\mathcal{R}\cos(\phi - \Phi)}\theta\,d\theta\,d\phi.
$$

If the emitted electric field has a harmonic ϕ-dependence of the form

$$
\tilde{E}(\theta, \phi) = \tilde{E}_1(\theta)\cos(2n\phi) \quad \text{or} \quad \tilde{E}(\theta, \phi) = \tilde{E}_1(\theta)\sin(2n\phi),
$$

we can use the integral representation of Bessel functions given in (9.1.21) of [41],

$$J_n(z) = \frac{i^{-n}}{\pi} \int_0^\pi e^{iz \cos\phi} \cos(n\phi) \, d\phi,$$

to express the integral over ϕ:

$$\int_0^{2\pi} e^{ik\theta\mathcal{R}\cos(\phi-\Phi)} \cos(2n\phi) \, d\phi = (-1)^n 2\pi J_{2n}(k\theta\mathcal{R}) \cos(2n\Phi)$$

or

$$\int_0^{2\pi} e^{ik\theta\mathcal{R}\cos(\phi-\Phi)} \sin(2n\phi) \, d\phi = (-1)^n 2\pi J_{2n}(k\theta\mathcal{R}) \sin(2n\Phi).$$

This gives an image with the same azimuthal dependence and symmetry as the emitted radiation. If it is independent of ϕ, we obtain an image that is a function of $\mathcal{R}$ only:

$$\tilde{E}(\mathcal{R}) = -ir_p k \int_0^\infty \tilde{E}(\theta) J_0(k\theta\mathcal{R}) \, \theta \, d\theta. \tag{12.6}$$

In all these calculations of the diffraction we assumed that we have a point source located at the distance r_p from the lens. This is an approximation for the case of synchrotron radiation. In the treatment of the depth-of-field effect we said that the length of the source is $\pm 2\sigma'_\gamma \rho$ and used for the RMS opening angle $\sigma'_\gamma \approx 0.41(\lambda/\rho)$. If we take the finite longitudinal extent of the source into account, the sphere of radius r_p around the image center is no longer an equi-phase surface. The calculation becomes more complicated and the exponent in the integral (12.4) will contain quadratic terms of coordinates x and y at the lens. This leads to the case of Fresnel diffraction, which rarely has a closed solution and will not be covered here but can be found in more extended treatments [69–71]. We expect the improved treatment to make a sizable correction if the observation of the radiation accepts a relatively large opening angle and is done from not too far a distance. However, in the following we will still use the Fraunhofer diffraction to illustrate some of the underlying physics.

12.2.2 The emittance of a photon beam

We consider now a general photon beam emitted from a point source with a symmetric Gaussian distribution with respect to the horizontal angle x' at frequency ω. A lens at a distance r_p focuses the source onto a plane with coordinates (X, Y) at the distance r_p from the lens, as illustrated in Fig. 12.3. The angular field distribution between the source and the lens is

$$\tilde{E}(x) \propto \exp\left(-\frac{x'^2}{4\sigma'^2}\right)$$

and the energy distribution is

$$\frac{dU}{dx'} \propto \exp\left(-\frac{x'^2}{2\sigma'^2}\right),$$

which has the RMS opening angle

$$\sqrt{\langle x'^2 \rangle} = \sigma'.$$

The field in the image plane is obtained from (12.4),

$$\tilde{E}(X) \propto \int_{-\infty}^{\infty} \exp\left(-\frac{x'^2}{4\sigma'^2}\right) e^{-ikx'X}\, dx' = 2\sqrt{\pi}\sigma' e^{-k^2 X^2 \sigma'^2},$$

and the energy distribution is

$$\frac{dU}{dX} \propto e^{-2k^2 \sigma'^2 X^2}$$

with the RMS width

$$\sqrt{\langle X^2 \rangle} = \sigma = \frac{1}{2k\sigma'}.$$

We call the product of the RMS emission angle and the RMS image size of the radiated energy the *emittance* ϵ_γ of the photon beam

$$\epsilon_\gamma = \sigma'\sigma = \frac{1}{2k} = \frac{\lambda}{4\pi}. \tag{12.7}$$

The emittance depends only on the wavelength, not on the opening angle. However, for a distribution different from the Gaussian our result will be multiplied by a factor larger than unity. In other words, a photon beam with a Gaussian distribution has the minimum emittance $\epsilon_\gamma = \lambda/(4\pi)$.

For a two-dimensional distribution the minimum-emittance condition applies to each direction:

$$\epsilon_{\gamma x} \geq \frac{\lambda}{4\pi}, \qquad \epsilon_{\gamma y} \geq \frac{\lambda}{4\pi}, \qquad \epsilon_{\gamma \theta} \geq \epsilon_{\gamma x} + \epsilon_{\gamma y} \geq \frac{\lambda}{2\pi}.$$

For a photon beam with rotational symmetry we have

$$\langle \theta^2 \rangle = 2\langle x'^2 \rangle = 2\langle y'^2 \rangle, \qquad \langle \mathcal{R}^2 \rangle = 2\langle X^2 \rangle = 2\langle Y^2 \rangle, \qquad \epsilon_{\gamma \theta} = 2\epsilon_{\gamma x} = 2\epsilon_{\gamma y}.$$

12.2.3 The diffraction of synchrotron radiation emitted in long magnets

We now use synchrotron radiation from a long magnet to image the cross section of the beam. The radiation depends only on the vertical emission angle y', which we used to call ψ. As mentioned before, this imaging is usually done with visible light having a frequency much smaller than ω_c and we can use an approximation. The Fourier-transformed electric field in the approximation of small frequencies $\omega \ll \omega_c$, (5.11), is

$$\tilde{\mathbf{E}}(\psi, \lambda) = \frac{e}{\sqrt{2\pi}\epsilon_0 cr} \left(\frac{\pi\rho}{\lambda}\right)^{1/3} \left[\mathrm{Ai}'\left(\left(\frac{\pi\rho}{\lambda}\right)^{2/3}\psi^2\right), \; i\psi\left(\frac{\pi\rho}{\lambda}\right)^{1/3} \mathrm{Ai}\left(\left(\frac{\pi\rho}{\lambda}\right)^{2/3}\psi^2\right)\right]$$

$$\tag{12.8}$$

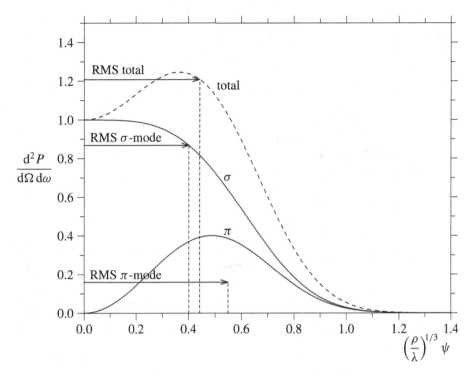

Fig. 12.4. The vertical distribution of synchrotron radiation from long magnets.

and the angular spectral power distribution is (5.12)

$$\frac{d^2 P_\sigma}{d\Omega\, d\omega} = \frac{2r_0 m_0 c^2}{\pi \rho} \left(\frac{\omega}{2\omega_0}\right)^{2/3} \mathrm{Ai}'^2\left(\left(\frac{\omega}{2\omega_0}\right)^{2/3}\psi^2\right)$$

$$\frac{d^2 P_\pi}{d\Omega\, d\omega} = \frac{2r_0 m_0 c^2}{\pi \rho} \left(\frac{\omega}{2\omega_0}\right)^{4/3} \psi^2 \mathrm{Ai}^2\left(\left(\frac{\omega}{2\omega_0}\right)^{2/3}\right).$$

The latter is plotted in Fig. 12.4. It has the RMS opening angles (5.21)

$$\sqrt{\langle\psi^2\rangle_\sigma} = 0.41\left(\frac{\lambda}{\rho}\right)^{1/3}, \qquad \sqrt{\langle\psi^2\rangle_\pi} = 0.55\left(\frac{\lambda}{\rho}\right)^{1/3}, \qquad \sqrt{\langle\psi^2\rangle} = 0.45\left(\frac{\lambda}{\rho}\right)^{1/3}.$$

$$(12.9)$$

The vertical image given by Fraunhofer diffraction is obtained by applying the transformation (12.4) to the vertical field distribution (12.8):

$$\tilde{\mathbf{E}}(Y, \lambda) \propto \int_{-\infty}^{\infty} \tilde{\mathbf{E}}(\psi, \lambda) e^{-i(\psi kY)}\, d\psi.$$

This integration has to be done numerically. The corresponding power density of the image is proportional to $|\tilde{\mathbf{E}}(Y, \lambda)|^2$ and is plotted in Fig. 12.5 for the horizontal and vertical

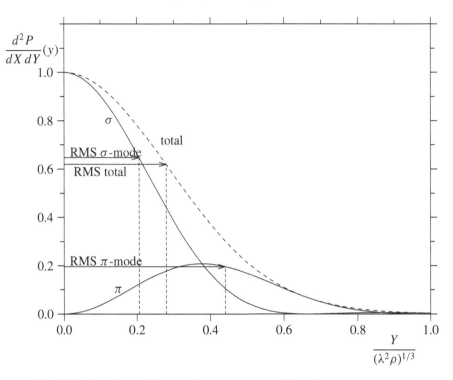

Fig. 12.5. Fraunhofer diffraction for synchrotron radiation from long magnets.

polarizations as well as for the total radiation. The RMS values of the image power are

$$\sigma_{Y\sigma} = 0.206(\lambda^2\rho)^{1/3}, \qquad \sigma_{Y\pi} = 0.429(\lambda^2\rho)^{1/3}, \qquad \sigma_{Y\,\text{total}} = 0.279(\lambda^2\rho)^{1/3}. \quad (12.10)$$

From Fig. 12.5 it is evident that the image is narrowest for the σ-mode of the radiation. Using a horizontal polarizing filter will therefore improve the resolution of the image by about 25%.

The product of the RMS image size (12.10) and the RMS opening angle (12.9) gives the emittances of the two modes of polarization and the total radiation:

$$\epsilon_{\gamma\sigma} = 1.06\frac{\lambda}{4\pi}, \qquad \epsilon_{\gamma\pi} = 2.96\frac{\lambda}{4\pi}, \qquad \epsilon_{\gamma\,\text{total}} = 1.57\frac{\lambda}{4\pi}.$$

Later we will also need the ratio between the size of the image and the opening angle:

$$\sqrt{\frac{\langle Y^2\rangle_\sigma}{\langle \psi^2\rangle_\sigma}} = 0.504(\lambda\rho^2)^{1/3}, \qquad \sqrt{\frac{\langle Y^2\rangle_\pi}{\langle \psi^2\rangle_\pi}} = 0.780(\lambda\rho^2)^{1/3}$$

$$\sqrt{\frac{\langle Y^2\rangle_\text{total}}{\langle \psi^2\rangle_\text{total}}} = 0.622(\lambda\rho^2)^{1/3}.$$

So far we have investigated only the vertical angular distribution and the resulting image. In the horizontal direction the field distribution is uniform in nature and will be determined by some aperture limitation due to a slit or lens. Since this aperture will determine the horizontal diffraction, we do not want to make it too small. On the other hand, by making it too large we will increase the magnitude of the depth-of-field effect. As a compromise we make the horizontal angular acceptance comparable to the natural vertical distribution. Considering that the horizontal limitation has a sharp edge, a value of the order $|x'| \leq 2\sigma'_\gamma$ is a reasonable compromise.

12.2.4 The diffraction of undulator radiation

We derived the radiation from a weak undulator in the frequency domain as

$$\tilde{E}(\omega) = \frac{4r_0 c B_0 \gamma^3}{\sqrt{2\pi} r_p} \frac{(1 - \gamma^2\theta^2 \cos(2\phi), \ \gamma^2\theta^2 \sin(2\phi)) \ \pi N_u}{(1 + \gamma^2\theta^2)^3} \frac{\sin\left(\dfrac{(\omega - \omega_1)\pi N_u}{\omega_1}\right)}{\dfrac{(\omega - \omega_1)\pi N_u}{\omega_1}},$$

(12.11)

with

$$\omega_1 = \frac{\omega_{10}}{1 + \gamma^2\theta^2} = \frac{2\gamma^2 k_u c}{1 + \gamma^2\theta^2}.$$

We filter out the frequency $\omega = \omega_{10}$ to use it for imaging the cross section of the beam and obtain from (7.41) the field containing only an x-component,

$$\tilde{E}_x(\omega_{10}) \approx \frac{4r_0 c B_0 \gamma^3}{\sqrt{2\pi} r_p} \frac{\pi N_u}{\omega_{10}} \frac{\sin(\gamma^2\theta^2\pi N_u)}{\gamma^2\theta^2\pi N_u},$$

(12.12)

and, from the angular spectral power distribution (7.42),

$$\frac{d^2 P(\omega_{10})}{d\Omega \, d\omega} = P_0 \gamma^2 \frac{3}{\pi} \frac{N_u}{\omega_{10}} \left(\frac{\sin(\gamma^2\theta^2\pi N_u)}{\gamma^2\theta^2\pi N_u}\right)^2.$$

This function is plotted in Fig. 12.6. Integrating it over the solid angle gives the spectral power density at the fundamental frequency:

$$\frac{dP}{d\omega}(\omega_{10}) = P_u \frac{3}{2\omega_{10}}.$$

(12.13)

As discussed before, this angular distribution at ω_{10} has a diverging variance caused by the unphysical abrupt termination of the magnetic field at $\pm L_u/2$ which enhances the high-frequency spectrum. Since a realistic undulator has a smooth field termination this problem is an artifact of our simplified description of the field. To avoid it, we fit the central part of

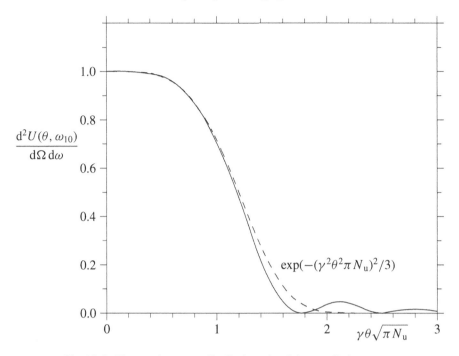

Fig. 12.6. The angular power distribution of undulator radiation at ω_{10}.

the distribution with an exponential, (7.44), as shown in Fig. 7.9,

$$\frac{\mathrm{d}^2 P(\omega_{10})}{\mathrm{d}\Omega\,\mathrm{d}\omega} \approx P_\mathrm{u}\gamma^2\frac{3}{\pi}\frac{N_\mathrm{u}}{\omega_{10}}e^{-(\pi N_\mathrm{u}\gamma^2\theta^2)^2/3},$$

which is also plotted in Fig. 12.6. It has the RMS value, (7.44),

$$\langle\gamma^2\theta^2\rangle = \frac{\sqrt{3\pi}}{\pi^2 N_\mathrm{u}} = 0.31\frac{1}{N_\mathrm{u}}, \tag{12.14}$$

which we will use as an approximation for the actual distribution.

The field in the focal plane is obtained from (12.12) with (12.6),

$$\tilde{E}_x(\mathcal{R},\omega_{10}) = \mathrm{i}\frac{\sqrt{2\pi}}{\lambda}\frac{2r_0 c B_0\gamma}{\omega_{10}}\,\mathrm{si}\!\left(\frac{k_\mathrm{u}^2\mathcal{R}^2}{\pi N_\mathrm{u}\gamma^2}\right),$$

where we used the sine integral function si z and the integral [72]

$$\int_0^\infty \frac{\sin(bz^2)}{z}J_0(cz)\,\mathrm{d}z = -\frac{1}{2}\,\mathrm{si}\!\left(\frac{c^2}{4b}\right).$$

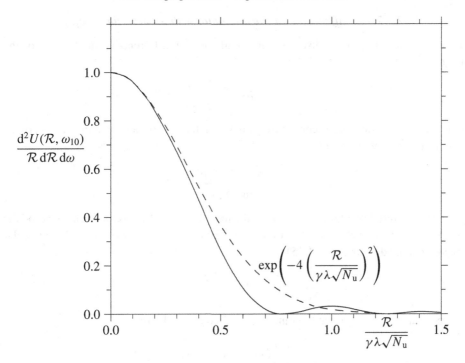

Fig. 12.7. Fraunhofer diffraction for imaging with undulator radiation at ω_{10}.

The spatial spectral power distribution of the image at the central frequency ω_{10} is

$$\frac{\mathrm{d}^3 P}{\mathcal{R}\,\mathrm{d}\mathcal{R}\,\mathrm{d}\Phi\,\mathrm{d}\omega}(\omega_{10}) = P_\mathrm{u}\frac{3k_\mathrm{u}^2\gamma^2}{\pi^3 N_\mathrm{u}\omega_{10}}\,\mathrm{si}^2\!\left(\frac{k_\mathrm{u}^2\mathcal{R}^2}{\pi N_\mathrm{u}\gamma^2}\right).$$

Integrating it over the area gives the same spectral density (12.13) as that which we found for the emitted radiation.

Since the above distribution has also a diverging variance, we perform again an exponential fit at the center:

$$\frac{\mathrm{d}^3 P}{\mathcal{R}\,\mathrm{d}\mathcal{R}\,\mathrm{d}\Phi\,\mathrm{d}\omega}(\omega_{10}) \approx P_\mathrm{u}\frac{3k_\mathrm{u}^2\gamma^2}{\pi^3 N_\mathrm{u}\omega_{10}}\,\exp\!\left(-\frac{4\mathcal{R}^2}{\lambda^2\gamma^2 N_\mathrm{u}}\right),$$

which is plotted in Fig. 12.7. Its variance

$$\langle \mathcal{R}^2\rangle = \frac{\gamma^2\lambda^2 N_\mathrm{u}}{4} \tag{12.15}$$

will be used as an approximation. With the RMS angle (12.14) and beam size (12.15) we find the emittance and the ratio:

$$\epsilon_{\gamma x} = \epsilon_{\gamma_y} \approx \sqrt[4]{3\pi}\,\frac{\lambda}{4\pi} = 1.75\frac{\lambda}{4\pi}, \qquad \sqrt{\frac{\langle Y^2\rangle_\sigma}{\langle \psi^2\rangle_\sigma}} \approx \frac{\pi}{4\sqrt[4]{3\pi}}L_\mathrm{u} = 0.448 L_\mathrm{u}.$$

12.2.5 The diffraction for the undulator with a Lorentzian profile

In the previous chapter we discussed an undulator with a Lorentzian modulation of the magnetic field,

$$B(z) = B_0 \frac{\cos(k_u z)}{1 + (z/z_0)^2},$$

giving a radiation field at the central frequency ω_{10} that has, for a large number of periods, only an x-component, (11.15),

$$\tilde{E}_x(\omega) = \frac{\sqrt{2\pi} e \gamma^3 k_u K_0 N_u}{4\epsilon_0 r_p \omega_{10}} e^{-\pi N_u \gamma^2 \theta^2}.$$

Owing to the variation of the amplitude of the field the radiated power also changes and the radiated energy is a better quantity to discuss. For this undulator we obtained the angular spectral energy distribution (11.15),

$$\frac{d^2 U}{d\Omega \, d\omega}(\omega_{10}) \approx U_0 \frac{3\gamma^2 N_u}{\omega_{10}} e^{-2\pi N_u \gamma^2 \theta^2},$$

the spectral energy density at the fundamental frequency (11.16),

$$\frac{dU}{d\omega}(\omega_{10}) = U_0 \frac{3}{2\omega_{10}},$$

and the RMS opening angle

$$\theta_{\text{RMS}} = \frac{1}{\gamma \sqrt{2\pi N_u}}. \tag{12.16}$$

To calculate the diffraction for the field distribution we make use of its azimuthal symmetry and use the relation (12.6):

$$\tilde{E}(\mathcal{R}, \Phi, \omega_{10}) = -\frac{i}{4\sqrt{2\pi}} \frac{e\gamma^3 k k_u K_0 N_u}{\epsilon_0 \omega_{10}} 2\pi \int_0^\infty e^{-\pi N_u \gamma^2 \theta^2} J_0(k\mathcal{R}\theta)\theta \, d\theta$$

$$= -\frac{i}{4\sqrt{2\pi}} \frac{e\gamma k k_u K_0}{\epsilon_0 \omega_{10}} \exp\left(-\frac{k^2 \mathcal{R}^2}{4\gamma^2 \pi N_u}\right).$$

The integral appearing above can be found as 11.4.29 in [41]. The spatial energy distribution of the diffraction pattern has the form

$$\frac{dU}{\mathcal{R} \, d\mathcal{R} \, d\Phi \, d\omega}(\omega_{10}) = \frac{2}{\mu_0 c} |\tilde{E}_x(\mathcal{R}, \Phi, \omega_{10})|^2 = U_0 \frac{3k^2}{4\pi \gamma^2 N_u \omega_{10}} \exp\left(-\frac{k^2 \mathcal{R}^2}{2\gamma^2 \pi N_u}\right).$$

Since it is the Fourier transform of the original Gaussian angular distribution of the radiation, the diffraction pattern at the fundamental frequency is also Gaussian. It has the RMS width

$$\sigma_{\mathcal{R}} = \frac{\sqrt{2\pi N_u} \gamma \lambda}{2\pi}, \qquad \sigma_X = \sigma_Y = \frac{\sqrt{\pi N_u} \gamma \lambda}{2\pi}. \tag{12.17}$$

Table 12.1. *RMS values of the emission angle and diffraction pattern,*
emittance and size-to-angle ratio of a 1 : 1 image formed with the
σ-mode for various sources

Source	Direction	σ'_γ	σ_γ	$\epsilon_\gamma \dfrac{4\pi}{\lambda}$	$\dfrac{\sigma_\gamma}{\sigma'_\gamma}$
Long magnet, $\omega \ll \omega_c$	y	$0.41\sqrt[3]{\lambda/\rho}$	$0.21\sqrt[3]{\rho\lambda^2}$	1.06	$0.50\sqrt[3]{\lambda\rho^2}$
Hard-edge undulator, $\omega = \omega_{10}$	x, y	$\approx\dfrac{0.40}{\gamma\sqrt{N_{\mathrm u}}}$	$\approx\dfrac{\gamma\lambda\sqrt{N_{\mathrm u}}}{2}$	≈ 2	$\approx 0.5 L_{\mathrm u}$
Lorentzian undulator, $\omega = \omega_{10}$	x, y	$\dfrac{1}{2\gamma\sqrt{\pi N_{\mathrm u}}}$	$\dfrac{\gamma\lambda\sqrt{\pi N_{\mathrm u}}}{2\pi}$	1.0	z_0

Integrating this over the area gives the spectral energy distribution,

$$\frac{\mathrm{d}U}{\mathrm{d}\omega}(\omega_{10}) = U_0 \frac{3}{2\omega_{10}},$$

which is the same as that of the original radiation.

From the RMS angle (12.16) and image size (12.17) we obtain the emittance

$$\epsilon_{\gamma x} = \epsilon_{\gamma y} = \frac{\lambda}{4\pi},$$

which is the minimum possible value because we have a Gaussian distribution. For the ratio
of the size and the angle we have

$$\sqrt{\frac{\langle \mathcal{R}^2 \rangle_\sigma}{\langle \theta^2 \rangle_\sigma}} = z_0 = \frac{N_{\mathrm u}\lambda_{\mathrm u}}{2}.$$

12.2.6 A comparison of the properties of beams from various sources

In Table 12.1 we list the types of radiation from the various sources we investigated and
give the RMS opening angle σ' and the 1 : 1 image size σ as well as the emittance ϵ_γ.
Furthermore, we list the ratio of the image size and opening angle $\sigma_\gamma/\sigma'_\gamma$, which we will
need later. The quality of a photon beam is higher the closer its emittance approaches the
minimum value $\lambda/(4\pi)$.

13

Electron-storage rings

13.1 Introduction

An electron-storage ring consists of an assembly of deflecting bending magnets and focusing quadrupole magnets, called a lattice (Fig. 13.1). The bending magnets determine a closed equilibrium orbit for a particle having the nominal values for energy, initial position, and angle. Particles with small deviations from these nominal parameters are kept in the vicinity of the nominal orbit by the focusing quadrupoles. In addition a storage ring has magnet-free straight sections. These contain an acceleration system, which is usually called the radio-frequency system or just the RF system. It is used to replace the energy lost by the particles due to the emission of synchrotron radiation or to accelerate the electron beam to higher energy.

The straight sections also house insertion devices (undulator and wiggler magnets) and other equipment such as instruments necessary for the operation of the storage ring. The particles are made to circulate inside a vacuum chamber with a very low pressure of a few nTorr (0.13 μPa) in order to minimize the interaction between the particles and the residual gas.

We restrict ourselves to a plane storage ring and describe the beam dynamics in a coordinate system that follows the nominal equilibrium orbit as shown in Fig. 13.2. The longitudinal coordinate s is the path-length along this nominal equilibrium orbit. The horizontal x-coordinate is perpendicular to the nominal orbit and points to the outside while the vertical y-coordinate is perpendicular to the (x, z) orbit plane and points upwards. Furthermore, to describe the emitted synchrotron radiation we use z as the coordinate along the tangent from a source point s. We also use the local curvature ρ, which is positive if it bends the particle to the inside. The nominal energy and momentum of the particles are E_e and p_0 with the deviations ΔE and Δp.

The derivatives with respect to s are indicated with a prime, e.g. the vertical angle of the trajectory is y'. Sometimes also the derivative with respect to the relative momentum is indicated in the same way, e.g. $v' = \mathrm{d}v/\mathrm{d}(\Delta p/p_0)$. We assume that there are small deviations from the nominal conditions and use paraxial approximations with

$$\frac{x}{\rho} \ll 1, \qquad x' \ll 1, \qquad \frac{y}{\rho} \ll 1, \qquad y' \ll 1, \qquad \frac{1}{\rho} = \frac{x''}{\sqrt{1 + x'^2}} \approx x'', \qquad \frac{\Delta E}{E_e} \ll 1.$$

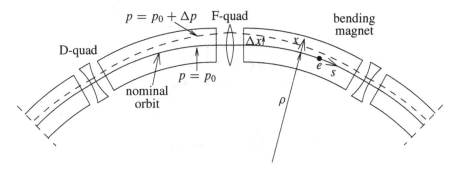

Fig. 13.1. Storage rings consisting of deflecting, bending, and focusing quadrupole lenses.

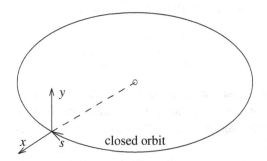

Fig. 13.2. Coordinates of a particle in a storage ring.

We will take these quantities only to first order. We deal with ultra-relativistic particles and approximate

$$\gamma = \frac{E_e}{m_0 c^2} \gg 1, \qquad \beta \approx 1, \qquad \frac{\Delta p}{p_0} = \frac{1}{\beta^2} \frac{\Delta E}{E_e} \approx \frac{\Delta E}{E_e}.$$

Many aspects of the beam dynamics are completely symmetric in the horizontal x- and vertical y-directions and we use the letter w to represent either of them:

$$w = x \ \text{ or } \ w = y. \tag{13.1}$$

Storage rings are described in many books and numerous reports. Many emphasize their application as sources of synchrotron radiation [14, 15, 18, 24] and cover in more detail the material presented here.

13.1.1 Lattice magnets

To describe the elements which make up the lattice we start with the bending magnets, which usually have a homogeneous dipole field B as shown in Fig. 13.3, but in some cases

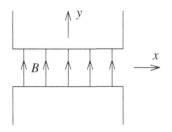

Fig. 13.3. A dipole bending magnet.

may also include a transverse gradient. The dipole magnet is characterized by the curvature $1/\rho$, the deflecting angle φ_B, and length $\rho\varphi_B$. A charged particle of momentum $p = \beta\gamma m_0 c$ traverses the magnet on a circular orbit with curvature

$$\frac{1}{\rho} = \frac{eB}{p} \approx \frac{eBc}{\gamma m_0 c^2}, \tag{13.2}$$

where we use the ultra-relativistic approximation, which is usually justified for electrons. These bending magnets determine the nominal equilibrium orbit. In most machines they are homogenous dipole magnets. If they all have the same field strength, the ring is called isomagnetic.

The quadrupole magnets, often called 'quads,' provide focusing and deflect particles with deviations in position back towards the nominal orbit. Furthermore, they provide a bending correction for particles that have a positive or negative momentum deviation and whose paths are bent less or more in the dipole magnets. Such particles are on a different equilibrium orbit, which is displaced to the outside or inside with respect to the nominal orbit. They traverse the quadrupoles off center and acquire some additional bending to the inside or outside.

A quadrupole has a transverse magnetic field with a linear gradient, resulting in a field strength proportional to the distance from the axis as indicated in Fig. 13.4:

$$B_x = \frac{\partial B_x}{\partial y} y = ky, \qquad B_y = \frac{\partial B_y}{\partial x} x = kx$$

$$B^2 = k^2(x^2 + y^2) \quad \text{with} \quad k = \frac{\partial B_x}{\partial y} = \frac{\partial B_y}{\partial x}.$$

The constant k gives the two cross gradients of the field, which are both the same due to the condition curl $\mathbf{B} = [\nabla \times \mathbf{B}] = 0$. In the horizontal median plane the field is vertical, resulting in a horizontal deflection proportional to $\pm kx$, whereas in the vertical median plane the deflection is proportional to $\mp ky$. In each of these two planes the quadrupole provides the same focusing strength but of opposite sign. The deflection of a particle with

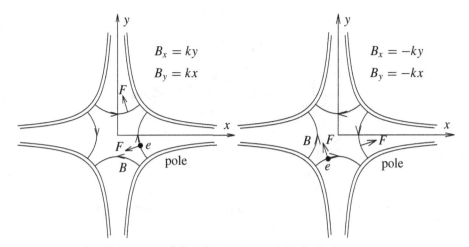

Fig. 13.4. Quadrupole magnets with deflecting forces for a positive charge moving out of the plane of the figure indicated. The magnet on the left-hand side is horizontally focusing and vertically defocusing, whereas the one on the right-hand side has the opposite focusing properties.

momentum $\mathbf{p}_0$ traversing the quadrupole on a paraxial trajectory is given by

$$\frac{1}{\rho_x} \approx -\frac{d^2x}{ds^2} = -x'' = -\frac{eB_y}{p} = \frac{ek}{p}x = K_f x, \qquad y'' = K_f y.$$

The focusing parameter

$$K_f = \frac{ek}{p} = \frac{e}{p}\frac{\partial B_y}{\partial x} \approx \frac{ec}{\gamma m_0 c^2}\frac{\partial B_y}{\partial x} = \frac{1}{\rho B_0}\frac{\partial B_y}{\partial x} \qquad (13.3)$$

is by convention positive in a horizontally focusing quadrupole, called an F-quad, and negative in a horizontally defocusing D-quad. The particle trajectory for $K_f > 0$ is given by

$$x'' + K_f x = 0, \qquad y'' - K_f y = 0$$

with solutions

$$x = \hat{x}\cos(\sqrt{|K_f|}s - \varphi_0), \quad y = \hat{y}\cosh(\sqrt{|K_f|}s - \varphi_0).$$

In many cases the quadrupole length ℓ is relatively short, resulting in an angular deflection but negligible change in excursion, and we can use a so-called short-lens approximation and obtain for the change in trajectory

$$\Delta x = 0, \qquad \Delta x' \approx -K_f \ell x = -\frac{x}{f} \quad \text{and} \quad \Delta y = 0, \qquad \Delta y' \approx K_f \ell y = \frac{y}{f}$$

with f being the focal length of the short lens replacing the quadrupole.

To provide overall focusing in both planes we have to use focusing and defocusing quadrupoles in combinations similar to that of the two lenses shown in Fig. 13.5. This is

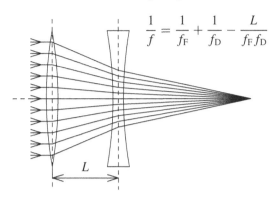

$$\frac{1}{f} = \frac{1}{f_{\mathrm{F}}} + \frac{1}{f_{\mathrm{D}}} - \frac{L}{f_{\mathrm{F}} f_{\mathrm{D}}}$$

Fig. 13.5. The overall focusing provided by an F-quad and a D-quad.

based on the principle that the particles have a larger transverse excursion in the F-quads and acquire a stronger deflection towards the axis, whereas in the D-quads the excursion is smaller, resulting in a weaker deflection away from the axis. This principle is well known in light optics.

13.2 The transverse particle dynamics in a storage ring

13.2.1 The particle dynamics over many revolutions

To study the transverse particle dynamics in a ring, we carry out a thought experiment. We start a single particle through the storage ring and observe its position and angle x_k and x_k' during successive turns k at selected locations, see Fig. 13.6. Since the focusing is linear, the observed points lie on an ellipse in this phase-space (x, x'). At any location around the ring the area of this ellipse is the same (Liouville's theorem), but its shape and orientation can be different. In simple lattices the centers of the quadrupoles are often symmetry points where the beam is neither converging nor diverging but about parallel. As a result the phase-space ellipse is upright. In the F-quads the excursions x_k are large but the angles x_k' small, which gives an ellipse with a large horizontal–vertical aspect ratio. In a D-quad the situation is reversed, with an ellipse being narrow and high. In between quads, there is a correlation between displacement and angle that is negative, or positive at locations where the beam is converging or diverging. The corresponding phase-space ellipses are tilted forwards after a D-quad and backwards following an F-quad.

Since these magnets are linear, the beam coordinates (x_1, x_1') at the exit from the magnet are linear functions of the input parameters (x_0, x_0') which we can express by writing a matrix equation:

$$\begin{matrix} x_1 = m_{11}x_0 + m_{12}x_0' \\ x_1' = m_{21}x_0 + m_{22}x_0' \end{matrix} \rightarrow \begin{pmatrix} x_1 \\ x_1' \end{pmatrix} = \begin{pmatrix} m_{11} & m_{12} \\ m_{21} & m_{22} \end{pmatrix} \begin{pmatrix} x_0 \\ x_0' \end{pmatrix} = M_x \begin{pmatrix} x_0 \\ x_0' \end{pmatrix}.$$

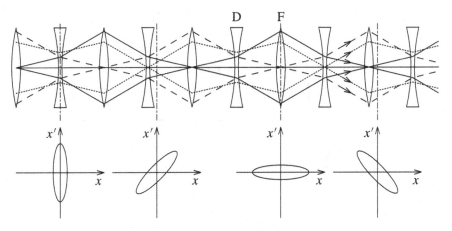

Fig. 13.6. The trajectory of a single particle making many revolutions in a storage ring.

This gives, for a straight section with length L, a short bending magnet of length $\rho\varphi_B$, and a short horizontally focusing quadrupole of focal length f, the following expressions:

$$\text{straight section:} \quad \begin{matrix} x_1 = x_0 + Lx_0' \\ x_1' = x_0' \end{matrix} \quad \rightarrow M_x = M_y = \begin{pmatrix} 1 & L \\ 0 & 1 \end{pmatrix}$$

$$\text{dipole:} \quad \begin{matrix} x_1 = x_0 + \rho\varphi_B x_0' \\ x_1' = x_0' \end{matrix} \quad \rightarrow M_x = M_y = \begin{pmatrix} 1 & \rho\varphi_B \\ 0 & 1 \end{pmatrix}$$

$$\text{quadrupole:} \quad \begin{matrix} x_1 = x_0 \\ x_1' = -x_0/f + x_0' \end{matrix} \quad \rightarrow M_x = \begin{pmatrix} 1 & 0 \\ -1/f & 1 \end{pmatrix}$$

$$\begin{matrix} y_1 = y_0 \\ y_1' = y_0/f + y_0' \end{matrix} \quad \rightarrow M_y = \begin{pmatrix} 1 & 0 \\ 1/f & 1 \end{pmatrix}.$$

(13.4)

The matrices of these elements have some general common properties.

- The determinant is unity, $|M| = m_{11}m_{22} - m_{12}m_{21} = 1$.
- The term m_{12} gives an indication for the length of the element.
- The element m_{21} indicates the dependence of the deflection on the initial displacement, which is characteristic for focusing if $m_{21} < 0$ and defocusing if $m_{21} > 0$.

The following investigation of the transverse beam dynamics is equivalent for the horizontal and vertical planes and hence we use the letter w to stand for x or y.

The calculation of the trajectory around the whole ring is complicated and can rarely be done in a closed form. However, the transfer through individual elements can be built up to form the whole ring. The transport of the particle through two elements is obtained by

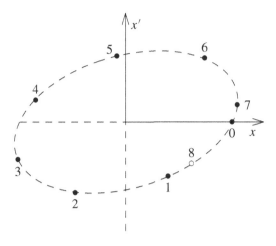

Fig. 13.7. The position and angle of a single particle making many revolutions observed at a given location in a storage ring.

multiplying the matrices of the individual elements in inverse order

$$\begin{pmatrix} w_1 \\ w_1' \end{pmatrix} = M_1 \begin{pmatrix} w_0 \\ w_0' \end{pmatrix}, \qquad \begin{pmatrix} w_2 \\ w_2' \end{pmatrix} = M_2 \begin{pmatrix} w_1 \\ w_1' \end{pmatrix} = M_2 \cdot M_1 \begin{pmatrix} w_0 \\ w_0' \end{pmatrix}.$$

By going through all the elements of a ring, we can find the one-turn matrix which gives the position and angle after one revolution as a linear expression of the initial value:

$$\begin{matrix} w_1 = m_{11}w_0 + m_{12}w_0' \\ w_1' = m_{21}w_0 + m_{22}w_0' \end{matrix} \rightarrow \begin{pmatrix} w_1 \\ w_1' \end{pmatrix} = \begin{pmatrix} m_{11} & m_{12} \\ m_{21} & m_{22} \end{pmatrix} \begin{pmatrix} w_0 \\ w_0' \end{pmatrix}.$$

This one-turn matrix has a unit determinant and represents a linear mapping like the individual element matrices. The values (x_k, x_k') after k turns are obtained from the multiplication (Fig. 13.7)

$$\begin{pmatrix} w_1 \\ w_1' \end{pmatrix} = \begin{pmatrix} m_{11} & m_{12} \\ m_{21} & m_{22} \end{pmatrix}^k \begin{pmatrix} w_0 \\ w_0' \end{pmatrix}.$$

With all elements being linear, the one-turn matrix represents a linear mapping of (w_0, w_0') onto points w_k, w_k' after k turns, which lie on an ellipse as will become clear later.

In the first step we assume that the particle is observed at a symmetry point where the ellipse is upright. We take a location with a straight ellipse, giving orthogonal mapping over k turns as shown in Fig. 13.8.

For a pair (u, v) of variables that are measured in the same units the general transformation matrix over one turn,

$$\begin{pmatrix} u_1 \\ v_1 \end{pmatrix} \begin{pmatrix} \cos\mu & \sin\mu \\ -\sin\mu & \cos\mu \end{pmatrix} \begin{pmatrix} u_0 \\ v_0 \end{pmatrix},$$

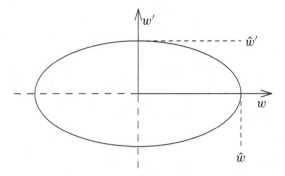

Fig. 13.8. A phase-space ellipse obtained by mapping a circulating particle at a symmetry point over many turns.

or a number n of turns,

$$
\begin{pmatrix} u_n \\ v_n \end{pmatrix} = \begin{pmatrix} \cos\mu & \sin\mu \\ -\sin\mu & \cos\mu \end{pmatrix}^n \begin{pmatrix} u_0 \\ v_0 \end{pmatrix} = \begin{pmatrix} \cos(n\mu) & \sin(n\mu) \\ -\sin(n\mu) & \cos(n\mu) \end{pmatrix} \begin{pmatrix} u_0 \\ v_0 \end{pmatrix},
$$

represents a rotation of the vector $[u, v]$ by an angle of μ_w each turn with the points u_n, v_n lying on a circle $u_n^2 + v_n^2 = u_0^2 + v_0^2$.

Since w and w' have different units, we need a scaling factor β_w having the dimension of a length, giving for the matrix

$$
\begin{pmatrix} w_1 \\ w_1' \end{pmatrix} = \begin{pmatrix} \cos\mu_w & \beta_w \sin\mu_w \\ -\sin\mu_w/\beta_w & \cos\mu_w \end{pmatrix} \begin{pmatrix} w_0 \\ w_0' \end{pmatrix}.
$$

To illustrate the meaning of μ_w and β_w we consider k turns:

$$
\begin{pmatrix} \cos\mu_w & \beta_w \sin\mu_w \\ -\sin\mu_w/\beta_w & \cos\mu_w \end{pmatrix}^k = \begin{pmatrix} \cos(k\mu_w) & \beta_w \sin(k\mu_w) \\ -\sin(k\mu_w)/\beta_w & \cos(k\mu_w) \end{pmatrix}.
$$

The points w_k, w_k' now lie on an ellipse and the parameter $\mu_w = 2\pi\nu$ represents a phase advance per turn of the trajectory. The particle motion observed each turn represents an oscillation around the nominal value called *betatron oscillation*. Correspondingly μ_w is the betatron phase advance per turn and ν_w is the betatron tune. We start the particle with $w_0 = 0$, $w_0' \neq 0$ and obtain after k turns $w_k = \sin(k\mu_w)\beta_w w_0'$. Therefore, $\beta_w = \hat{w}/\hat{w}'$ gives the ratio between the maximum displacement and the angle at a given location. The area of the phase-space ellipse mapped by a particle is the same everywhere, $A = \pi\epsilon_w = \pi\hat{w}\hat{w}'$, where ϵ_w is called the emittance of a trajectory. We have

$$
\epsilon_w = \hat{w}\hat{w}' = \frac{\hat{w}^2}{\beta_w} = \hat{w}'^2\beta_w, \qquad \hat{w} = \sqrt{\epsilon_w\beta_w}, \qquad \hat{w}' = \sqrt{\frac{\epsilon_w}{\beta_w}}. \tag{13.5}
$$

The equation for the phase-space ellipse is

$$
\frac{w^2}{\beta_w} + \beta_w w'^2 = \epsilon_w.
$$

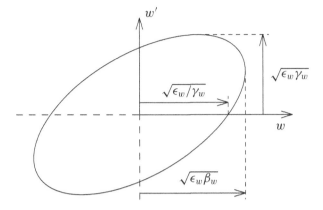

Fig. 13.9. The phase-space ellipse for the general case.

We can now express the mapping over many turns for a symmetry point with the phase advance per turn μ_w and the amplitude function β_w,

$$w_k = \cos(k\mu_w)\, w_0 + \beta_w \sin(k\mu_w)\, w_0' = C \cos(k\mu_w - \varphi_0)$$

$$C = \sqrt{w_0^2 + \beta_w^2 w_0'^2} = \sqrt{\beta_w \epsilon_w}, \qquad \tan\varphi_0 = \frac{\beta_w w_0'}{w_0}$$

$$w_k = \sqrt{\epsilon_w \beta_w} \cos(k\mu_w - \varphi_0), \qquad w_k' = -\sqrt{\frac{\epsilon}{\beta_w}} \sin(k\mu_w - \varphi_0)$$

with ϕ_0 being a constant phase given by the initial conditions.

In the general case the particle is not observed at a symmetry point but at a location where the beam is converging or diverging. The resulting phase-space ellipse is tilted (Fig. 13.9). However, since the one-turn matrix still has Det $M = 1$, the area of the ellipse stays the same.

To understand the general situation, we consider a location at $-\Delta s$ upstream of the symmetry point with coordinates $\bar{w}$, $\bar{w}'$. The optical properties of the structure between the two locations are described by a matrix T with elements t_{ij}, giving the relations

$$\begin{pmatrix} w_0 \\ w_0' \end{pmatrix} = \begin{pmatrix} t_{11} & t_{12} \\ t_{21} & t_{22} \end{pmatrix} \begin{pmatrix} \bar{w}_0 \\ \bar{w}_0' \end{pmatrix}, \qquad \begin{pmatrix} w_1 \\ w_1' \end{pmatrix} = \begin{pmatrix} t_{11} & t_{12} \\ t_{21} & t_{22} \end{pmatrix} \begin{pmatrix} \bar{w}_1 \\ \bar{w}_1' \end{pmatrix},$$

which we use to transform the one-turn matrix for the symmetry point,

$$\begin{pmatrix} w_1 \\ w_1' \end{pmatrix} = \begin{pmatrix} m_{11} & m_{12} \\ m_{21} & m_{22} \end{pmatrix} \begin{pmatrix} w_0 \\ w_0' \end{pmatrix},$$

into one for a general location:

$$\begin{pmatrix} t_{11} & t_{12} \\ t_{21} & t_{22} \end{pmatrix} \begin{pmatrix} \bar{w}_1 \\ \bar{w}_1' \end{pmatrix} = \begin{pmatrix} m_{11} & m_{12} \\ m_{21} & m_{22} \end{pmatrix} \begin{pmatrix} t_{11} & t_{12} \\ t_{21} & t_{22} \end{pmatrix} \begin{pmatrix} \bar{w}_0 \\ \bar{w}_0' \end{pmatrix}.$$

We multiply this by the inverse matrix

$$T^{-1} = \begin{pmatrix} t_{22} & -t_{12} \\ -t_{21} & t_{11} \end{pmatrix} \tag{13.6}$$

and obtain the desired one-turn matrix:

$$\begin{pmatrix} \bar{w}_1 \\ \bar{w}'_1 \end{pmatrix} = \begin{pmatrix} t_{22} & -t_{12} \\ -t_{21} & t_{11} \end{pmatrix} \begin{pmatrix} m_{11} & m_{12} \\ m_{21} & m_{22} \end{pmatrix} \begin{pmatrix} t_{11} & t_{12} \\ t_{21} & t_{22} \end{pmatrix} \begin{pmatrix} \bar{w}_0 \\ \bar{w}'_0 \end{pmatrix}.$$

Using the one-turn matrix with β_{wc} for a symmetry point,

$$M_s = \begin{pmatrix} \cos\mu_w & \beta_{wc}\sin\mu_w \\ -\sin\mu_w/\beta_{wc} & \cos\mu_w \end{pmatrix},$$

we find for the new one-turn matrix for a general location

$$M = T^{-1}M_s T$$

$$= \begin{pmatrix} \cos\mu_w + \left(\dfrac{t_{11}t_{12}}{\beta_{wc}} + t_{21}t_{22}\beta_{wc}\right)\sin\mu_w & \left(\dfrac{t_{12}^2}{\beta_{wc}} + t_{22}^2\beta_{wc}\right)\sin\mu_w \\ -\left(\dfrac{t_{11}^2}{\beta_{wc}} + t_{21}^2\beta_{wc}\right)\sin\mu_w & \cos\mu_w - \left(\dfrac{t_{11}t_{12}}{\beta_{wc}} + t_{21}t_{22}\beta_{wc}\right)\sin\mu_w \end{pmatrix}.$$

To express this in a more compact form, we introduce the so-called Twiss parameters :

$$\beta_w = \frac{t_{12}^2}{\beta_{wc}} + t_{22}^2\beta_{wc}$$

$$\alpha_w = \frac{t_{11}t_{12}}{\beta_{wc}} + t_{21}t_{22}\beta_{wc} \tag{13.7}$$

$$\gamma_w = \frac{t_{11}^2}{\beta_{wc}} + t_{21}^2\beta_{wc} = \frac{1+\alpha_w^2}{\beta_w}.$$

They can be calculated from the lattice elements and their physical meanings will become clear later. From the fact that all our matrices have unit determinants we obtain the relation

$$\beta_w\gamma_w = 1 + \alpha_w^2.$$

We obtain the general one-turn matrix starting at a point with β_x, α_w, γ_w,

$$M = \begin{pmatrix} \cos\mu_w + \alpha_w\sin\mu_w & \beta_w\sin\mu_w \\ -\gamma_w\sin\mu_w & \cos\mu_w - \alpha_w\sin\mu_w \end{pmatrix} = \begin{pmatrix} m_{11} & m_{12} \\ m_{21} & m_{22} \end{pmatrix},$$

and the matrix describing k turns,

$$M_k = M^k = \begin{pmatrix} \cos(k\mu_w) + \alpha_w\sin(k\mu_w) & \beta_w\sin(k\mu_w) \\ -\gamma\sin(k\mu_w) & \cos(k\mu_w) - \alpha_w\sin(k\mu_w) \end{pmatrix}. \tag{13.8}$$

This illustrates that μ_w still has the meaning of a phase advance per turn. It can be calculated from the trace of the matrix:

$$\cos \mu_w = \frac{1}{2} \, \text{Trace} \, M = \frac{m_{11} + m_{22}}{2}.$$

From the computed matrix element we can calculate the Twiss parameters:

$$\beta_w^2 = \frac{m_{12}^2}{1 - \left(\dfrac{m_{11} + m_{22}}{2} \right)^2}, \qquad \alpha_w^2 = \frac{\left(\dfrac{m_{11} - m_{22}}{2} \right)^2}{1 - \left(\dfrac{m_{11} + m_{22}}{2} \right)^2}.$$

The general phase-space ellipse is given by

$$\gamma_w w^2 + 2\alpha_w w w' + \beta_w w'^2 = \frac{1}{\beta_w}(w^2 + (\alpha_w w + \beta_w w')^2) = \epsilon_w.$$

Compared with the orthogonal case, the relations between maximum excursions and emittance are the same for $\hat{w}$ but different for $\hat{w}'$:

$$\hat{w} = \sqrt{\epsilon_w \beta_w}, \qquad \hat{w}' = \sqrt{\epsilon_w \gamma_w}, \qquad \epsilon_w = \frac{\hat{w}^2}{\beta_w} = \frac{\hat{w}'^2}{\gamma_w} = \frac{\hat{w}\hat{w}'}{\sqrt{\beta_w \gamma_w}} = \frac{\hat{w}\hat{w}'}{\sqrt{1 + \alpha_w^2}}.$$

Using (13.8), we can write the coordinates turn by turn in the forms

$$w_k = (\cos(k\mu_w) + \alpha_w \sin(k\mu_w))w_0 + \beta_w \sin(k\mu_w)w_0' = C \cos(k\mu_w - \varphi_0)$$

$$C = \sqrt{w_0^2 + (\alpha_w w_0 + \beta_w w_0')^2} = \sqrt{\epsilon_w \beta_w}$$

and

$$w_k' = -\gamma_w \sin(k\mu_w) w_0 + (\cos(k\mu_w) - \alpha_w \sin(k\mu_w))w_0' = C' \cos(k\mu_w - \varphi_0')$$

$$C' = \sqrt{(\gamma_w w_0 + \alpha_w w_0')^2 + w_0'^2} = \sqrt{\epsilon_w \gamma_w},$$

giving

$$w_k = \sqrt{\epsilon_w \beta_w} \cos(k\mu_w - \varphi_0), \qquad w_k' = \sqrt{\epsilon_w \gamma_w} \sin(k\mu_w - \varphi_1). \tag{13.9}$$

To obtain the lattice functions β_w and α_w at a certain location s around the ring, we have to start at this point and multiply the matrices of the elements of the ring. To obtain these functions for another location s_1 (Fig. 13.10) it is useful to find their dependences on s directly. This is particularly important for a straight section, where the coordinates (w_0, w_0') at $s = 0$ change as follows:

$$w_1 = w_0 + w_0's, \qquad w_1' = w_0'.$$

We introduce this relation into the expression for the phase-space ellipse:

$$\epsilon_w = \gamma_{w0} w_0^2 + 2\alpha_{w0} w_0 w_0' + \beta_{w0} w_0'^2$$

$$\epsilon_w = \gamma_{w0} w_1^2 + 2(\alpha_{w0} - \gamma_{w0}s)w_1 w_1' + (\beta_{w0} - 2\alpha_{w0}s + \gamma_{w0}s^2)w_1'^2. \tag{13.10}$$

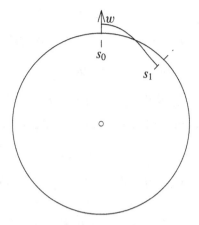

Fig. 13.10. Transfer between two points along the ring.

From this we obtain the propagation of the lattice functions in a straight section:

$$\beta_w(s) = \beta_w(0) - 2\alpha_w(0)s + \gamma_w(0)s^2 \tag{13.11}$$
$$\alpha_w(s) = \alpha_w(0) - \gamma_w(0)s, \qquad \gamma_w(s) = \gamma_w(0).$$

This equation becomes especially simple if we express it around the location s_0 where $\beta_w(s_0)$ has a minimum, i.e. $\alpha(s_0) = 0$:

$$\beta_w(s - s_0) = \beta_w(s_0)\left(1 + \left(\frac{s - s_0}{\beta_w(s_0)}\right)^2\right).$$

Next we investigate the effect of a short lens of focusing strength $1/f$ on the Twiss parameters. The change in trajectory at the lens is

$$w_1 = w_0, \qquad w_1' = w_0' - \frac{w_0}{f},$$

which we use to substitute for (w_0, w_0') in (13.10), giving

$$\epsilon_w = w_1^2\left(\gamma_{w0} + \frac{2\alpha_{w0}}{f} + \frac{\beta_{w0}}{f^2}\right) + 2w_1 w_1'\left(\alpha_{w0} + \frac{\beta_{w0}}{f}\right) + w_1'^2 \beta_{w0},$$

which gives the changes of the Twiss parameter in a short focusing lens:

$$\Delta\beta_w = 0, \qquad \Delta\alpha_w = \frac{\beta_w}{f}, \qquad \Delta\gamma_w = \frac{2\alpha_w}{f} + \frac{\beta_w}{f^2}. \tag{13.12}$$

Finally we consider continuous focusing in a long quadrupole with focusing parameter K_f and take a short element of length ds and focusing strength $K\,ds$, (13.3), which makes the first-order trajectory changes

$$w_1 = w_0 + w_0'\,ds, \qquad w_1' = w_0' - w_0 K_f\,ds$$

or

$$w_0 = w_1 - w_1' \, ds, \qquad w_0' = w_1' + w_1 K_f \, ds,$$

which we use to substitute for (w_0, w_0') in (13.10):

$$\epsilon_w = w_1^2(\gamma_{w0} + 2\alpha_{w0} K_f \, ds) + 2w_1 w_1'(\alpha_{w0} - \gamma_{w0} \, ds + \beta_{w0} K_f \, ds) + w_1'^2(\beta_{w0} - 2\alpha_{w0} \, ds).$$

This gives for the derivatives of the Twiss parameter in a general linearly focusing section

$$\gamma_w' = 2\alpha_w K_f, \qquad \alpha_w' = \beta_w K_f - \gamma_w, \qquad \beta_w' = -2\alpha_w.$$

On making the substitutions $\alpha = -\beta_w'/2$ and $\gamma_w = (1 + \alpha_w^2)/\beta_w$ we obtain a non-linear differential equation for $\beta_w(s)$:

$$\tfrac{1}{2}\beta_w \beta_w'' - \tfrac{1}{4}\beta_w'^2 + \beta_w^2 K_f = 1.$$

This general equation rarely has an analytic solution and the Twiss parameters are usually computed by matrix multiplication.

We try to find a convenient expression for the trajectory as a function of the longitudinal coordinate s within one revolution similar to the expression (13.9) which gives the trajectory coordinates for each turn k at a certain location s_0. It can be shown that the latter expression can be generalized by replacing the phase $k\mu_w$ per turn (13.4) by the continuous betatron phase $\varphi_w(s)$, leading to the expressions

$$w(s) = \sqrt{\epsilon_w \beta_w(s)} \cos(\varphi_w(s) - \varphi_0)$$
$$w'(s) = -\sqrt{\epsilon_w/\beta_w(s)} \, (\alpha_w(s) \cos(\varphi_w(s) - \varphi_0) + \sin(\varphi_w(s) - \varphi_0))$$
$$= -\sqrt{\epsilon_w \gamma_w} \cos(\varphi_w(s) - \varphi_0')$$

and the betatron phase is related to the beta function by

$$\frac{d\varphi_w(s)}{ds} = \frac{1}{\beta_w(s)}. \tag{13.13}$$

The phase $\varphi_w(s)$ advances fast where β_w is small and slowly where it is large.

It is interesting to form the variances of the coordinates at a location s by averaging over many turns:

$$\langle w^2 \rangle = \tfrac{1}{2}\epsilon\beta_w, \qquad \langle ww' \rangle = -\tfrac{1}{2}\epsilon\alpha_w, \qquad \langle w'^2 \rangle = \tfrac{1}{2}\epsilon\gamma_w. \tag{13.14}$$

The envelope of the multi-turn trajectory is proportional to $\sqrt{\beta_w}$ (Fig. 13.11).

13.2.2 The beam with many particles

The large number of particles in a beam execute betatron oscillations of different phases and amplitudes. They describe ellipses that have different areas but, at a given location, have the same orientation and aspect ratio. Therefore, the beam as a whole covers a phase-space

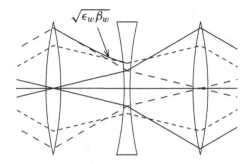

Fig. 13.11. The multi-turn trajectory and its envelope $\propto \sqrt{\epsilon_w \beta_w(s)}$.

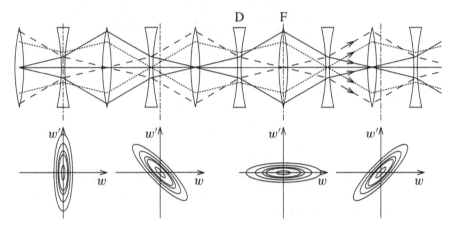

Fig. 13.12. The phase-space ellipses of different particles have the same orientation and aspect ratio but different sizes.

distribution Ψ of the same form as the particles, with the beam emittance being the average of the individual particle ellipses $E_w = \langle \epsilon_w \rangle$ (Fig. 13.12).

The phase-space distribution $\Psi(w, w')$ is shown in Fig. 13.13. Usually it can not be observed directly; however, we can measure its projections $f(w)$ and $g(w')$ on the planes of position or angle,

$$f(w) = \int \Psi(w, w') \, dw', \qquad g(w') = \int \Psi(w, w') \, dw,$$

the average (center-of-charge) position and angle,

$$\langle w \rangle = \frac{\displaystyle\int f(w) w \, dw}{\displaystyle\int f(w) \, dw}, \qquad \langle w' \rangle = \frac{\displaystyle\int g(w') w' \, dw'}{\displaystyle\int g(w') \, dw'},$$

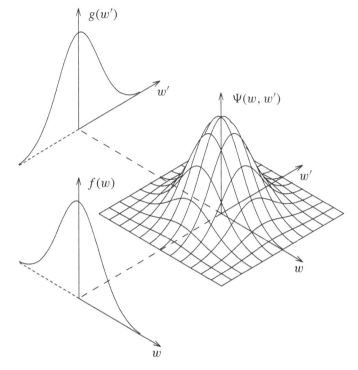

Fig. 13.13. The phase-space distribution and projections of a beam with many particles.

and its variances in position and angle,

$$\sigma_w^2 = \frac{\int f(w) w^2 \, dw}{\int f(w) \, dw}, \qquad \sigma_w'^2 = \frac{\int g(w') w'^2 \, dw'}{\int g(w) \, dw'}. \qquad (13.15)$$

The emittance of the beam containing many particles is

$$\langle \epsilon_w \rangle = \sigma_w \sigma_w'.$$

In electron machines the horizontal emittance of a single particle is given by the quantum excitation due to emission of synchrotron radiation, as will be shown in Chapter 14. Since this is the case for each particle, the emittance of the whole beam is usually the same. However, instabilities or scattering between particles can lead to a blow-up of this natural emittance at high intensities.

13.2.3 The dispersion

A particle with a momentum deviation $\Delta p / p_0$ has a different bending angle in a dipole magnet. For a plane ring this results in a new equilibrium orbit at a distance $\Delta x = D_x \, \Delta p / p_0$

and angle $\Delta x' = D'_x \, \Delta p/p_0$ with respect to the nominal orbit, as indicated in Fig. 13.1. This deviation, normalized by $\Delta p/p_0$, is called the dispersion D_x and the derivative is denoted by D'_x. The curvature $1/\rho$ of the particle trajectory in a bending magnet of field B depends on the momentum p:

$$\frac{1}{\rho} = \frac{eB}{p} \approx \frac{eBc}{m_0 c^2 \gamma}.$$

The angular deflection in a bending element of length ds is therefore

$$\Delta\varphi_B = \frac{ds}{\rho} = \frac{ds}{\rho_0(1 + \Delta p/p_0)} \approx \frac{ds}{\rho_0}(1 - \Delta p/p_0)$$

with ρ_0 being the bending radius of the particle with nominal energy. A particle with an excess of energy is deflected less and will obtain a positive angular deviation to the outside:

$$dx' = \frac{ds}{\rho_0}\frac{\Delta p}{p_0}, \qquad dD'_x = \frac{dx'}{\Delta p/p_0} = \frac{ds}{\rho_0}.$$

In a bending magnet the dispersion will therefore propagate like

$$D'_x(s) = D'_x(0) + \frac{s}{\rho_0}, \qquad D_x(s) = D_x(0) + D'_x(0)s + \frac{s^2}{2\rho_0}. \tag{13.16}$$

Since the off-momentum orbit is displaced with respect to the nominal one, the particle goes off center through the quadrupoles and acquires some extra deflection. For a short quadrupole lens of focusing strength $1/f$ this leads to an angular deflection of

$$\Delta x' = -\frac{x}{f}, \qquad \Delta D'_x = -\frac{D_x}{f}, \qquad \Delta D_x = 0. \tag{13.17}$$

For a long quadrupole lens with focusing parameter K we obtain

$$D''_x = -D_x K,$$

which indicates the focusing effects of the quadrupoles on the off-energy orbit. Finally, in a straight section the dispersion propagates like

$$D'_x(s) = D'_x(0), \qquad D_x(s) = D_x(0) + D'_x(0)s. \tag{13.18}$$

With the above relations we can evaluate the dispersion function $D_x(s)$ and its derivative $D'_x(s)$.

13.2.4 The chromatic aberrations and their correction with sextupoles

The focusing strength of a quadrupole depends on the momentum of the particle, $1/f \propto 1/p$. As a result the tune depends on the momentum deviation, which is expressed as the chromaticity, in terms of either the absolute tune ν'_x or the ratio ξ_x of the relative changes

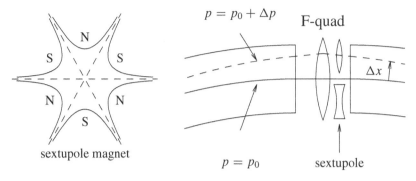

Fig. 13.14. Chromatic correction with sextupole magnets.

in tune and momentum

$$v_x' = \frac{\Delta v_x}{\Delta p / p_0}, \qquad \xi_x = \frac{\Delta v_y / v_y}{\Delta p / p_0} \quad \text{and} \quad v_y' = \frac{\Delta v_y}{\Delta p / p_0}, \qquad \xi_y = \frac{\Delta v_y / v_y}{\Delta p / p_0}.$$

This can be corrected with sextupole magnets located at finite dispersion. They have the geometry shown in Fig. 13.14 and a field of the form

$$B_x = \frac{\partial^2 B_y}{\partial x^2} xy, \qquad B_y = \frac{1}{2} \frac{\partial^2 B_y}{\partial x^2} (x^2 - y^2), \qquad |B| = \frac{1}{2} \left| \frac{\partial^2 B_y}{\partial x^2} \right| (x^2 + y^2).$$

Developing this field around a point $x = x_0, \ y = 0$ on the horizontal axis,

$$B_x = \frac{\partial^2 B_y}{\partial x^2} x_0 \, \Delta y, \qquad B_y = \frac{\partial^2 B_y}{\partial x^2} \left(\frac{x_0^2}{2} + x_0 \, \Delta x \right),$$

results in a dipole field combined with a quadrupole field of a strength proportional to the displacement x_0.

A horizontally focusing sextupole is placed close to an F-quad where particles with excess energy have an orbit displaced to the outside and receive some extra focusing, as shown in Fig. 13.14. Conversely, vertically focusing sextupoles located close to D-quads provide correction in the vertical plane.

We develop the sextupole field around a point $x = 0, \ y = y_0$ on the vertical axis,

$$B_x = \frac{\partial^2 B_y}{\partial x^2} y_0 \, \Delta x, \qquad B_y = -\frac{\partial^2 B_y}{\partial x^2} \left(\frac{y_0^2}{2} + y_0 \, \Delta y \right),$$

and obtain a field consisting of a dipole and a rotated quadrupole component, which creates coupling between the horizontal and vertical motions. The strength of the latter is proportional to the vertical distance y_0.

Sextupoles are non-linear elements that can not be treated with simple matrices. They have to be rather strong in highly focusing lattices and can have adverse effects, e.g. they often limit the so-called dynamic aperture which is the maximum betatron amplitude accepted by the ring.

In the absence of sextupoles the horizontal and vertical chromaticities have the 'natural' values, which are always negative. With uncorrected chromaticities the betatron tune ν_x and ν_y depend on the momentum of the particle, which may result in a large tune spread. It can lead also to beam instabilities.

13.2.5 Coupling and vertical dispersion

Rotated elements and misalignment of magnets can lead to a situation in which horizontal and vertical betatron oscillations are coupled. In storage rings mainly rotated quadrupole fields cause coupling. A small rotation of a quadrupole magnet can be regarded as a normal quadrupole field with a weaker one that is rotated by 45°. The latter is of the form

$$B_x = \frac{\partial B_x}{\partial y} x, \qquad B_y = \frac{\partial B_x}{\partial y} y$$

and results in a vertical deflection for a particle traversing it with a horizontal displacement from the axis. Vertical displacements in sextupole magnets also result in a rotated quadrupole field acting as a source of coupling. As will be seen in the next chapter, the horizontal emittance of an electron beam has a finite size due to emission of synchrotron radiation, while the natural vertical emittance is much smaller. Coupling will usually determine the vertical beam size.

In an ideal plane storage ring there is only a horizontal dispersion D_x while the vertical one vanishes, $D_y = 0$. However, errors such as rotational misalignment of bending magnets or of quadrupoles at locations of horizontal dispersion can produce vertical dispersion. Owing to quantum excitation during emission of synchrotron radiation, this will lead to a larger vertical emittance, as will be discussed in Chapter 14.

The vertical emittance is mainly caused by coupling or vertical dispersion. To keep it small, we have to keep rotational misalignment of dipoles and quadrupoles small, reduce vertical orbit distortion in quadrupoles and sextupoles, and avoid vertical position errors of these elements.

13.2.6 An example: The FODO lattice

To illustrate the calculation of the lattice parameters we take as an example a storage ring with N identical cells of length L, each having only a thin F-quad and a D-quad of equal strength, with bending magnets filling the space between them, as shown in Fig. 13.15. This beam optics is called the FODO lattice and is used in many machines. To calculate the beta function we start from a symmetry point in the center of the F-quad. Since each cell has

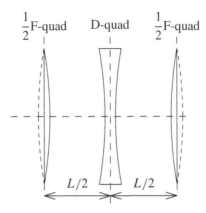

$\frac{1}{2}$F-quad   D-quad $\frac{1}{2}$F-quad

$L/2$ $L/2$

Fig. 13.15. A FODO lattice consisting of equidistant F- and D-quads.

the same configuration and the same lattice functions at both ends, we can treat it as a ring
with a single cell of phase advance μ_x. Later we take the N_cth power of the transfer matrix
for a cell in order to obtain the full ring. We obtain the horizontal matrices of the elements
from (13.4) and multiply them:

$$
M_x = \begin{pmatrix} 1 & 0 \\ -\dfrac{1}{2f} & 1 \end{pmatrix} \begin{pmatrix} 1 & \dfrac{L}{2} \\ 0 & 1 \end{pmatrix} \begin{pmatrix} 1 & 0 \\ \dfrac{1}{f} & 1 \end{pmatrix} \begin{pmatrix} 1 & \dfrac{L}{2} \\ 0 & 1 \end{pmatrix} \begin{pmatrix} 1 & 0 \\ -\dfrac{1}{2f} & 1 \end{pmatrix}
$$

$$
= \begin{pmatrix} 1 - \dfrac{L^2}{8f^2} & L\left(1 + \dfrac{L}{4f}\right) \\ -\dfrac{L}{4f^2}\left(1 - \dfrac{L}{4f}\right) & 1 - \dfrac{L^2}{8f^2} \end{pmatrix} = \begin{pmatrix} \cos\mu_x & \beta_{xF}\sin\mu_x \\ -\dfrac{1}{\beta_{xF}}\sin\mu_x & \cos\mu_x \end{pmatrix}.
$$

On comparing the two matrices in the lower line, we obtain

$$
\cos\mu_x = 1 - 2\sin^2(\mu_x/2) = 1 - \frac{L^2}{8f^2} \;\rightarrow\; \sin(\mu_x/2) = \frac{L}{4f}
$$

$$
\beta_{xF}^2 = -\frac{m_{12}}{m_{21}} = L^2\left(\frac{2f}{L}\right)^2 \frac{1 + \dfrac{L}{4f}}{1 - \dfrac{L}{4f}} \;\rightarrow\; \beta_{xF} = L\frac{1 + \sin(\mu_x/2)}{2\cos(\mu_x/2)\sin(\mu_x/2)}.
$$

For the vertical matrix the starting quadrupole is defocusing and we could again multiply
the corresponding matrices. However, we can just change the sign of $1/f$ in the horizontal

expression and obtain

$$\cos\mu_y = 1 - 2\sin^2(\mu_y/2) = 1 - \frac{L^2}{8f^2} = \sin\mu_x \rightarrow \sin(\mu_y/2) = \frac{L}{4f}$$

$$\beta_{yF}^2 = -\frac{m_{12}}{m_{21}} = L^2\left(\frac{2f}{L}\right)^2 \frac{1 - \dfrac{L}{4f}}{1 + \dfrac{L}{4f}} \rightarrow \beta_{yF} = L\frac{1 - \sin(\mu_y/2)}{2\cos(\mu_y/2)\sin(\mu_y/2)}.$$

The FODO lattice has perfect symmetry between horizontal focusing and vertical defocusing and vice versa, resulting in the same phase advance per cell, and a horizontal beta function at the D-quad that is equal to the vertical one at the F-quad:

$$\mu_x = \mu_y, \qquad \beta_{xF} = \beta_{yD}, \qquad \beta_{xD} = \beta_{yF}.$$

The beta functions are largest at the centers of the quadrupoles. The maximum value depends on the length L and the phase advance μ of the cell and is smallest around $\mu_x \approx 76°$. It is interesting to note that β_x diverges as the phase advance per cell μ_x approaches π, where the lattice becomes unstable.

Next we calculate α_{xF+} at the exit of the F-quad and α_{xD-} at the entrance of the D-quad. Using the fact that $\alpha = 0$ at the symmetry points represented by the centers of the quadrupoles and the relation (13.12) for the change of α_x in the half quads:

$$\alpha_{xF+} = \frac{\beta_{xF}}{2f} = \frac{1 + \sin(\mu_x/2)}{\cos(\mu_x/2)}, \qquad \alpha_{xD-} = \frac{\beta_{xD}}{2f} = \frac{1 - \sin(\mu_x/2)}{\cos(\mu_x/2)}.$$

Using (13.11), we obtain the lattice functions between the F- and D-quads:

$$\beta_x(s) = \beta_{xF}\left(1 - 2\sin(\mu_x/2)\frac{2s}{L} + \frac{2\sin^2(\mu_x/2)}{1 + \sin(\mu_x/2)}\left(\frac{2s}{L}\right)^2\right)$$

$$\alpha_x(s) = \alpha_{xF+}\left(1 - \frac{2\sin(\mu_x/2)}{1 + \sin(\mu_x/2)}\frac{2s}{L}\right) \tag{13.19}$$

$$\gamma_x(s) = \frac{4\tan(\mu_x/2)}{L} = \text{constant}$$

with

$$\beta_{xF} = L\frac{1 + \sin(\mu_x/2)}{2\cos(\mu_x/2)\sin(\mu_x/2)}, \qquad \alpha_{xF+} = \frac{1 + \sin(\mu_x/2)}{\cos(\mu_x/2)}. \tag{13.20}$$

To calculate the dispersion we start in the center of the F-quad, where $D'_x = 0$ and the dispersion has the unknown value D_{xF}. Using (13.17), we obtain the derivative at the exit of the F-quad,

$$D'_{xF+} = -\frac{D_{xF}}{2f},$$

and we use (13.16) to propagate this to the entrance of the D-quad:

$$D'_{xD-} = -\frac{D_{xF}}{2f} + \frac{L}{2\rho}, \qquad D_{xD} = D_{xF} - \frac{D_{xF}}{2f}\frac{L}{2} + \frac{L^2}{8\rho}.$$

Finally, going through the half D-quad must lead to a vanishing dispersion for symmetry reasons,

$$D'_{xD} = D'_{xD-} + \frac{D_{xD}}{2f} = 0,$$

which gives

$$D_{xF} = \frac{4f^2}{\rho}\left(1 + \frac{L}{8f}\right) = \frac{L^2}{4\rho\,\sin^2(\mu_x/2)}\left(1 + \frac{1}{2}\sin(\mu_x/2)\right)$$

$$D_{xD} = \frac{4f^2}{\rho}\left(1 - \frac{L}{8f}\right) = \frac{L^2}{4\rho\,\sin^2(\mu_x/2)}\left(1 - \frac{1}{2}\sin(\mu_x/2)\right).$$

The propagation of the dispersion between the quads is, according to (13.16),

$$D_x(s) = D_{xF}\left(1 - \sin(\mu_x/2)\frac{2s}{L} + \frac{\sin^2(\mu_x/2)}{2\left(1 + \frac{1}{2}\sin(\mu_x/2)\right)}\left(\frac{2s}{L}\right)^2\right)$$

$$D'_x(s) = D_{xF+}\left(1 - \frac{\sin(\mu_x/2)}{1 + \frac{1}{2}\sin(\mu_x/2)}\frac{2s}{L}\right) \tag{13.21}$$

with

$$D_{xF} = \frac{L^2}{\rho}\frac{1 + \frac{1}{2}\sin(\mu_x/2)}{4\sin^2(\mu_x/2)}, \qquad D'_{xF+} = -\frac{L}{\rho}\frac{1 + \frac{1}{2}\sin(\mu_x/2)}{2\sin(\mu_x/2)}. \tag{13.22}$$

In Fig. 13.16 the beta functions and the dispersion are plotted as functions of s for a FODO lattice with $\mu_x = 90°$. Figure 13.16 shows also the emittance function $\mathcal{H}$, which will be discussed later.

13.3 The longitudinal particle dynamics

13.3.1 Introduction

A particle with a momentum deviation moves on a new equilibrium orbit that differs from the nominal one by a distance $\Delta x = D_x \Delta p/p_0$ and angle $\Delta x' = D'_x \Delta p/p_0$ as shown in Fig. 13.1, where $D_x(s)$ is the dispersion and $D'_x(s)$ its derivative. As a consequence this new orbit has also a different circumference C. These deviations are normalized and described in a linear approximation by the dispersion D_x and the so-called momentum

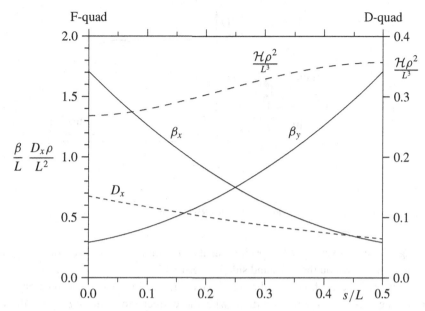

Fig. 13.16. Beta functions, dispersion, and $\mathcal{H}$ in a FODO lattice with $\mu_x = 90°$.

compaction α_c:

$$\Delta x = D_x \frac{\Delta p}{p_0}, \qquad \frac{\Delta C}{C_0} = \alpha_c \frac{\Delta p}{p_0}, \qquad \alpha_c = \frac{1}{C_0} \oint \frac{D_x(s)}{\rho} \, ds. \qquad (13.23)$$

In general both have to be computed for a given beam optics. For the FODO lattice, consisting of thin lenses with bending magnets filling the space between them, we have calculated the dispersion (13.21) and integrate over it to obtain the momentum compaction:

$$\alpha_c(\text{FODO}) = \left(\frac{L}{\rho}\right)^2 \frac{1 - \frac{1}{12} \sin^2(\mu_x/2)}{4 \sin^2(\mu_x/2)}.$$

The different path lengths and velocities of an off-momentum particle lead to changes in revolution time T_{rev} and frequency ω_{rev}. In most rings $\alpha_c > 0$, which results in a longer orbit for particles of higher energy:

$$\frac{\Delta T}{T_{\text{rev}}} = \left(\alpha_c - \frac{1}{\gamma^2}\right) \frac{\Delta p}{p_0} = \eta_c \frac{\Delta p}{p_0} \approx \alpha_c \frac{\Delta p}{p_0}, \qquad \frac{\Delta \omega_{\text{rev}}}{\omega_{\text{rev}}} \approx -\alpha_c \frac{\Delta p}{p_0}.$$

For ultra-relativistic electrons usually $1/\gamma^2 \ll \alpha_c$. Particles of higher energy take longer to complete a turn. Hence their frequencies of revolution are smaller than that of a nominal particle. However, proton rings do not always operate in an ultra-relativistic regime and there exists an energy with Lorentz factor $\gamma_T = 1/\sqrt{\alpha_c}$ for which these two terms cancel out, which is called the transition energy.

To replace the energy U_s lost by a circulating particle due to the emission of synchrotron radiation, one uses a longitudinal field provided by an RF (radio-frequency) system

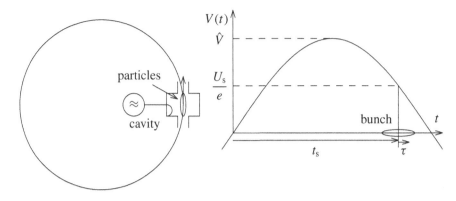

Fig. 13.17. An RF system.

consisting of one or several cavities oscillating at a harmonic h of the revolution frequency $\omega_{RF} = h\omega_{rev}$ as shown on the left-hand side in Fig. 13.17.

The cavity voltage is shown on the right-hand side in Fig. 13.17 as a function of the traversal time $t_s + \tau$. A nominal particle with the correct momentum p_0 goes through the cavity at the synchronous time t_s with respect to the zero crossing of the voltage. It receives a gain in energy that compensates for the loss due to synchrotron radiation $\Delta E = U_s = e\hat{V}\sin(\omega_{RF}t_s)$.

A particle arriving later than the synchronous time receives a smaller gain in energy. As a result the energy will become too small and the related revolution time shorter, which corrects the original delay. For an early particle the situation is reversed. A particle with too much energy takes longer to go around and arrives late next time when it receives less energy and vice versa. This creates longitudinal focusing and leads to an oscillation with frequency ω_s around the synchronous arrival time and the nominal energy, which is called the synchrotron or phase oscillation and also energy oscillation.

13.3.2 The longitudinal focusing – small amplitudes

We consider a particle of nominal momentum $p_0 \approx E_e/c$, revolution frequency ω_{rev}, synchronous arrival time t_s, and phase $\varphi_s = h\omega_{rev}t_s$. The deviations from these values are

$$\frac{\Delta T}{T_{rev}} = \alpha_c \frac{\Delta p}{p_0} \approx \alpha_c \frac{\Delta E}{E_e} = \alpha_c \epsilon, \qquad \tau = t - t_s.$$

The change in energy per turn δE is

$$\delta E = e\hat{V}\sin(h\omega_{rev}(t_s + \tau)) - U_s \approx e\hat{V}(\sin\varphi_s + \cos\varphi_s\, h\omega_{rev}\tau) - U_s, \qquad (13.24)$$

where we assumed that there is a small deviation $h\omega\tau \ll 1$ and developed the trigonometric functions. We call the relative energy deviation $\epsilon = \Delta E/E_e$, assume that there are small changes per turn, and make a smooth approximation,

$$\frac{d(\Delta E/E_e)}{dt} = \dot{\epsilon} \approx \frac{\omega_{rev}}{2\pi}\frac{\delta E}{E_e}, \qquad \dot{\tau} \approx \frac{\Delta T}{T_{rev}} = \alpha_c \epsilon,$$

giving

$$\dot{\epsilon} \approx \frac{\omega_{rev}e\hat{V}}{2\pi E_e}(\sin\varphi_s + \cos\varphi_s\, h\omega_{rev}\tau) - \frac{\omega_{rev}}{2\pi}\frac{U_s(\epsilon)}{E_e}.$$

The energy loss U_s due to the emission of synchrotron radiation depends on the particle energy. We linearize this dependence and determine the synchronous phase for the nominal energy:

$$U_s = U_{s0} + \frac{dU_s}{dE}\Delta E, \qquad e\hat{V}\sin\varphi_s = U_{s0}.$$

This leads to the two first-order differential equations

$$\dot{\epsilon} = \omega_{rev}^2\frac{he\hat{V}\cos\varphi_s}{2\pi E_e}\tau - \frac{\omega_{rev}}{2\pi}\frac{dU_s}{dE}\epsilon, \qquad \dot{\tau} = \alpha_c\epsilon,$$

which can be combined in

$$\ddot{\epsilon} + \frac{\omega_{rev}}{2\pi}\frac{dU_s}{dE}\dot{\epsilon} - \omega_{rev}^2\frac{h\alpha_c e\hat{V}\cos\varphi_s}{2\pi E_e}\epsilon = 0.$$

The solution is a damped oscillation of the form

$$\epsilon(t) = \hat{\epsilon}e^{-\alpha_\epsilon t}\cos\left(\sqrt{\omega_s^2 - \alpha_\epsilon^2}\,t + \varphi_0\right) \approx e^{-\alpha_\epsilon t}\cos(\omega_s t + \varphi_0).$$

Here, ω_s is the synchrotron frequency without damping, which we assume to be much larger than the damping rate α_ϵ:

$$\omega_s = \omega_{rev}\sqrt{\frac{h\alpha_c e\hat{V}\cos\varphi_s}{2\pi E_e}}, \qquad \alpha_\epsilon = \frac{1}{2}\frac{\omega_{rev}}{2\pi}\frac{dU_s}{dE}.$$

For $\alpha_\epsilon > 0$ the emission of synchrotron radiation provides damping for the longitudinal oscillation.

To have a gain in energy and longitudinal focusing the synchronous phase has to satisfy $\pi/2 < \varphi_s < \pi$ for $\alpha_c > 0$ and $0 < \varphi_s < \pi/2$ for the more exotic case of $\alpha_c < 0$.

The damping of the synchrotron oscillations is weak and can be neglected for certain investigations, giving

$$\epsilon(t) = \hat{\epsilon}\cos(\omega_s t + \varphi_0), \qquad \tau(t) = \hat{\tau}\sin(\omega_s t + \varphi_0)$$

with the constant of motion

$$\frac{\alpha_c\epsilon^2}{2} + \frac{\omega_s^2\tau^2}{2\alpha_c} = \frac{\alpha_c\hat{\epsilon}^2}{2} = \frac{\omega_s^2\hat{\tau}^2}{2\alpha_c} = H' = \text{constant},$$

which can be used to relate the maximum excursions in energy and arrival time. The electrons execute synchrotron oscillation around the nominal arrival time t_s and energy E_e.

They group together and form a so-called bunch. Their distribution is usually Gaussian, of the form

$$\Psi(\epsilon, \tau) = \frac{1}{2\pi \sigma_\epsilon \sigma_\tau} \exp\left(-\frac{\epsilon^2}{2\sigma_\epsilon^2}\right) \exp\left(-\frac{\tau^2}{2\sigma_\tau^2}\right), \qquad \iint \Psi(\epsilon, \tau)\, d\epsilon\, d\tau = 1$$

with the RMS values $\sigma_\epsilon = \sigma_E / E_e$ and $\sigma_\tau = \sigma_t - t_s$ related by

$$\sigma_\tau = \frac{\alpha_c}{\omega_s} \sigma_\epsilon = \frac{1}{\omega_{rev}} \sqrt{\frac{2\pi E_e \alpha_c}{h e \hat{V} \cos \varphi_s}} \, \sigma_\epsilon. \tag{13.25}$$

It is interesting to note that, for a given relative energy spread σ_ϵ, the bunch length σ_τ is proportional to $\sqrt{\alpha_c}$. Operating a ring in a low-α_c mode is a method by which to obtain short bunches.

The phase-space distribution $\Psi(\epsilon, \tau)$ can usually not be observed directly, but its projection onto the time axis multiplied by the total charge $q = Ne$ of the N particles represents the instantaneous beam current

$$I(t) = Ne \int_{-\infty}^{\infty} \Psi(\epsilon, \tau)\, d\epsilon = \frac{Ne}{\sqrt{2\pi}\sigma_\tau} \exp\left(-\frac{\tau^2}{2\sigma_\tau^2}\right) = \hat{I} \exp\left(-\frac{t^2}{2\sigma_\tau^2}\right)$$

with the peak current $I_p = Ne/(\sqrt{2\pi}\sigma_\tau)$. In a machine with an RF system oscillating at the harmonic number h of the frequency of revolution $\omega_{RF} = h\omega_{rev}$ there are h synchronous time locations t_s around which the particles can oscillate in a stable fashion. The system supports a maximum of h bunches.

13.3.3 The longitudinal focusing – large amplitudes

We go back to the derivation of the synchrotron oscillation and use the longitudinal coordinates and the energy loss per turn:

$$h\omega_{rev}\tau = \varphi, \qquad h\omega_{rev}t_s = \varphi_s, \qquad U_s = e\hat{V} \sin \varphi_s.$$

We will no longer assume that we have small amplitudes but take the change in energy per turn

$$\delta E = e\hat{V} \sin(h\omega_{rev}(t_s + \tau)) - U_s$$

and use a smooth approximation and replace the localized change in energy in the cavity by one that is distributed around the ring. We obtain a system of non-linear differential equations:

$$\dot{\epsilon} \approx \frac{\omega_{rev}}{2\pi} \frac{\delta E}{E_e} = \frac{\omega_{rev} e \hat{V} \left(\sin(h\omega_{rev}(t_s + \tau)) - \sin \varphi_s\right)}{2\pi E_e}, \qquad \dot{t} = \alpha_c \epsilon.$$

We put $\varphi = h\omega_{\mathrm{rev}}\tau$, express $\dot{\epsilon}$ with φ, multiply the above equation by $\dot{\varphi}$, and integrate once over t to obtain

$$\frac{1}{2}\dot{\varphi}^2 - \omega_{\mathrm{rev}}^2 \frac{\alpha_c h e \hat{V}}{2\pi E_e}\left(\cos(\varphi_s + \varphi) - \cos\varphi_s + \varphi\sin\varphi_s\right) = \frac{1}{2}\hat{\dot{\varphi}}^2 = \text{constant}.$$

This expression gives the phase-space trajectories $\epsilon(\varphi)$ of the particles executing a phase oscillation. We can use it to determine the maximum amplitude of the deviation in energy which still gives stable oscillations. The function $f(\varphi) = -(\cos(\varphi_s + \varphi) - \cos\varphi_s + \varphi\sin\varphi_s)$ gives the potential provided by the RF voltage for the particles shown in the central part of Fig. 13.18. It has a maximum at $\varphi_1 = (\pi - 2\varphi_s)$ that determines the energy acceptance of

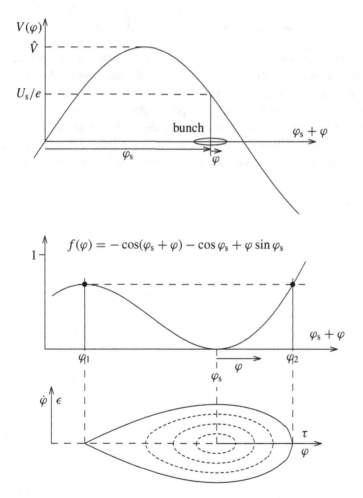

Fig. 13.18. Large amplitude oscillation and energy acceptance.

the RF system using $\sin \varphi_s = U_s/(e\hat{V})$:

$$f(\varphi)_{\max} = 2\left(\sqrt{1 - \left(\frac{U_s}{e\hat{V}}\right)^2} - \frac{U_s}{e\hat{V}}\arccos\left(\frac{U_s}{e\hat{V}}\right)\right)$$

$$\epsilon_{\max}^2 = \left(\frac{\Delta E}{E_e}\right)_{\max}^2 = \frac{e\hat{V}}{\pi\alpha_c h E_e}2(\cos\varphi_s - \sin\varphi_s \arccos(\sin\varphi_s))$$

$$= 2\frac{e\hat{V}}{\pi\alpha_c h E_e}\left(\sqrt{1 - \left(\frac{U_s}{e\hat{V}}\right)^2} - \frac{U_s}{e\hat{V}}\arccos\left(\frac{U_s}{e\hat{V}}\right)\right).$$

For a small energy loss $U_s/(e\hat{V}) = \sin\varphi_s \ll 1$ we can approximate

$$\epsilon_{\max}^2 \approx \frac{2e\hat{V}}{\pi\alpha_c h E_e}\left(1 - \frac{\pi}{2}\frac{U_s}{e\hat{V}} + \frac{1}{2}\left(\frac{U_s}{e\hat{V}}\right)^2\cdots\right).$$

The RF voltage, the potential function $f(\varphi)$, and the phase-space trajectories are shown in Fig. 13.18. The latter are shown as solid lines for the maximum stable energy accepted by the system and as dashed lines for smaller deviations. Their form approaches an ellipse for smaller amplitudes, for which a linear approximation can be used.

The electrons in a ring have a Gaussian energy distribution with RMS value σ_ϵ. The RF bucket has to contain this distribution and the energy acceptance has to be larger, $\epsilon_{\max} \geq 7\sigma_\epsilon$, in order to minimize losses of the particles in the tails and to assure that there is a good beam life time.

14

Effects of radiation on the electron beam

14.1 The energy loss

In the bending magnet of field B an ultra-relativistic particle of energy E_e has a trajectory of curvature $1/\rho = eB/p$ and emits synchrotron radiation of power (3.13):

$$P_s = \frac{2cr_0m_0c^2\gamma^4}{3\rho^2} = \frac{2c^3r_0e^2E_e^2B^2}{3(m_0c^2)^3}. \tag{14.1}$$

To obtain the energy loss U_s per turn we have to integrate this over the time. We assume that the relative change in particle energy during one revolution is small. However, the radiated power varies around the ring since it depends on the local field. For this reason the energy loss U_s during one turn is often the relevant quantity. Using an ultra-relativistic approximation with $dt \approx ds/c$,

$$U_s = \int P(t)\,dt \approx \frac{1}{c}\oint P(s)\,ds = \frac{2c^2r_0e^2E_e^2}{3(m_0c^2)^3}\oint B^2(s)\,ds \tag{14.2}$$

$$= \frac{2r_0m_0c^2\gamma^4}{3}\oint \frac{ds}{\rho^2(s)} = \frac{2r_0m_0c^2\gamma^4}{3}I_{s2},$$

where we use the synchrotron radiation integral I_{s2}, which was introduced in Chapter 5, (5.6),

$$I_{s2} = \oint \frac{ds}{\rho^2(s)} = \oint \left(\frac{eB_0(s)c}{m_0c^2\gamma}\right)^2 ds, \tag{14.3}$$

which represents an integration over the square of the curvature or of the field *on the nominal orbit*. In a lattice consisting of dipole and quadrupole magnets, the circulating electrons are usually very close to axes of the latter, where the magnetic field is small. Therefore we can neglect the radiation produced in the quadrupoles. Assuming that there are identical bending magnets producing a curvature $1/\rho$, we have

$$I_{s2} = \frac{2\pi}{\rho} \quad \text{and} \quad U_s = \frac{4\pi r_0m_0c^2\gamma^4}{3\rho} = \frac{4\pi r_0ceE_e^3B}{3(m_0c^2)^3}.$$

In this approach, a magnetic field is assumed to have a constant value over the full length with an abrupt drop to zero at the end, neglecting the effects of fringe fields. This tends to overestimate the energy loss. On the basis of measurements of the magnetic field $B(s)$ along the orbit, a maximum value $\hat{B}$ in the center and an effective length ℓ_1 are defined and used for lattice calculations:

$$\hat{B}\ell_1 = \int B(s)\,ds.$$

Using these parameters also to determine the energy loss with $\ell_1 \hat{B}^2$ in the presence of fringe fields results in a value for U_s that is too large, since

$$\ell_1 \hat{B}^2 \geq \int B^2(s)\,ds = \ell_2 B_2^2. \tag{14.4}$$

At this point we would like to clarify the difference between the power P_s given by (14.1), which is radiated by a particle while it traverses a bending magnet with field B or curvature $1/\rho$, and the power averaged over one turn. In a ring that has only bending magnets of the same curvature $1/\rho$ and thin quadrupoles without any field-free straight sections the angular velocity is everywhere ω_0 and the time $T_{rev} = T_0 = 2\pi/\omega_0$. A ring that has bending magnets of the same field strength and also field-free straight sections, called an isomagnetic ring, will have a longer revolution time $T_{rev} > T_0$ and a smaller frequency of revolution $\omega_{rev} = 2\pi/T_{rev} < \omega_0$. Therefore we obtain the following relations for the instantaneous and averaged power:

$$P_s = \frac{2cr_0m_0c^2\gamma^4}{3\rho^2} = \frac{\omega_0}{2\pi}U_s, \qquad \langle P \rangle = \frac{\omega_{rev}}{2\pi}U_s = \frac{\omega_{rev}}{\omega_0}P_s \leq P_s.$$

We can easily generalize this for non-isomagnetic rings.

14.2 The radiation damping

14.2.1 Introduction

The power radiated as synchrotron radiation reduces not only the energy of the nominal electron motion but also that of the synchrotron and betatron oscillations, while the RF system restores the energy of the former. This produces damping. In the following treatment we include the effect of the RF system which replaces the lost energy implicitly by keeping the average longitudinal particle momentum p_0 or the energy E_e constant. This method is based on a smooth replacement of the energy, which is justified as long as the energy loss in one revolution or between cavities is small compared with the energy itself.

In calculating the radiation damping we need the change of the energy loss U_s per turn due to a deviation dE of the particle energy. In the calculation of the energy loss carried out in the previous section we assumed that the particle has the nominal energy and circulates on the nominal orbit, which usually goes through the centers of the quadrupoles. Now we need the change of this loss with an energy deviation, and later, also with a betatron displacement. The energy loss is proportional to E_e^2, which leads to a direct effect. An energy deviation also changes the orbit at locations of finite dispersion. In cases in which the magnets have a gradient this leads directly to a change of the field seen by the particle:

$$dB = \frac{dB}{dx} dx = \frac{dB}{dx} \frac{dE}{E_e} D_x.$$

We obtain the change of the radiated power (14.1) with the energy deviation

$$dP = 2P_s \left(\frac{dE}{E_e} + \frac{dB}{B} \right) = 2P_s \left(1 + \frac{D_x}{\rho} \frac{\rho}{B} \frac{dB}{dx} \right) \frac{dE}{E_e}.$$

This is often written in a more compact form by introducing a normalized gradient, called the field index n_B,

$$n_B = -\frac{\rho}{B} \frac{dB}{dx} = -\rho^2 K_f, \tag{14.5}$$

where K_f is the focusing parameter (13.3). We obtain for the power deviation

$$dP_s = 2P_s \left(1 - \frac{D_x n_B}{\rho} \right) \frac{dE}{E_e}.$$

In this change of the radiated power with energy deviation we include the effect $dP \propto 2B\,dB$ due to the change of the magnetic field seen by the electron in dipole magnets with a gradient. However, we do not include quadrupole magnets because they have a vanishing field on the axis and an orbit deviation would lead only to a second-order effect.

To obtain the resulting change dU_s of the energy loss per turn we have to integrate the power over time. However, this integration follows the displaced orbit having a path-length element ds' different from the nominal ds. Inside a magnet with bending radius ρ the relation between them is

$$ds' = \left(1 + \frac{x}{\rho} \right) ds = \left(1 + \frac{D_x}{\rho} \frac{dE}{E_e} \right) ds.$$

However, for the integration around the ring, the shape of the ends of the magnet has an important effect. In a sector magnet of nominal deflecting angle $d\varphi$ the full path-length is $d\varphi\,(\rho + x)$, whereas for a rectangular magnet of length L the path-length is independent of the radial displacement. To accommodate this difference, we write the path-length element with a factor κ being about 1 for a sector and 0 for a rectangular magnet,

$$ds' = \left(1 + \kappa \frac{D_x}{\rho} \frac{dE}{E_e} \right) ds. \tag{14.6}$$

We obtain the change in U_s by integrating over $dt = ds/c$:

$$dU_s = \frac{1}{c} \oint P_s(s) \left(2 + \frac{(\kappa - 2n_B)D_x}{\rho} \right) \frac{dE}{E_e} ds. \tag{14.7}$$

We express the power in the form

$$P_s = \frac{2cr_0 m_0 c^2 \gamma^4}{3\rho^2(s)}$$

and obtain for the change of energy loss with energy

$$\frac{dU_s}{dE} = \frac{2r_0 m_0 c^2 \gamma^4}{3E_e} \oint \left(\frac{2}{\rho^2(s)} + \frac{(\kappa - 2n_B(s))D_x(s)}{\rho^3(s)} \right) ds.$$

We use the second synchrotron-radiation integral I_{s2} and introduce a new one [42],

$$I_{s4} = \oint \frac{(1 - 2n_B)D_x}{\rho^3} ds.$$

We set $\kappa = 1$ to obtain

$$\frac{dU_s}{dE} = \frac{2r_0 m_0 c^2 \gamma^4}{3E_e}(2I_{s2} + I_{s4}) = \frac{U_{s0}}{E_e}\left(2 + \frac{I_{s4}}{I_{s2}}\right). \tag{14.8}$$

14.2.2 The damping of synchrotron oscillations

The damping rate for synchrotron oscillation was derived in the previous chapter:

$$\omega_s = \omega_{rev}\sqrt{\frac{h\alpha_c e\hat{V}\cos\phi_s}{2\pi E_e}}, \qquad \alpha_\epsilon = \frac{1}{2}\frac{\omega_{rev}}{2\pi}\frac{dU_s}{dE}. \tag{14.9}$$

For a homogeneous magnet and neglecting path-length effects, we find

$$\alpha_\epsilon = \frac{\omega_{rev}}{2\pi}\frac{U_{s0}}{E_e} = \frac{U_{s0}}{T_{rev}E_e} = \frac{\langle P \rangle}{E_e}. \tag{14.10}$$

The damping time $1/\alpha_\epsilon$ is just the time it takes to radiate the total particle energy.

For the general case involving gradient magnets and path-length effects we use the more accurate derivative of the energy loss (14.8) in (14.9) and obtain the longitudinal damping rate

$$\alpha_\epsilon = \frac{U_{s0}}{2T_{rev}E_e}(2 + I_{s4}/I_{s2}) = \frac{U_{s0}}{2T_{rev}E_e}(2 + \mathcal{D}) = \frac{r_0\gamma^3}{3T_{rev}}(2I_{s2} + I_{s4}), \tag{14.11}$$

where the parameter $\mathcal{D} = I_{s4}/I_{s2}$ is often used in the literature.

14.2.3 The damping of vertical betatron oscillations

We assume that we have a plane storage ring without vertical bending and no coupling. The damping of vertical betatron oscillations is caused by the fact that the synchrotron radiation is emitted in the direction of the particle momentum $\mathbf{p}$ and reduces both its components p_s and p_x while the RF-cavity voltage has a longitudinal field and increases only the longitudinal momentum component p_s, as illustrated in Fig. 14.1. In this process we have a reduction of the transverse momentum:

$$\frac{\delta p_y}{p_0} = \delta y' = \frac{\delta p_s}{p_0}y' = -\frac{\delta E_\gamma}{E_e}y'.$$

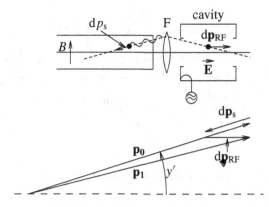

Fig. 14.1. Damping of vertical betatron oscillation.

Since the change in energy of the particle is always negative, representing a loss, we give also the positive energy δE_γ of the photon in order to clarify the situation.

We consider now a short element ds of the trajectory, where an energy element $\delta E = P\,\delta s/c$ is lost due to synchrotron radiation, and replaced in the forward direction by the RF system. Using a paraxial approximation, we have $p_s \approx |\mathbf{p}| \approx E_e/c$ and $p_y \approx y'|\mathbf{p}| \approx y'E_e/c$, from which we obtain for the change of the electron coordinates y and y'

$$\delta y = 0, \qquad \delta y' = -\frac{\delta E_\gamma}{E_e}y',$$

which results in a first-order change of the emittance:

$$\epsilon_y = \gamma_y y^2 + 2\alpha_y yy' + \beta_y y'^2$$
$$\delta\epsilon_y = 2\alpha_y y\,\delta y' + \beta_y 2y'\,\delta y' = -2\frac{\delta E_\gamma}{E_e}(\alpha_y yy' + \beta_y y'^2).$$

To find the average change in emittance over many turns, we use the expressions (13.14),

$$\langle w^2 \rangle = \tfrac{1}{2}\epsilon\beta_w, \qquad \langle ww' \rangle = -\tfrac{1}{2}\epsilon\alpha_w, \qquad \langle w'^2 \rangle = \tfrac{1}{2}\epsilon\gamma_w, \qquad (14.12)$$

and obtain

$$\langle\delta\epsilon_y\rangle = \epsilon_y\big(\alpha_y^2 - \beta_y\gamma_y\big)\frac{\delta E_\gamma}{E_e} = -\epsilon_y\frac{\delta E_\gamma}{E_e}.$$

This differential loss has to be integrated over one turn, resulting in

$$\frac{d\epsilon_y}{dt} = -\epsilon_y\frac{U_s}{T_{rev}E_e} \quad \text{or} \quad \epsilon_y(t) = \epsilon_{y0}\exp\left(-\frac{U_s}{T_{rev}E_e}t\right).$$

The amplitude of the betatron oscillation is proportional to $\sqrt{\epsilon_y}$ and has therefore the damping rate

$$\alpha_y = \frac{U_s}{2T_{rev}E_e}. \qquad (14.13)$$

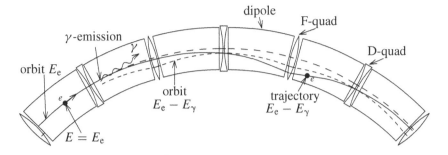

Fig. 14.2. Horizontal damping and quantum excitation

14.2.4 The damping of horizontal betatron oscillations

The case of horizontal betatron oscillations is more complicated due to the effect of the dispersion. An electron that radiates a photon loses a certain amount of energy δE and has a new equilibrium orbit displaced by $\delta x_\beta = D_x \, \delta E / E_e$. Although the actual radial position of the electron has not changed, it will start a betatron oscillation around its new, displaced equilibrium orbit. If there is also a derivative of the dispersion, the electron trajectory will have an angle with respect to the new off-energy orbit and oscillate around it. This mechanism is illustrated in Fig. 14.2.

To calculate this effect we take a location on the orbit where the electron has the horizontal displacement due to a betatron and an energy motion

$$x = x_\beta + x_\epsilon, \qquad x' = x'_\beta + x'_\epsilon.$$

In a path element $\mathrm{d}s'$ along its trajectory the electron radiates the relative energy

$$\frac{\delta E}{E} = P(s')\frac{\delta s'}{c} = -\frac{\delta E_\gamma}{E_e},$$

resulting in change of the coordinates:

$$\delta x = \delta x_\beta + \delta x_\epsilon = 0, \qquad \delta x' = \delta x'_\beta + \delta x'_\epsilon = -x'_\beta \frac{\delta E_\gamma}{E_e}.$$

We assume that we have a dispersion D_x and derivative D'_x at the location of emission, resulting in a change of the energy-oscillation coordinates:

$$\delta x_\epsilon = D_x \frac{\delta E}{E_e} = -D_x \frac{\delta E_\gamma}{E_e}, \qquad \delta x'_\epsilon = D'_x \frac{\delta E}{E_e} = -D'_x \frac{\delta E_\gamma}{E_e}.$$

This gives a change of the betatron coordinates, assuming that the electron had the nominal energy before:

$$\delta x_\beta = -D_x \frac{\delta E}{E_e} = D_x \frac{\delta E_\gamma}{E_e}, \qquad \delta x'_\beta = (x'_\beta - D'_x)\frac{\delta E}{E_e} = -\left(x_\beta - D'_x\right)\frac{\delta E_\gamma}{E_e}.$$

The resulting change of the electron emittance is

$$\delta\epsilon_x = 2(\gamma_x x_\beta \, \delta x_\beta + \alpha_x(x'_\beta \, \delta x_\beta + x_\beta \, \delta x'_\beta) + \beta_x x'_\beta \, \delta x'_\beta)$$
$$= -2\frac{\delta E_\gamma}{E_e}\left((\alpha_x x_\beta x'_\beta + \beta_x x'^2_\beta) - (\gamma_x D_x x_\beta + \alpha_x(D_x x'_\beta + D'_x x_\beta) + \beta_x D'_x x'_\beta)\right).$$

$$(14.14)$$

Next we discuss the energy-loss element δE. On the nominal orbit it would just be $\delta E = P_s(s) \, ds/c$; however, for the trajectory including a betatron-oscillation excursion x_β there is a correction for a possible magnetic-field gradient and one for the path-length (14.6) and hence we have to take the expression

$$dE = P_s\left(1 + \frac{1}{B}\frac{dB}{dx}x_\beta\right)\frac{\delta s'}{c} = \frac{P_s}{c}\left(1 - n_B\frac{x_\beta}{\rho}\right)\left(1 + \kappa\frac{x_\beta}{\rho}\right)\delta s$$
$$\approx \frac{P_s}{c}\left(2 + \frac{\kappa - 2n_B}{\rho}x_\beta\right)ds,$$

where P_s is now the power corresponding to the nominal orbit. The above expression gives the dependence of the energy loss on the betatron excursion x_β and has a similar form to (14.7), where this dependence on a deviation in energy was investigated.

To obtain the change in emittance due to this energy loss occurring in a nominal path element ds we substitute it into (14.14):

$$\delta\epsilon_x = \frac{2P_s}{cE_e}\left[\left(2 + \frac{\kappa - 2n_B}{\rho}x_\beta\right)\right.$$
$$\left. \times \left((\alpha_x x_\beta x'_\beta + \beta_x x'^2_\beta) - (\gamma_x D_x x_\beta + \alpha_x(D_x x'_\beta + D'_x x_\beta) + \beta_x D'_x x'_\beta)\right)\right]ds.$$

The phase of the betatron oscillation will be different in successive revolutions and we average over the square terms in x_β and x'_β using the expressions (14.12) and the relation $\beta_x\gamma_x - \alpha_x^2 = 1$ and obtain vanishing averages for the third powers:

$$\delta\epsilon_x = -\epsilon_x\frac{P_s}{cE_e}\left(1 - (\kappa - 2n_B)\frac{D_x}{\rho}\right)\delta s.$$

To obtain the total change in emittance during one revolution we have to integrate over ds. We separate in $P_s(s)$ the s-dependent parameter $\rho(s)$,

$$\frac{P_s}{E_e} = \frac{2r_0 c\gamma^3}{3}\frac{1}{\rho^2(s)},$$

and obtain for $\kappa = 1$

$$\frac{d\epsilon_x}{T_{\text{rev}}} = -\epsilon_x\frac{2r_0\gamma^3}{3T_{\text{rev}}}\oint\left(\frac{1}{\rho^2} - \frac{(1 - 2n_B)D_x}{\rho^3}\right)ds = -\epsilon_x\frac{2r_0\gamma^3}{3T_{\text{rev}}}(I_{s2} - I_{s4}).$$

Assuming that there is a small relative change of the emittance per turn, we make a smooth approximation:

$$\frac{d\epsilon_x}{dt} = -\epsilon_x \frac{2r_0\gamma^3}{3T_{\mathrm{rev}}}(I_{s2} - I_{s4}).$$

The emittance is exponentially damped at a rate $2r_0\gamma^3(I_{s2} + I_{s4})/(3T_0)$. The betatron amplitude is proportional to $\sqrt{\epsilon_s}$ and has the damping rate

$$\alpha_{\mathrm{h}} = \frac{r_0\gamma^3 I_{s2}}{3T_{\mathrm{rev}}}\left(1 - \frac{I_{s4}}{I_{s2}}\right) \stackrel{*}{=} \frac{U_{s0}}{2T_0 E_{\mathrm{e}}}(1 - \mathcal{D}).$$

14.2.5 The sum of the damping rates

We sum the damping rates of the three modes of electron oscillation and find

$$\alpha_\epsilon + \alpha_v + \alpha_{\mathrm{h}} = 2\frac{U_{s0}}{T_{\mathrm{rev}}E_{\mathrm{e}}}. \tag{14.15}$$

This sum depends only on the energy loss per turn, the energy, and the revolution time and is not affected by the details of the ring lattice. This result is valid for a very general ring, which can also contain vertical deflections and dispersion, as shown in [73]. Since the damping is based on the fact that the synchrotron radiation reduces on average the nominal energy as well as the one contained in the oscillations while the RF system only restores the first one, the above sum rule agrees with intuition. Owing to the importance of this, one introduced in the literature damping partition numbers D_i for the three modes of oscillation. For a plane ring they are

$$J_\epsilon = 2 + I_{s4}/I_{s2}, \qquad J_y = 1, \qquad J_x = 1 - I_{s4}/I_{s2}, \qquad \sum_i J_i = 4.$$

We treated here a ring lying in the horizontal plane without any vertical bending. In this case already the sum of two damping rates is constant, $D_s + D_x = 3$. For a ring having only homogeneous dipole magnets and quadrupoles the synchrotron integral I_{s4} is

$$I_{s4} = \oint \frac{D_x}{\rho^3}\,ds,$$

neglecting path-length effects. Since, in a strong focusing ring, $D_x \ll \rho$, we have $I_{s4} \ll I_{s2}$ and obtain for the damping partition numbers

$$J_\epsilon \approx 2, \qquad J_y \approx J_x \approx 1.$$

In bending magnets with a positive gradient $dB/dx > 0$, i.e. $n_B < 0$, the field is larger at positive deviations x from the central orbit, leading to more power being radiated. Since most rings have positive dispersion, a particle with a positive energy deviation radiates more and the damping of the oscillation in energy is increased. For the horizontal betatron oscillation an energy loss occurring on the outside, $x_\beta > 0$, will move the off-energy equilibrium orbit to the inside by $-D_x \frac{\delta E_\gamma}{E_e}$ and the betatron amplitude is increased. For a loss occurring on the inside, $x_\beta < 0$, the situation is reversed and the betatron amplitude decreases.

14.3 The quantum excitation of oscillations

14.3.1 Introduction

We have calculated the damping of particle oscillations in storage rings by evaluating the effect of an infinitesimal energy loss on these oscillations. In reality the energy is emitted in the form of photons of finite energies $\hbar\omega$. This will shock-excite oscillations, which are damped but excited again in a statistical manner. As a result there are oscillations in energy and betatron oscillations of finite amplitudes in a storage ring, which manifest themselves as non-vanishing energy spread and emittance.

To calculate the resulting energy spread and emittance we use Campbell's theorem [74], which applies to a statistical excitation of a damped oscillation. We consider a weakly damped oscillator that has a response to a pulse excitation $a\,\delta(t)$ at the time $t = 0$ of the form

$$x(t) = a e^{-\alpha t} \cos(\omega_r t)$$

and calculate the integral over the square of the normalized response,

$$g^2 = \frac{1}{a^2} \int_0^\infty x^2(t)\,dt = \int_0^\infty e^{-2\alpha t} \cos^2(\omega_r t) \approx \frac{1}{4\alpha},$$

for $\alpha \ll \omega_r$.

We assume now that the pulse has a variance $\langle a^2 \rangle$ in size and occurs randomly at an average rate $\dot{n}$. According to Campbell's theorem, the resulting variance of the response is

$$\langle x^2 \rangle = \dot{n}\langle a^2 \rangle g^2 = \frac{\dot{n}\langle a^2 \rangle}{4\alpha}. \tag{14.16}$$

This result looks like a reasonable intuition. A detailed discussion of it is given in [73].

In the following we apply this to a storage ring where synchrotron radiation is emitted in long magnets and calculate the energy spread and emittance. Such a ring often also contains undulators but they usually do not contribute significantly to the energy loss and therefore to quantum excitation. The synchrotron-radiation spectrum for long magnets has been evaluated in Chapter 5. The number of photons $\dot{n}$ radiated per unit time and the variance of their energy are expressed in (5.45) by the critical energy ϵ_c and the total power

$$\dot{n} = \frac{15\sqrt{3}}{8}\frac{P_s}{E_{\gamma c}}, \qquad \langle E_\gamma^2 \rangle = \frac{11}{27}E_{\gamma c}^2, \qquad E_{\gamma c} = \frac{3ch\gamma^3}{4\pi|\rho|} = \frac{3m_0c^2\lambda_{\text{Comp}}\gamma^3}{4\pi|\rho|}, \tag{14.17}$$

where $\lambda_{\text{Comp}} = h/(m_0 c) = 2.426 \times 10^{-12}$ m is the Compton wavelength (5.43). Since $E_{\gamma c}$ is always positive while the curvature $1/\rho$ can change sign in a reversed bend, we use the absolute sign for the latter.

In calculating the effect of quantum excitation we assume that the particle has no initial deviation in energy, displacement, and angle from the nominal values and calculate the effect of the photon emission on these parameters. This is different from the investigation of the radiation damping, for which an initial oscillation was assumed and its increase or decrease due to the radiation was investigated.

14.3.2 The energy spread

Synchrotron oscillations are directly excited by the energy loss due to photon emissions, and we obtain the resulting energy spread of the electron beam from (14.16) on the basis of Campbell's theorem [75]. The number of photons and the variance of the photon energy depend on the local magnetic field. We average their product over one revolution and divide the result by the longitudinal damping rate α_ϵ, which is itself an average over the ring. With (14.11) for the damping rate, (14.1) for the radiated power, and (14.17) for the photon flux and RMS energy we obtain

$$\frac{\langle dE^2 \rangle}{E_e^2} = \frac{\langle \dot{n}\langle \epsilon^2 \rangle \rangle}{4\alpha\epsilon E_e^2} = \frac{55\sqrt{3}}{288} \frac{1}{E_e^2 \alpha_\epsilon c T_{\text{rev}}} \oint P_s E_{\gamma c} \, ds = \frac{55\sqrt{3}}{96} \frac{\lambda_{\text{Comp}} \gamma^2 I_{s3}}{2\pi(2I_{s2} + I_{s4})},$$

where we introduce another synchrotron radiation integral:

$$I_{s3} = \oint \frac{1}{|\rho|^3} \, ds.$$

It should be noted that field-free straight sections do reduce the damping rate and the average number of photons emitted in the same way and have therefore no effect on the energy spread. The photon emission produces directly an energy spread, which results, through the longitudinal focusing, also in a finite bunch length σ_τ. It is given by (13.25):

$$\sigma_\tau = \frac{\alpha_\epsilon}{\omega_s} \sigma_\epsilon = \frac{1}{\omega_{\text{rev}}} \sqrt{\frac{2\pi E_e \alpha_c}{h e \hat{V} \cos \phi_s}} \sigma_\epsilon.$$

One sometimes calls the product of RMS values of the electron excursion in time σ_τ and energy $\sqrt{\langle dE^2 \rangle}$ the longitudinal emittance:

$$\epsilon_s = \sigma_\tau \sqrt{\langle dE^2 \rangle} = E_e \sigma_\tau \sigma_\epsilon = E_e \frac{\alpha_c}{\omega_s} \sigma_\epsilon^2 = \frac{E_e}{\omega_{\text{rev}}} \sqrt{\frac{2\pi \alpha_c E_e}{h e \hat{V} \cos \phi_s}} \sigma_\epsilon^2.$$

14.3.3 The horizontal emittance

Emission of a photon of energy δE at a location of dispersion D_x and derivative D'_s results in a horizontal betatron oscillation, with initial parameters for an electron originally on the nominal orbit of

$$\delta x_\beta = D_x \frac{\delta E_\gamma}{E}, \qquad \delta x' = D'_x \frac{\delta E_\gamma}{E},$$

resulting in an emittance given by

$$\delta\epsilon_x = \gamma_x\,\delta x^2 + 2\alpha_x\,\delta x\,\delta x' + \beta x'^2 = \left(\gamma_x D_x^2 + 2\alpha_x D_x D_x' + \beta_x D_x'^2\right)\left(\frac{\delta E_\gamma}{E}\right)^2.$$

The expression in the parentheses,

$$\mathcal{H} = \gamma_x D_x^2 + 2\alpha_x D_x D_x' + \beta_x D_x'^2, \tag{14.18}$$

is usually called the emittance function $\mathcal{H}$ and can be calculated from the beam optics.

This emittance is proportional to the square of the betatron amplitude and we can apply Campbell's theorem directly. The number of photons emitted per unit time and the variance of their energy are the same as for the previous case of the energy spread and we obtain

$$\begin{aligned}
\epsilon_x &= \frac{55\sqrt{3}}{288}\frac{1}{cE_e^2\alpha_x T_{\text{rev}}}\oint P_s\left(\gamma_x D_x^2 + 2\alpha_h D_x D_x' + \beta_x D_x'^2\right)E_{\gamma c}\,ds \\
&= \frac{55\sqrt{3}}{96}\frac{\lambda_{\text{Comp}}\gamma^2}{2\pi(I_{s2}-I_{s4})}\oint\frac{\gamma_x D_x^2 + 2\alpha_x D_x D_x' + \beta_x D_x'^2}{|\rho|^3}\,ds.
\end{aligned}$$

We introduce one more synchrotron-radiation integral [42],

$$I_{s5} = \oint\frac{\gamma_x D_x^2 + 2\alpha_x D_x D_x' + \beta_x D_x'^2}{|\rho|^3}\,ds = \oint\frac{\mathcal{H}}{|\rho|^3}\,ds,$$

and obtain for the equilibrium emittance

$$\epsilon_x = \frac{55\sqrt{3}}{96}\frac{\hbar}{m_0 c}\gamma^2\frac{I_{s5}}{I_{s2}-I_{s4}} = \frac{55\sqrt{3}}{96}\frac{\lambda_{\text{Comp}}\gamma^2}{2\pi}\frac{I_{s5}}{I_{s2}-I_{s4}}. \tag{14.19}$$

14.3.4 The vertical emittance

In an ideal flat ring there is no vertical dispersion and a photon emitted in the forward direction does not excite any vertical betatron oscillations. However, the photons have a finite vertical opening angle, which can give rise to a small excitation and a resulting vertical emittance. We estimate the magnitude of this effect. A photon emitted at a vertical angle ψ gives the electron a deflection

$$\delta y' = -\frac{\delta p_{\gamma y}}{p_0} = -\frac{\hbar\omega\psi}{E_e},$$

which results in a change in emittance of

$$\delta\epsilon_y = \beta_y\,\delta y'.$$

This occurs statistically and results, together with the radiation damping, in a vertical equilibrium emittance, which is calculated using Campbell's theorem:

$$\epsilon_y = \frac{\beta_y\dot{n}\langle\delta y'^2\rangle}{4\alpha_v}.$$

We assume here that we have a ring without straight sections having a constant average beta function. For the number of photons per second (5.45) and the damping rate (14.13) we obtain

$$\dot{n} = \frac{5\sqrt{3}}{12} \frac{e^2 \gamma}{\epsilon_0 h \rho}, \qquad \alpha_v = \frac{U_s}{2 T_0 E_e}$$

and the variance of the vertical deflection (5.24) is

$$\langle \delta y'^2 \rangle = \frac{8}{27} \frac{(\hbar \omega_c)^2}{\gamma^2 E_e^2} = \frac{1}{6\pi^2} \frac{\lambda_{\text{Comp}}^2 \gamma^2}{\rho^2}.$$

This gives for the vertical emittance

$$\epsilon_y = \frac{5\sqrt{3}}{12} \frac{h}{2\pi m_0 c} \frac{\beta_y}{\rho} = \frac{5\sqrt{3}}{24\pi} \frac{\beta_y}{\rho} \lambda_{\text{Comp}}.$$

With $\lambda_{\text{Comp}} = 2.436 \times 10^{-12}$ m and β_y of the same order as ρ this emittance is extremely small.

In a practical ring, magnetic and alignment errors lead to some vertical dispersion and to coupling between horizontal and vertical betatron oscillations. The resulting vertical emittance is typically less than 1% of the horizontal one in a well-corrected ring. This is small, but still much larger than the emittance created by the vertical opening angle of the radiation.

14.4 A summary of the effects of radiation on the electron beam

We summarize the effects of synchrotron radiation on the damping rate of betatron and synchrotron oscillations, on the energy spread, and on the horizontal emittance in a plane ring. We introduced convenient synchrotron-radiation integrals that can be calculated from the ring lattice. They represent an integration on the nominal orbit around the ring, involving the curvature $1/\rho(s)$ or its absolute value, the horizontal dispersion $D_x(s)$ and its derivative $D_x'(s)$, the horizontal beta function $\beta_x(s)$, its normalized derivative $\alpha_x(s) = -\beta_x'(s)/2$, and the function $\gamma_x(s) = (1 + \alpha_x^2(s))/\beta_x(s)$:

$$\boxed{\begin{aligned} I_{s2} &= \oint \frac{1}{\rho^2(s)}\, ds, \qquad I_{s3} = \oint \frac{1}{|\rho(s)|^3}\, ds \\[2mm] I_{s4} &= \oint \frac{(1 - 2 n_B(s)) D_x(s)}{|\rho(s)|^3}\, ds, \qquad I_{s5} = \oint \frac{\mathcal{H}(s)}{|\rho(s)|^3}\, ds. \end{aligned}}$$

$$(14.20)$$

Here we also use the field index n_B to describe a gradient in the dipole magnets and the emittance function $\mathcal{H}$:

$$n_B(s) = -\frac{\rho}{B}\frac{dB}{dx}$$

$$\mathcal{H}(s) = \gamma_x(s)D_x^2(s) + 2\alpha_x(s)D_x(s)D_x'(s) + \beta_x(s)D_x'^2(s).$$

The energy loss U_s per turn of one particle on the nominal orbit is

$$U_{s0} = \frac{2r_0 m_0 c^2 \gamma^4}{3} I_{s2}.$$

In a plane ring the damping rates of the longitudinal, vertical, and horizontal oscillations are

$$\begin{aligned}
\alpha_\epsilon &= \frac{r_0\gamma^3}{3T_{\mathrm{rev}}}(2I_{s2} + I_{s4}) = \frac{U_{s0}}{2E_e\gamma T_{\mathrm{rev}}}J_\epsilon \\[4pt]
\alpha_y &= \frac{r_0\gamma^3}{3T_{\mathrm{rev}}}I_{s2} = \frac{U_{s0}}{2m_0c^2\gamma T_{\mathrm{rev}}}J_y \\[4pt]
\alpha_x &= \frac{r_0\gamma^3}{3T_{\mathrm{rev}}}(I_{s2} - I_{s4}) = \frac{U_{s0}}{2m_0c^2\gamma T_{\mathrm{rev}}}J_x \\[4pt]
\sum_i \alpha_i &= \frac{r_0\gamma^3}{3T_{\mathrm{rev}}}4I_{s2}\frac{2U_{s0}}{m_0c^2 T_{\mathrm{rev}}},
\end{aligned}$$

(14.21)

where we use the damping partition numbers $D_i = 4\alpha_i / \sum_i \alpha_i$:

$$J_\epsilon = 2 + I_{s4}/I_{s2}, \qquad J_y = 1, \qquad J_x = 1 - I_{s4}/I_{s2}, \qquad \sum_i J_i = 4. \qquad (14.22)$$

The square of the RMS (variance) energy spread and the horizontal emittance are

$$\begin{aligned}
\frac{\langle dE^2\rangle}{E_e^2} &= \frac{55\sqrt{3}}{96}\frac{\lambda_{\mathrm{Comp}}\gamma^2}{2\pi}\frac{I_{s3}}{2I_{s2} + I_{s4}} \\[4pt]
\epsilon_x &= \frac{55\sqrt{3}}{96}\frac{\lambda_{\mathrm{Comp}}\gamma^2}{2\pi}\frac{I_{s5}}{I_{s2} + I_{s4}}
\end{aligned}$$

(14.23)

with the Compton wavelength

$$\lambda_{\mathrm{Comp}} = \frac{h}{m_0 c} = 2.426 \times 10^{-12}\,\mathrm{m}.$$

To write the above equations more compactly, some authors [18] use the abbreviation

$$C_q = \frac{55\sqrt{3}}{96} \frac{\lambda_{\text{Comp}}}{2\pi} = 3.84 \times 10^{-13} \, \text{m}. \qquad (14.24)$$

14.5 Changing effects of radiation with wiggler magnets

In some machines wiggler magnets are installed in a straight section in order to change the damping rates, energy spread, and emittance [76]. The effect of such wigglers can be calculated with the expressions given in the previous section by including the wiggler fields in the synchrotron-radiation integrals. However, we would like to give some explicit expressions for practical situations. We consider a ring with all bending magnets having the same field and curvature B_r, $1/\rho_r$ being much smaller than the corresponding parameters of the wigglers B_w and $1/\rho_w$. On the other hand, the effective length of the wiggler is much smaller than that of a ring magnet, $L_w \ll 2\pi\rho_r$. The values of the two lengths should be properly defined by an integral over the squared field, as explained before in (14.4), but we ignore here this relatively small difference. We obtain for the synchrotron-radiation integrals

$$I_{s2} = \oint \frac{1}{\rho^2(s)} \, ds = \frac{2\pi\rho_r}{\rho_r^2} + \frac{L_w}{\rho_w^2} = I_{s2r}\left(1 + \frac{\rho_r^2}{\rho_w^2} \frac{L_w}{2\pi\rho}\right)$$

$$I_{s3} = \oint \frac{1}{|\rho_r|^3(s)} \, ds = \frac{2\pi\rho_r}{\rho_r^3} + \frac{L_w}{|\rho_w|^3} = I_{s3r}\left(1 + \frac{|\rho_r|^3}{|\rho_w|^2} \frac{L_w}{2\pi|\rho_r|}\right).$$

This gives the ratio of the total value and that of the ring alone for the energy loss per turn or the sum of the damping rates and also of the energy-spread variance,

$$\frac{U_s}{U_{sr}} = \frac{\Sigma_i \alpha_i}{\Sigma_i \alpha_{ir}} = 1 + \frac{I_{s2w}}{I_{s2r}} = 1 + \frac{B_w^2}{B_r^2} \frac{L_w}{2\pi\rho_r}$$

$$\frac{\langle dE^2 \rangle}{\langle dE^2 \rangle_r} = \frac{1 + \left|\frac{B_w}{B_r}\right|^3 \frac{L_w}{2\pi r}}{1 + \left(\frac{B_w}{B_r}\right)^2 \frac{L_w}{2\pi r}} \approx 1 + \left(\frac{B_w}{B_r}\right)^2 \frac{L_w}{2\pi r}\left(\left|\frac{B_w}{B_r}\right| - 1\right),$$

where I_{s2r}, I_{s3r}, U_{sr}, $\Sigma_i \alpha_i$, and $\langle dE^2 \rangle_r$ are the values for the ring alone. The approximation made for the energy spread is valid as long as the effect of the wiggler is not too large.

Of some interest are wigglers with a magnetic-field gradient, also called Robinson wigglers [73], to change the damping distribution. In a ring with a gradient wiggler the longitudinal damping partition number (14.22) is

$$J_\epsilon = 2 + \frac{I_{s4r} + I_{s4w}}{I_{s2r} + I_{s2w}} \approx 2 + \frac{I_{s4r}}{I_{s2r}} + \frac{I_{s4w}}{I_{s2r}}.$$

We neglect here the influence of this gradient wiggler on the total energy loss given by the integral I_{s2} but concentrate on the energy-dependent change in loss produced due to its strong gradient and its location at a large dispersion. The change of the damping partition number due to this gradient wiggler becomes

$$
\mathrm{d}J_\epsilon = \frac{I_{\mathrm{s4w}}}{I_{\mathrm{s2r}}} = \frac{(1 - 2n_b)D_x \rho_\mathrm{r}^2}{|\rho_\mathrm{w}|^3} \frac{L_\mathrm{w}}{2\pi \rho_\mathrm{r}} = \left(1 + 2\frac{\rho_\mathrm{w}}{B_\mathrm{w}} \frac{\mathrm{d}B_\mathrm{w}}{\mathrm{d}x}\right) \frac{D_x \rho_\mathrm{r}^2}{|\rho_\mathrm{w}|^3} \frac{L_\mathrm{w}}{2\pi \rho_\mathrm{r}}.
$$

The gradient of such a wiggler is usually made very large and we can approximate $\kappa - n_B \approx n_B$. We express the curvature in terms of the field and use the fact that, for a plane machine, the sum of the longitudinal and horizontal damping partition numbers is constant, $J_\epsilon + J_x = 3 \gg 1$, resulting in

$$
\mathrm{d}J_\epsilon = -\mathrm{d}J_x = \frac{D_x}{B_\mathrm{r}^2} \frac{\mathrm{d}B_\mathrm{w}^2}{\mathrm{d}x} \frac{L_\mathrm{w}}{2\pi \rho_\mathrm{r}}.
$$

A wiggler of modest length with a high field gradient located at a large dispersion can bring about a significant change of the damping partition numbers.

A wiggler magnet without gradient located at a large dispersion D_x, i.e. a large value of the function $\mathcal{H}$, will increase the emittance. This is of little interest for synchrotron-radiation sources, but is sometimes used to adjust the beam–beam interaction in colliding-beam facilities. On the other hand, we can locate this wiggler in a dispersion-free straight section with $\mathcal{H} \approx 0$, where it does not produce quantum excitations but increases the damping rate and thereby decreases the emittance. However, with increasing strength this wiggler creates its own dispersion D_x and derivative D_x', resulting in an increasing value of $\mathcal{H}$ and an increase of the emittance. This effect can be kept small by using a wiggler with many short periods.

15

Radiation emitted by many particles

15.1 Effects of the electron distribution on the radiation

15.1.1 Introduction

We calculated the synchrotron radiation emitted by a single particle on the nominal orbit. In other words, the position and angular direction of the moving electron were identical to the origin and axis we used to describe the synchrotron radiation. For an electron beam of finite emittance there is a spread in position and angle, with RMS values σ_x, σ_x' and σ_y, σ_y' around the nominal values. As a result, we observe radiation emitted at different angles and positions. As a first consequence, the emitted photon beam has an increased emittance, at least equal to the sum of the electron emittance and the natural photon emittance. Furthermore, since the frequency emitted by an electron depends on the angle of observation, the angular spread of the electrons tends to broaden the spectrum observed on the axis. This effect is negligible for standard synchrotron radiation but can be pronounced in the case of undulators with many periods.

Apart from a spread in angle and position, the electrons have also a spread in energy. The synchrotron-radiation frequency depends strongly on the electron energy. The spread of the latter leads again to a broadening of the synchrotron-radiation spectrum. This is again negligible for the radiation from long magnets but very pronounced for higher harmonics emitted in strong undulators.

We will discuss the above limitations imposed by having many electrons with different angles and positions. The same effects appear with a single electron being observed over many turns in a storage ring. Quantum excitation changes the electron parameters randomly over time, resulting in finite energy spread and emittance.

15.1.2 The radiation geometry in the case of a large electron emittance

We consider a situation in which the electron emittance is much larger than the natural one of the radiation. The photons are emitted basically in the direction of the electron momentum. They will therefore have the same angular spread but it will be slightly broadened by their natural opening angle, Fig. 15.1.

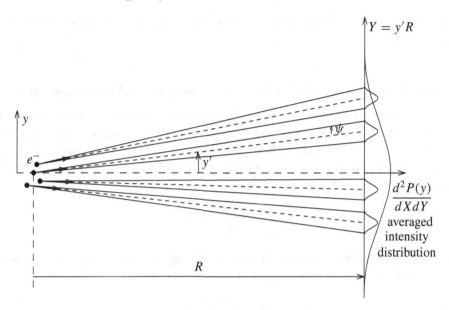

Fig. 15.1. The distribution of the radiation emitted by electrons with an angular spread.

A convenient method in which one uses the lattice functions $\beta_x(0)$, $\alpha_x(0)$, and $\gamma_x(0)$ at the source to optimize the photon-beam distribution for an experiment at a distance $s = R$ has been developed [77]. If these functions are known at the source, $s = 0$, we can calculate their propagation in a drift space free of focusing elements at a distance R (13.11):

$$\beta(s) = \beta_x(0) - 2\alpha_x(0)R + \gamma_x(0)R^2.$$

Apart from the finite opening angle we can treat the synchrotron radiation emitted in the forward direction like particles in a straight section and describe their propagation by using the same lattice functions. We can therefore define a beta function for the photon beam at the screen $\beta_\gamma(R) = \beta(R)$. Neglecting the finite opening angle of the radiation, the size of the photon beam σ_R on the screen is determined by the emittance of the electron beam ϵ and the beta function $\beta(R)$:

$$\sigma_\gamma = \sqrt{\epsilon \beta_\gamma(R)}.$$

For example, in order to minimize the size of the photon beam at a given point, we select a magnet serving as source at a location where the beam is converging and choose the lattice function such that the lens-free propagation has minimum values for the two beta functions at the detector.

To correct for the finite natural photon opening angle, we convolute the two distributions. Using Gaussian distributions for both gives approximately

$$\sigma^2 \approx \sigma_\gamma^2 + \sigma_\epsilon^2.$$

If the source is located at a finite dispersion the corresponding contribution has to be included too.

15.1.3 The electron and natural photon emittances are of the same magnitude

If the electron-beam emittance is comparable to the photon emittance, we should match the two angular distributions. As shown in Chapter 11 and in particular in Table 12.1, the ratio between the size of the photon beam and its angular spread is about half the formation length ℓ_γ of the radiation. The corresponding ratio for the electron beam is given by the beta function and should be about the same:

$$\beta_{\gamma x} \approx \frac{\sigma_{\gamma x}}{\sigma'_{\gamma x}} \approx \frac{\ell_\gamma}{2}.$$

For the vertical distribution of the radiation emitted in long magnets at low frequencies $\omega \ll \omega_c$, this leads to

$$\beta_{\gamma y} \approx \sqrt[3]{\lambda \rho^2}/2,$$

which often leads to rather small values that are difficult to realize in practical cases. For undulators of length L_u the matching gives

$$\beta_{\gamma x} \approx L_u/2,$$

which can be achieved more easily. It is interesting to note that minimizing the necessary vertical aperture in an undulator, i.e. minimizing the beta function at the ends, leads to the same condition.

15.2 The spatial coherence

15.2.1 The diffraction limit

Once the electron-beam emittance is smaller than the photon-beam emittance,

$$\epsilon_x < \epsilon_{x\gamma} \approx \frac{\lambda}{4\pi},$$

and the beta functions are approximately matched, we have a situation in which the size and angular spread of the electron beam no longer influence the photon beam. This condition is called the diffraction limit for a given wavelength λ. As a consequence, the electron-beam emittance can not be resolved using this photon beam of wavelength λ for imaging and direct observation. The radiation at any angle or location contains contributions from all the electrons. For this reason this situation is also called *spatial coherence* [78].

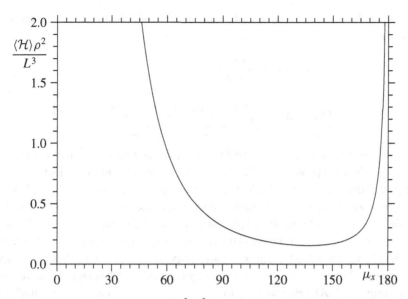

Fig. 15.2. The average emittance function $\mathcal{H}\rho^2/L^3$ normalized with the bending radius ρ and the full cell length L versus the phase advance μ_x per cell.

15.2.2 Small-emittance rings

It is desirable for a synchrotron-radiation storage ring to have a small emittance. This can be achieved with optics that keeps the function $\mathcal{H}(s)$ (14.18) small in the bending magnets:

$$\mathcal{H} = \gamma_x D_x^2 + 2\alpha_x D_x D_x' + \beta_x D_x'^2.$$

To illustrate this we consider the FODO lattice discussed in Chapter 12 and use the lattice functions (13.19)–(13.21) to construct the function $\mathcal{H}(s)$. It is plotted in Fig. 13.16 as a function of s and its average value is

$$\langle \mathcal{H} \rangle = \frac{L^3}{8\rho^2 \cos(\mu_x/2)\sin^3(\mu_x/2)} \left(1 - \frac{3}{4}\sin^2(\mu_x/2) + \frac{1}{60}\sin^4(\mu_x/2) \right).$$

The normalized $\langle \mathcal{H} \rangle$ is plotted in Fig. 15.2 as a function of the phase advance per cell μ_x. It has a broad minimum at $\mu_x \approx 137°$.

The emittance of the ring is given by (14.23), which involves the synchrotron-radiation integrals I_{s2}, I_{s4}, and I_{s5}. For our idealized ring consisting of N_c thin-lens FODO cells of length L with the space between the quadrupoles filled with dipole magnets, we have

$$I_{s2} = \oint \frac{ds}{\rho^2} = N_c \frac{L}{\rho^2}, \qquad I_{s4} \approx 0, \qquad I_{s5} = N_c \frac{L\langle\mathcal{H}\rangle}{|\rho^3|}.$$

We introduce the bending angle $\varphi_c = L/\rho$ of one cell and obtain for the emittance of the FODO ring

$$\epsilon_x = \frac{55\sqrt{3}}{96} \frac{\hbar}{m_0 c} \gamma^2 \frac{I_{s5}}{I_{s2} - I_{s4}}$$

$$= \frac{55\sqrt{3}}{96} \frac{\hbar}{m_0 c} \gamma^2 \varphi_c^3 \frac{\left(1 - \frac{3}{4} \sin^2(\mu_x/2) + \frac{1}{60} \sin^4(\mu_x/2)\right)}{8 \cos(\mu_x/2) \sin^3(\mu_x/2)}.$$

Apart from some natural constants, the ring emittance is proportional to the square of the beam energy, the third power of the bending angle φ_c per cell, and a form factor determined by the phase advance μ_x per cell. The latter can be optimized with $\mu_x \approx 137°$. For a given energy and size of the ring, the emittance can be reduced by having more but shorter cells. This approach is basically simple but has some technical limitations. To have the same phase advance μ_x in a shorter cell requires stronger quadrupoles, which will have a larger size. Using shorter cells leads to a small dispersion, which actually keeps $\mathcal{H}$ and the emittance small but also makes the chromaticity correction with sextupoles more difficult. In optimizing of a FODO ring we have to take these factors into account and might end up with a smaller phase advance per cell than 137°.

The thin-lens FODO lattice is very useful for understanding the basic possibilities of and limitations on obtaining a small emittance. However, it is idealized and may require unrealistic components. Furthermore, a ring needs straight sections for the RF system and injection components for undulators. The optics of the regular lattice has to be matched to that of the straight sections, which has different optimization conditions. As a result the ring is larger and less regular than the simple FODO lattice. However, the scaling of the emittance with the bending angle per cell to the third power is still approximately correct and can serve as for guidance optimizing the size of a storage ring.

Small-emittance rings are usually not based on a FODO structure but on more complicated lattices, which are not described here. Practically all existing rings are listed and their lattices shown in [79].

15.3 The temporal coherence

In most practical cases the observed wavelength is smaller than the bunch length and there is no time correlation among the different particles in a bunch. Even if the distances between them are smaller than the observed wavelength, they change in a random fashion due to the quantum excitation and smear out any constructive interference among field contributions from different electrons. Therefore, the sum of the radiation fields produced by different electrons vanishes on average and only those of each individual particle contribute to the emitted power. Calling this power emitted by a single electron P_{0i}, we find the total power P_{0t} radiated by all N_b particles in the bunch:

$$P_{0t} = \sum_{1}^{N_b} P_{0i} = N_b P_0.$$

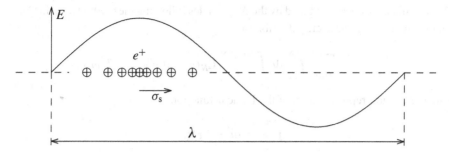

Fig. 15.3. Coherent radiation emitted by the particles in a bunch shorter than the wavelength.

This total power is often called the incoherent power P_{incoh} since there is no time correlation among the particles.

However, if the wavelength of the observed radiation is longer than the bunch length, $\lambda > \sigma_s$, as shown in Fig. 15.3, all electrons emit radiation with individual phases ϕ_i being about the same on the scale of λ, i.e. $\phi_i \ll 2\pi$. For this wavelength the bunch represents a macro-particle with charge $N_b e$ and the fields E_i emitted by individual electrons are added [36, 80–84],

$$E_t = \sum_1^{N_b} E_i = N_b E_i,$$

resulting in a coherent power

$$P_{coh} \propto E_t^2 = N_b^2 E_i^2 \quad \text{or} \quad P_{coh} = N_b P_{incoh}.$$

The coherent power emitted at $\lambda > \sigma_s$ is enhanced by a factor of N_b.

In all practical cases bunches are much longer than the critical wavelength λ_c and the condition for coherent emission is fulfilled only for a small part of the spectrum at very low frequencies. It turns out that this part is actually often greatly reduced by the presence of the conduction vacuum chamber, but we will neglect this suppression effect for the moment.

A single electron gives a radiation pulse $\mathbf{E}(t)$ with Fourier transform $\tilde{\mathbf{E}}(\omega)$ related by

$$\tilde{\mathbf{E}}(\omega) = \frac{1}{\sqrt{2\pi}} \int_{-\infty}^{\infty} \mathbf{E}(t) e^{-i\omega t} \, dt, \qquad \mathbf{E}(t) = \frac{1}{\sqrt{2\pi}} \int_{-\infty}^{\infty} \tilde{\mathbf{E}}(\omega) e^{i\omega t} \, d\omega.$$

The field from N_b electrons having statistically distributed delays τ_k relative to a reference time is

$$\mathbf{E}_N(t) = \sum_{k=1}^{N_b} \mathbf{E}(t - \tau_k)$$

and its Fourier transform is

$$\tilde{\mathbf{E}}_N(\omega) = \sum_{i=1}^{N_b} e^{-i\omega \tau_k} \tilde{\mathbf{E}}(\omega). \tag{15.1}$$

We calculate the energy radiated by the N_b particles following the method used in Chapter 3 and obtain the angular energy distribution

$$\frac{dU_N}{d\Omega} = \frac{r^2}{2\pi\mu_0 c} \int_{-\infty}^{\infty} dt \int_{-\infty}^{\infty} \int_{-\infty}^{\infty} \tilde{\mathbf{E}}_N(\omega)\tilde{\mathbf{E}}_N(\omega')e^{i(\omega+\omega')t} \, d\omega \, d\omega'.$$

Using the integral representation of the Dirac δ-function,

$$\int_{-\infty}^{\infty} e^{iat} \, dt = 2\pi\delta(a),$$

we obtain for the integration over t

$$\frac{dU}{d\Omega} = \frac{r^2}{\mu_0 c} \int_{-\infty}^{\infty} \int_{-\infty}^{\infty} \tilde{\mathbf{E}}_N \tilde{\mathbf{E}}_N(\omega')\delta(\omega+\omega') \, d\omega \, d\omega'$$

$$= \frac{r^2}{\mu_0 c} \int_{-\infty}^{\infty} \tilde{\mathbf{E}}_N(\omega)\tilde{\mathbf{E}}_N(-\omega) \, d\omega.$$

Since the field $\mathbf{E}(t)$ is a real function, its Fourier transform has the symmetry property

$$\tilde{\mathbf{E}}_N(-\omega) = \tilde{\mathbf{E}}_N^*(\omega).$$

This gives for the angular distribution of the radiation energy

$$\frac{dU}{d\Omega} = \frac{r^2}{\mu_0 c} \int_{-\infty}^{\infty} |\tilde{\mathbf{E}}_N(\omega)|^2 \, d\omega = \frac{2r^2}{\mu_0 c} \int_0^{\infty} |\tilde{\mathbf{E}}_N(\omega)|^2 \, d\omega.$$

With the expression (15.1) for $\tilde{\mathbf{E}}_N(\omega)$ we obtain

$$\frac{dU_N}{d\Omega} = \frac{2r^2}{\mu_0 c} \sum_{k=1}^{N_b} \sum_{l=1}^{N_b} \int_0^{\infty} |\tilde{\mathbf{E}}(\omega)|^2 e^{-i\omega(\tau_k - \tau_l)} \, d\omega. \tag{15.2}$$

We assume first that there is a uniform statistical time distribution of the particles. Averaging over the delays τ_k and τ_l gives zero except if $k = l$, in which case we obtain a single sum:

$$\frac{dU_N}{d\Omega} = \frac{2r^2}{\mu_0 c} \sum_{k=1}^{N_b} \int_0^{\infty} |\tilde{\mathbf{E}}(\omega)|^2 \, d\omega = N_b \frac{2r^2}{\mu_0 c} \int_0^{\infty} |\tilde{\mathbf{E}}(\omega)|^2 \, d\omega.$$

The total angular energy density is just N_b times that of a single electron;

$$\frac{dU_N}{d\omega} = N_b \frac{dU}{d\Omega} = \frac{dU_{\text{incoh}}}{d\Omega},$$

which is the incoherent case. Differentiating this with respect to ω gives the spectral angular energy distribution of the radiation:

$$\frac{d^2 U_{\text{incoh}}}{d\Omega \, d\omega} = N_b \frac{2r^2 |\tilde{\mathbf{E}}(\omega)|^2}{\mu_0 c} = N_b \frac{d^2 U}{d\Omega \, d\omega}.$$

In the next step we still assume that the time distribution of the electrons is statistical on a small scale but that the global electron density has a Gaussian distribution:

$$\frac{dN}{d\tau} = \frac{N_b}{\sqrt{2\pi}\,\sigma_t} \exp\left(-\frac{\tau^2}{2\sigma_t^2}\right), \qquad \int_{-\infty}^{\infty} \frac{dN}{d\tau}\, d\tau = N_b.$$

We replace now the sums in (15.2) by integrals. In doing so we lose the incoherent part of the radiated energy, which is based on the fact that the field of each particle is multiplied by itself. Therefore we have only the coherent part, to which the incoherent one has to be added:

$$\frac{dU_{coh}}{d\Omega} = \frac{2r^2}{\mu_0 c 2}\frac{N_b^2}{\pi\sigma_t^2} \int_0^\infty d\omega\, |\tilde{E}(\omega)|^2 \int d\tau_k$$

$$\times \int d\tau_l \exp\left(-\frac{\tau_k^2}{2\sigma_t^2}\right) e^{-\omega\tau_k} \exp\left(-\frac{\tau_l^2}{2\sigma_t^2}\right) e^{\omega\tau_l}$$

$$= N_b^2 \frac{2r^2}{\mu_0 c} \int_0^\infty d\omega\, |\tilde{E}(\omega)|^2 e^{-\omega^2\sigma_t^2}.$$

We obtain for the coherent angular spectral energy distribution

$$\frac{d^2 U_{coh}}{d\Omega\,d\omega} = \frac{d^2 U}{d\Omega\,d\omega} N_b^2 e^{-\omega^2\sigma_t^2} = \frac{d^2 U_{incoh}}{d\Omega\,d\omega} N_b e^{-\omega^2\sigma_t^2},$$

or, expressed in terms of the power,

$$\frac{d^2 P_{coh}}{d\Omega\,d\omega} = \frac{d^2 P}{d\Omega\,d\omega} N_b^2 e^{-\omega^2\sigma_t^2} = \frac{d^2 P_{incoh}}{d\Omega\,d\omega} N_b e^{-\omega^2\sigma_t^2},$$

where $d^2 P/(d\Omega\,d\omega)$ is the single-particle angular spectral power distribution. We integrate this over the solid angle and obtain the spectral distribution from the expression (5.15) in Chapter 5,

$$\frac{dP_{coh}}{d\omega} = \frac{dP}{d\omega} N_b^2 e^{-\omega^2\sigma_t^2} = \frac{P_0}{\omega_c} S_s\left(\frac{\omega}{\omega_c}\right) N_b^2 e^{-\omega^2\sigma_t^2}$$

with P_0 being the total power emitted by a single ultra-relativistic particle, (5.1);

$$P_0 = \frac{2r_0 c m_0 c^2 \gamma^4}{3\rho^2},$$

and $S_s(\omega/\omega_c)$ the normalized power spectral function (5.16) of the total radiation,

$$S_s\left(\frac{\omega}{\omega_c}\right) = \frac{54}{16}\frac{\omega}{\omega_c}\left(-2\frac{Ai'(z)}{z} - \frac{1}{3} + \int_0^z Ai(z')\,dz'\right) \qquad \text{with } z = \left(\frac{3\omega}{2\omega_c}\right)^{2/3}.$$

The coherent radiation becomes large only at low frequencies at which $\omega\sigma_t$ is not too large. These frequencies are in practical cases much smaller than the critical frequency

$\omega_c = 3c\gamma^3/(2\rho)$. We use the approximation (5.19) for the spectral function at $\omega \ll \omega_c$:

$$S_s\left(\frac{\omega}{\omega_c}\right) \approx \frac{27}{4}\left(\frac{2}{3}\right)^{2/3}\frac{3^{1/6}\Gamma(2/3)}{2\pi}\left(\frac{\omega}{\omega_c}\right)^{1/3} = 1.333\,23\left(\frac{\omega}{\omega_c}\right)^{1/3}.$$

With this we obtain for the spectral power distribution of the coherent radiation in the absence of shielding

$$\frac{\mathrm{d}P_{\mathrm{coh}}}{\mathrm{d}\omega} \approx \frac{P_0}{\omega_c}\frac{27}{4}\left(\frac{2}{3}\right)^{2/3}\frac{3^{1/6}\Gamma(2/3)}{2\pi}\left(\frac{\omega}{\omega_c}\right)^{1/3}N_b^2 e^{-\omega^2\sigma_t^2}.$$

We are interested in the total coherent power and integrate the above equation over frequency using the integral

$$\int_0^\infty x^a e^{-bx^2}\,\mathrm{d}x = \frac{\Gamma((1+a)/2)}{2b^{(1+a)/2}},$$

giving

$$P_{\mathrm{coh}} \approx N_b^2 P_0\frac{27}{8}\left(\frac{2}{3}\right)^{2/3}\frac{3^{1/6}(\Gamma(2/3))^2}{2\pi(\omega_c\sigma_t)^{4/3}} = N_b^2\frac{r_0 c m_0 c^2}{\rho^2}\frac{3^{1/6}(\Gamma(2/3))^2}{2\pi(\sigma_s/\rho)^{4/3}}.$$

We called here P_{incoh} and P_{coh} the total incoherent and coherent powers emitted by all particles. The corresponding power per particle is of course just divided by N_b. Using the ratio of the coherent and incoherent powers avoids this ambiguity:

$$\frac{P_{\mathrm{coh}}}{P_{\mathrm{incoh}}} = \frac{3\cdot 3^{1/6}(\Gamma(2/3))^2}{4\pi\gamma^4}\frac{N_b}{(\sigma_s/\rho)^{4/3}} = 0.5257\frac{N_b}{\gamma^4(\sigma_s/\rho)^{4/3}}.$$

The coherent radiation treated here appears only in the lowest part of the spectrum and is therefore independent of the particle energy. The ratio of the coherent and incoherent radiation decreases proportionally to γ^{-4} because the incoherent part increases as γ^4.

In this low part of the spectrum the wavelength becomes comparable to the height of the vacuum chamber and wave propagation is inhibited. This process is rather complicated for the case of a realistic geometry. However, the simplified case of an electron orbit between two parallel plates at distance $\pm a/2$ was treated rather early on and much work on more complicated geometries is now going on [36, 80, 85, 86]. It turns out that, in most practical cases, the suppression appears already at a wavelength shorter than that which could be radiated coherently. However, for very short bunches coherence can be achieved over a certain range of wavelengths that are longer than the bunch length, but shorter than the cut-off wavelength at which suppression occurs. The spectrum expected is shown qualitatively in Fig. 15.4. Special efforts have been made in various laboratories to produce bunches sufficiently short to provide coherent radiation that is not suppressed. This has been achieved in linear accelerators [87, 88]. In storage rings quantum excitation makes short bunches more difficult to obtain. Some sporadic coherent radiation in the far infrared has been observed in several machines, probably having been produced by some high-frequency bunch structures caused by instabilities. Steady-state coherent radiation was recently observed at the BESSY ring [89].

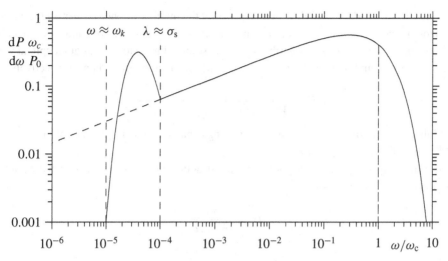

Fig. 15.4. Qualitative spectrum of the radiation from a short bunch containing a region of coherence.

15.4 Flux and brightness

We are looking for a parameter with which to evaluate the quality of a synchrotron-radiation source. Since the importance of some photon-beam parameters depends on the particular experiments, it is difficult to give a single quantity that has to be optimized. However, we will try to address two classes of measurements for which we give a photon-beam parameter that is important.

Most experiments involve the use of a monochromator that selects a narrow band of the spectrum. Some of them are mainly concerned with the number of these selected photons received per second on a relatively large target. Often bending-magnet or wiggler radiation is used and the full vertical distribution is accepted. In this case, a small emittance is of secondary significance, but the important quality is the photon flux per horizontal bending angle and relative frequency band. This parameter was calculated (5.49) under the idealized conditions of electrons with Lorentz factor γ and beam current I passing through a dipole magnet with bending radius ρ. This determines the critical frequency $\omega_c = 3c\gamma^3/(2\rho)$ which enters into the normalized spectrum $S(\omega/\omega_c)$ given by (5.16). We obtain for this flux

$$\frac{\mathrm{d}^2\dot{n}_I}{\mathrm{d}\phi\,\mathrm{d}\omega/\omega} \approx \frac{4\alpha_f\gamma\,I\,S_s(\omega/\omega_c)}{9e},$$

which is usually measured in units of $(\mathrm{s\,mrad\,0.1\%})^{-1}$. Taking $\omega = \omega_c$ where $S_s(1) = 0.404$, a typical electron energy of 1 GeV, and a current of 100 mA, we obtain for this flux per angle and relative frequency band about $1.6 \times 10^{12}\,(\mathrm{s\,mrad\,0.1\%})^{-1}$.

A second class of experiments profits from a small emittance to focus the beam onto a small spot, in order to have a very parallel beam, or sufficient intensity after selecting a very narrow frequency band giving a long coherence length. The parameter to optimize in this case is the *brightness* [90], which is basically the number of photons per unit time, per horizontal and vertical source size and angular spread, and per relative bandwidth given in units of 0.1%. For a Gaussian distribution of the source dimension and angular spread with RMS values σ_γ and σ'_γ (usually measured in millimeters and milliradians, respectively)

$$\mathcal{B} = \frac{\dot{n}}{4\pi^2 \sigma_{\gamma x}, \ \sigma_{\gamma y}, \ \sigma'_{\gamma x}, \ \sigma'_{\gamma y} \, \mathrm{d}\omega/\omega}.$$

Undulator beam lines at modern storage rings can reach a brightness of up to $\mathcal{B} \approx 10^{19} \, \mathrm{mm}^{-2} \, \mathrm{mrad}^{-2} \, \mathrm{s}^{-1} \, (0.1\%)^{-1}$.

It is interesting to give this brightness for a diffraction-limited beam for which the photon emittance is $\sigma_\gamma \sigma'_\gamma = \lambda/(4\pi)$:

$$\mathcal{B} = \frac{4\dot{n}}{\lambda^2 \, \mathrm{d}\omega/\omega}.$$

15.5 The synchrotron radiation emitted by protons and ions

15.5.1 Introduction

So far we have always considered electrons as sources of synchrotron radiation either as a single particle or, in this chapter, as a group of charges. However, radiation is emitted by any accelerated charge such as a proton or an ion. Its properties are completely determined by the treatment presented here so far if the different mass and charge of the particle are taken into account. However, the practical operating parameters of electron and proton rings are very different, which affects the emitted radiation. A quick inspection of the relevant equations shows that the intensity of the radiation emitted by protons or ions is much smaller and hence such rings are hardly of any interest as sources. The main purpose of this investigation is to illustrate the parameter dependence of synchrotron radiation and in particular the effect of coherent radiation emitted by the protons or ions. Furthermore, synchrotron radiation as a diagnostic tool has also been used for protons.

15.5.2 The radiation from protons

As an introduction we look at the radiation emitted by protons. This contains no new physics and all our previous investigation can be applied by taking the different rest mass and classical radius (3.8) of the proton into account:

$$m_0 c^2 = \begin{cases} 0.511 \, \mathrm{MeV} & \text{for electrons} \\ 938.27 \, \mathrm{MeV} & \text{for protons} \end{cases}$$

$$r_0 = \frac{e^2}{4\pi \epsilon_0 m_0 c^2} = \begin{cases} 2.818 \times 10^{-15} \, \mathrm{m} & \text{for electrons} \\ 1.535 \times 10^{-18} \, \mathrm{m} & \text{for protons.} \end{cases} \qquad (15.3)$$

The difference between the two cases lies in the machine parameters of electron and proton rings. To obtain radiation of any interest for diagnostics, the proton energy should exceed $100\,\text{GeV}$ or $\gamma > 100$, which allows us in most cases to use the ultra-relativistic approximation.

For ordinary synchrotron radiation one assumes that the dipole magnet is long and deflects the beam by an angle $\Delta\varphi > 2/\gamma$. This leads to a condition for the magnet length ℓ of

$$\Delta\varphi = \frac{\ell}{\rho} = \frac{eBc\ell}{m_0c^2\gamma} > \frac{2}{\gamma} \quad \text{or} \quad \ell > \frac{m_0c^2}{eBc}.$$

Taking a field of $B = 0.5\,\text{T}$, we find $\ell > 3\,\text{mm}$ for electrons, whereas a field of $1.5\,\text{T}$ gives $\ell > 2\,\text{m}$ for protons as the condition to be treated as a long magnet. For electrons the bending magnets can nearly always be considered as long and even fringe fields change little over the relevant length ℓ. For protons the situation is more restricted. Although magnets are usually several meters long and higher fields obtained in super-conducting machines make the limiting magnet length ℓ shorter, fringe fields can have a stronger influence. They can actually be used to shift the spectrum towards higher frequencies [60, 91].

To overcome problems associated with the small intensity emitted by protons, one can use undulators that are tuned to visible light [92, 93].

Since realistic magnets have to be relatively long in order to deflect protons by an angle larger than the natural opening angle of synchrotron radiation, undulators for protons will in most cases be weak. For the undulator parameter

$$K_\text{u} = \frac{eB_0}{m_0ck_\text{u}} = \frac{\lambda_\text{u}eB_0c}{2\pi m_0c^2}$$

to become unity for a typical period length of $\lambda_\text{u} = 0.2\,\text{m}$ one needs a field strength of about $100\,\text{T}$.

The effects of synchrotron radiation on the proton beam are very small. Calculated damping times range between years for older, iron-magnet synchrotrons and about a day for the new super-conducting rings. Consequently also quantum excitation is negligible and the calculated equilibrium energy spread and emittance are extremely small if other effects are neglected. In proton rings these parameters are given by the source and the adiabatic damping during acceleration. The emittance obtained is usually very small and can in some cases reach the diffraction limit. This restricts the diagnostics applications of proton synchrotron radiation.

15.5.3 The radiation from ions

A fully ionized atom consists of a nucleus of mass Am_P and charge Ze with Z being the atomic number, A the atomic weight (a dimensionless number), and m_P the mass of the proton. The small effect of the binding energy is neglected here. If the atom is not fully ionized but has only Z' electrons removed, we replace Z by $Z' < Z$. From a particle point of view this ion is not a point particle, but consists of N protons and $A - Z$ neutrons. For the wavelength contained in the synchrotron-radiation spectrum, however, this ion is

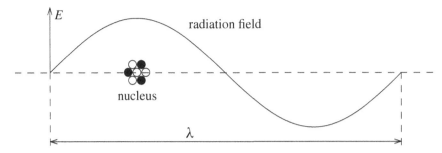

Fig. 15.5. Coherent radiation from the protons inside a nucleus.

sufficiently small that it can be considered a single charge Ze, as indicated in Fig. 15.5 and described in [94].

The total radiated power P_Z emitted by this nucleus, circulating in a storage ring with curvature $1/\rho$, is obtained from (3.9) by substituting the expression for the classical radius and replacing the elementary charge e by $q = Ze$,

$$P_Z = \frac{(eZ)^2 c \beta^4 \gamma^4}{6\pi \epsilon_0 \rho^2} \approx \frac{(eZ)^2 \gamma^4}{6\pi \epsilon_0 \rho^2}, \tag{15.4}$$

where we used the ultra-relativistic approximation, although this might introduce some inaccuracy for heavy ions. The Lorentz factor applies here to the whole ion.

It is interesting to compare the energy loss U_Z of the ion with the corresponding loss U_P of a proton circulating in the same ring with the same curvature $1/\rho$ in the same magnetic field B. Owing to the relation between field and curvature,

$$\frac{1}{\rho} = \frac{qB}{p} = \frac{qB}{m_0 c \beta \gamma},$$

the two particles must have different momenta p,

$$\frac{p_Z}{p_P} = Z = \frac{A m_P \beta_Z \gamma_Z}{m_P \beta_P \gamma_P} \approx A \frac{\gamma_Z}{\gamma_P},$$

which gives for the ratio between the Lorentz factors and energies

$$\frac{\gamma_Z}{\gamma_P} = \frac{Z}{A}, \qquad \frac{E_Z}{E_P} = \frac{A m_P c^2}{m_P c^2} \frac{\gamma_Z}{\gamma_P} = Z.$$

The whole nucleus of the ion has more energy than the proton due to its higher charge, but its energy per nucleon is smaller:

$$\frac{E_Z/A}{E_P} = \frac{Z}{A}.$$

This gives a ratio of the emitted powers of

$$\frac{P_Z}{P_P} = \frac{Z^2 \gamma_Z^4}{\gamma_P^4} = \frac{Z^6}{A^4}.$$

More relevant is the ratio of the powers normalized for the same charge, i.e. for the same beam current:

$$\frac{P_Z/Z}{P_P} = \frac{Z^5}{A^4}.$$

For a lead ion with $Z_{Pb} = 82$ and $A_{Pb} = 208$ we find for this ratio 1.98. For a ring with a given radius and magnetic field a lead beam radiates about twice the power of a proton beam with the same total current. This is due to the coherent radiation of the protons in the lead nucleus, which more than compensates for the lower value of γ_{Pb} compared with the value for protons. It is also interesting to consider the relative energy loss per unit time for ions and protons, which is related to the sum of the damping rates (14.21),

$$-\frac{1}{E_e}\frac{dE_e}{dt} = \frac{P}{E_e} = \frac{1}{2}\sum_i \alpha_i,$$

with the ratio

$$\frac{\left(\dfrac{1}{E_e}\dfrac{dE_e}{dt}\right)_Z}{\left(\dfrac{1}{E_e}\dfrac{dE_e}{dt}\right)_P} = \frac{Z^5}{A^4}$$

being the same as that for the radiated power per unit charge and hence we find that lead ions are damped faster than protons are.

Appendix A

Airy functions

A.1 Definitions and developments

The Airy functions are special cases of Bessel functions and are described in the standard mathematics books [41, 95]. We summarize here the basic properties relevant for our application, and give some integrals that are difficult to find elsewhere. The Airy functions can be defined by integrals similar to the ones we encountered in calculating the field components of synchrotron radiation:

$$
\text{Ai}(x) = \frac{1}{2\pi} \int_{-\infty}^{\infty} \cos\left(\frac{u^3}{3} + xu\right) du
$$

$$
\text{Ai}'(x) = \frac{d\text{Ai}(x)}{dx} = -\frac{1}{2\pi} \int_{-\infty}^{\infty} u \sin\left(\frac{u^3}{3} + xu\right) du.
$$

(A.1)

They satisfy the differential equation

$$
\text{Ai}''(x) - x\,\text{Ai}(x) = 0.
$$

(A.2)

At the origin $x = 0$ they take the values

$$
\text{Ai}(0) = \frac{1}{3^{2/3}\Gamma(2/3)} = \frac{\Gamma(1/3)}{3^{1/6}2\pi} = 0.355\,03
$$

$$
\text{Ai}'(0) = -\frac{1}{3^{1/3}\Gamma(1/3)} = -\frac{3^{1/6}\Gamma(2/3)}{2\pi} = -0.258\,82,
$$

where we used the relation $\Gamma(x)\Gamma(1 - x) = \pi/\sin(\pi x)$ to replace the Gamma function in the denominator.

We need the Airy functions for positive arguments only. They are plotted in Fig. A.1 together with the integral over the Airy function.

The Airy functions are closely related to the modified Bessel functions of order $n/3$:

$$
\text{Ai}(x) = \frac{1}{\pi}\sqrt{\frac{x}{3}}K_{1/3}\left(2x^{3/2}/3\right), \qquad \text{Ai}'(x) = -\frac{1}{\pi}\frac{x}{\sqrt{3}}K_{2/3}\left(2x^{3/2}/3\right).
$$

(A.3)

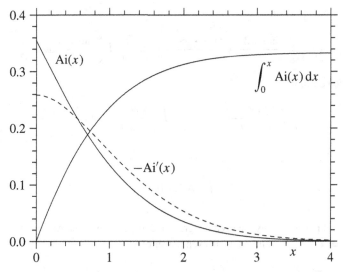

Fig. A.1. The Airy function $Ai(x)$, its derivative $Ai'(x)$, and the integral $\int_0^x Ai(x)\,dx$.

We will approximate these functions for small and large arguments and use the developments. For small arguments $x \ll 1$ we have

$$Ai(x) = Ai(0)\left(1 + \frac{1}{3!}x^3 + \cdots\right) + Ai'(0)\left(x + \frac{2}{4!}x^4 + \cdots\right) \tag{A.4}$$

$$Ai'(x) = Ai(0)\left(\frac{1}{2!}x^2 + \frac{1\cdot 4}{5!}x^5 + \cdots\right) + Ai'(0)\left(1 + \frac{2}{3!}x^3 + \cdots\right)$$

$$\int_0^x Ai(x)\,dx = Ai(0)\left(x + \frac{1}{4!}x^4 + \cdots\right) + Ai'(0)\left(\frac{1}{2!}x^2 + \frac{2}{5!}x^5 + \cdots\right)$$

and for large arguments

$$Ai(x) = \frac{e^{-z}}{2\sqrt{\pi}x^{1/4}}\left(1 - \frac{5}{72z} + \frac{385}{2\cdot 72^2 z^2}\cdots\right)$$

$$Ai'(x) = -\frac{x^{1/4}e^{-z}}{2\sqrt{\pi}}\left(1 + \frac{7}{72z} - \frac{455}{2\cdot 72^2 z^2}\cdots\right) \tag{A.5}$$

$$\int_x^\infty Ai(x')\,dx' = \frac{e^{-z}}{2\sqrt{\pi}x^{3/4}}\left(1 - \frac{41}{72z} + \frac{9241}{2\cdot 72^2 z^2}\cdots\right)$$

with $z = \frac{2}{3}x^{3/2} \gg 1$.

A.2 Integrals involving Airy functions

For several applications we need some integrals involving Airy functions.

- We start with the definite integral over the Airy function itself:

$$\int_0^\infty Ai(x)\,dx = \frac{1}{2\pi}\int_{-\infty}^\infty \frac{\sin(u^3 + xu)}{u}\,du\Big|_0^\infty.$$

We obtain for the upper limit

$$\frac{1}{2\pi} \int_{-\infty}^{\infty} \frac{\sin(u^3/3 + u(x \to \infty))}{u} \, du \approx \frac{1}{2\pi} \int_{-\infty}^{\infty} \frac{\sin(xu)}{u} \, du = \frac{1}{2}$$

and for the lower limit $x = 0$

$$\frac{1}{2\pi} \int_{-\infty}^{\infty} \frac{\sin(u^3/3)}{u} \, du = \frac{1}{6\pi} \int_{-\infty}^{\infty} \frac{\sin(u^3/3)}{u^3/3} \, d(u^3/3) = \frac{1}{6},$$

resulting in

$$\int_{0}^{\infty} \mathrm{Ai}(x) \, dx = \frac{1}{3}. \tag{A.6}$$

- For the spectral power density we need integrals over the square of the Airy functions with argument $x = a + by^2$ multiplied by an even power of y:

$$\int_{0}^{\infty} y^{2n} \, \mathrm{Ai}^2(a + by^2) \, dy. \tag{A.7}$$

First we express the square of the Airy function in terms of integrals,

$$\mathrm{Ai}^2(x) = \frac{1}{4\pi^2} \int_{-\infty}^{\infty} dt \int_{-\infty}^{\infty} ds \cos\left(\frac{t^3}{3} + xt\right) \cos\left(\frac{s^3}{3} + xs\right),$$

and substituting $s + t = u$ and $s - t = v$ gives

$$\mathrm{Ai}^2(x) = \frac{1}{16\pi^2} \int_{-\infty}^{\infty} du \int_{-\infty}^{\infty} dv \left[\cos\left(\frac{u^3}{12} + \frac{uv^2}{4} + xu\right) + \cos\left(\frac{v^3}{12} + \frac{vu^2}{4} + xv\right)\right].$$

The expression in the square brackets is symmetric in u and v and both terms give the same value after integration. Furthermore, it does not change with the sign of u and v:

$$\mathrm{Ai}^2(x) = \frac{1}{2\pi^2} \int_{0}^{\infty} \int_{0}^{\infty} \cos\left(\frac{u^3}{12} + \frac{uv^2}{4} + xu\right) du \, dv. \tag{A.8}$$

We substitute $x = a + by^2$ and integrate over y to obtain our first integral of the form (A.7):

$$\int_{0}^{\infty} \mathrm{Ai}^2(a + by^2) \, dy = \frac{1}{2\pi^2} \int_{0}^{\infty} du \int_{0}^{\infty} dv \int_{0}^{\infty} \cos\left[\frac{u^3}{12} + u\left(\frac{v^2}{4} + a + by^2\right)\right] dy.$$

We use

$$v = 2r \cos\phi, \qquad \sqrt{b}y = r \sin\phi, \qquad dv \, dy = 2r \, dr \, d\phi/\sqrt{b} \tag{A.9}$$

and integrate over ϕ from 0 to $\pi/2$:

$$\int_{0}^{\infty} \mathrm{Ai}^2(a + by^2) \, dy = \frac{1}{2\pi\sqrt{b}} \int_{0}^{\infty} du \int_{0}^{\infty} \cos\left(\frac{u^3}{12} + (a + r^2)u\right) r \, dr. \tag{A.10}$$

A further substitution $r^2 = w$, $u = 2^{2/3}u'$ leads to

$$\int_0^\infty \mathrm{Ai}^2(a + by^2)\,\mathrm{d}y = \frac{2^{2/3}}{4\pi\sqrt{b}} \int_0^\infty \mathrm{d}u' \int_0^\infty \cos\left(\frac{u'^3}{3} + 2^{2/3}(a + w)u'\right) \mathrm{d}w$$

$$= \frac{2^{2/3}}{4\sqrt{b}} \int_0^\infty \mathrm{Ai}\left(2^{2/3}(a + w)\right) \mathrm{d}w.$$

Putting $z' = 2^{2/3}(a + w)$ and $z = 2^{2/3}a$ gives the final integral

$$\int_0^\infty \mathrm{Ai}^2(a + by^2)\,\mathrm{d}y = \frac{1}{4\sqrt{b}} \int_z^\infty \mathrm{Ai}(z')\,\mathrm{d}z'. \tag{A.11}$$

For $a = 0$ this becomes

$$\int_0^\infty \mathrm{Ai}^2(by^2)\,\mathrm{d}y = \frac{1}{12\sqrt{b}}, \tag{A.12}$$

where the definite integral (A.6) over the Airy function has been used.

For the next integral, containing the factor y^2, we go back to the expression (A.10) with $\int_0^{\pi/2} \sin^2(\phi)\,\mathrm{d}\phi = \pi/4$ to obtain

$$\int_0^\infty y^2\,\mathrm{Ai}^2(a + by^2)\,\mathrm{d}y = \frac{1}{4\pi b\sqrt{b}} \int_0^\infty \mathrm{d}u \int_0^\infty \cos\left(\frac{u^3}{12} + (a + r^2)u\right) r^3\,\mathrm{d}r.$$

Using the same substitutions as before, we obtain

$$\int_0^\infty y^2\,\mathrm{Ai}^2(a + by^2)\,\mathrm{d}y = \frac{2^{2/3}}{8\pi b\sqrt{b}} \int_0^\infty \mathrm{d}u' \int_0^\infty \cos\left(\frac{u'^3}{12} + 2^{2/3}(a + w)u'\right) w\,\mathrm{d}w$$

$$= \frac{2^{2/3}}{8\sqrt{b}} \int_0^\infty \mathrm{Ai}\left(2^{2/3}(a + w)\right) w\,\mathrm{d}w.$$

Using again $z' = 2^{2/3}(a + w)$ and $z = 2^{2/3}a$ or $w = (z' - z)/2^{2/3}$, this becomes

$$\int_0^\infty y^2\,\mathrm{Ai}^2(a + by^2)\,\mathrm{d}y = \frac{1}{8 \cdot 2^{2/3}b\sqrt{b}} \int_z^\infty \mathrm{Ai}(z')\,(z' - z)\,\mathrm{d}z'$$

and, with the relation (A.2),

$$\int_0^\infty y^2\,\mathrm{Ai}^2(a + by^2)\,\mathrm{d}y = -\frac{a}{8b\sqrt{b}} \left(\frac{\mathrm{Ai}'(z)}{z} + \int_z^\infty \mathrm{Ai}(z')\,\mathrm{d}z'\right). \tag{A.13}$$

For the next power we have $\int_0^{\pi/2} \sin^4 \phi\,\mathrm{d}\pi = 3\pi/16$, giving, with the same method as before,

$$\int_0^\infty y^4\,\mathrm{Ai}^2(a + by^2)\,\mathrm{d}y = \frac{3}{32 \cdot 2^{4/3}b^{5/2}} \int_z^\infty \mathrm{Ai}(z')\,(z' - z)^2\,\mathrm{d}z'$$

$$= \frac{3a^2}{32b^{5/2}} \left(\frac{\mathrm{Ai}(z)}{z^2} + \frac{\mathrm{Ai}'(z)}{z} + \int_z^\infty \mathrm{Ai}(z')\,\mathrm{d}z'\right). \tag{A.14}$$

For the factor y^6 we have $\int_0^{\pi/2} \sin^6 \phi \, d\phi = 5\pi/32$ and obtain

$$\int_0^\infty y^6 \, \mathrm{Ai}^2(a + by^2) \, dy = \frac{5}{64 \cdot 2^2 b^{7/2}} \int_z^\infty \mathrm{Ai}(z')(z' - z)^3 \, dz'$$

$$= -\frac{5a^3}{64b^{7/2}} \left[\frac{\mathrm{Ai}(z)}{z^2} + \frac{\mathrm{Ai}'(z)}{z} + \left(1 - \frac{2}{z^3}\right) \int_z^\infty \mathrm{Ai}(z') \, dz' \right]. \quad (\mathrm{A}.15)$$

We need this integral also for $a = 0$:

$$\int_0^\infty y^6 \, \mathrm{Ai}^2(by^2) \, dy = \frac{5}{384 b^{7/2}}. \quad (\mathrm{A}.16)$$

As a further integral we need

$$\int_0^\infty y^{12} \, \mathrm{Ai}^2(by^2) \, dy, \quad (\mathrm{A}.17)$$

which we solve in the same way but set to begin with $a = 0$. We use the expression (A.8) for the square of the Airy function, make the substitutions (A.9), and apply the integral $\int_0^{\pi/2} \sin^{12} \phi \, d\phi = 231\pi/2048$, giving

$$\int_0^\infty y^{12} \, \mathrm{Ai}^2(by^2) \, dy = \frac{231 \cdot 2^{2/3}}{2048\pi b^{13/2}} \int_0^\infty \int_0^\infty du \cos\left(\frac{u^3}{12} + r^2 u\right) r^7 \, dr. \quad (\mathrm{A}.18)$$

A further substitution $z = 2^{2/3} r^2$, $dz = 2 \cdot 2^{2/3} r \, dr$, $u = 2^{2/3} u'$ leads to

$$\int_0^\infty y^{12} \, \mathrm{Ai}^2(by^2) \, dy = \frac{231}{2^{16} b^{13/2}} \int_0^\infty \mathrm{Ai}(z) \, z^6 \, dz.$$

We apply repetitive integrations by parts and applications of the relation (A.2) to obtain

$$\int_0^\infty \mathrm{Ai}(z) \, z^6 \, dz = 40 \int_0^\infty \mathrm{Ai}(z) \, dz = \frac{40}{3}$$

and

$$\int_0^\infty y^{12} \, \mathrm{Ai}^2(by^2) \, dy = \frac{5 \cdot 7 \cdot 11}{2^{13} b^{13/2}}. \quad (\mathrm{A}.19)$$

For its application to calculate the variance of the transverse photon momentum we modify this integral with the substitution

$$y = p^{1/3}, \qquad dy = \frac{1}{3 p^{2/3}} \, dp, \quad (\mathrm{A}.20)$$

giving

$$\int_0^\infty p^{10/3} \, \mathrm{Ai}^2\left(bp^{2/3}\right) dp = \frac{3 \cdot 5 \cdot 7 \cdot 11}{2^{13} b^{13/2}}. \quad (\mathrm{A}.21)$$

- Integrals of the form

$$\int_0^\infty y^{2n} \, \mathrm{Ai}'^2(a + by^2) \, dy.$$

We differentiate expression (A.11) twice with respect to a, use the relation (A.2),

$$\int_0^\infty \left[\mathrm{Ai}'^2(a + by^2) + (a + by^2) \, \mathrm{Ai}^2(a + by^2) \right] dy = -\frac{\mathrm{Ai}'(z)}{2 \cdot 2^{2/3} \sqrt{b}},$$

and combine it with (A.11) and (A.13) to obtain the integral

$$\int_0^\infty \text{Ai}'^2(a+by^2)\,dy = \frac{a}{8\sqrt{b}}\left(-3\frac{\text{Ai}'(z)}{z} - \int_z^\infty \text{Ai}(z')\,dz'\right).\tag{A.22}$$

Differentiating (A.11) once with respect to a and once with respect to b gives

$$\int_0^\infty y^2[\text{Ai}'^2(a+by^2) + (a+by^2)\,\text{Ai}^2(a+by^2)]\,dy = \frac{\text{Ai}(z)}{4 \cdot 2^{4/3}b\sqrt{b}},$$

which is combined with (A.13) and (A.14) to give

$$\int_0^\infty y^2\,\text{Ai}'^2(a+by^2)\,dy = \frac{a^2}{32b^{3/2}}\left(\frac{5\text{Ai}(z)}{z^2} + \frac{\text{Ai}'(z)}{z} + \int_z^\infty \text{Ai}(z')\,dz'\right).\tag{A.23}$$

We differentiate (A.12) twice with respect to b:

$$\int_0^\infty y^4[\text{Ai}'^2(by^2) + by^2\,\text{Ai}^2(by^2)]\,dy = \frac{1}{32b^{5/2}}.$$

Using (A.16), we obtain the integral

$$\int_0^\infty y^4\,\text{Ai}'^2(by^2)\,dy = \frac{7}{384b^{5/2}}.\tag{A.24}$$

For the variance of the transverse photon momentum we need the integral

$$\int_0^\infty y^{10}\,\text{Ai}'^2(by^2)\,dy,$$

which we obtain by differentiating the previous expression (A.24) twice with respect to b and using (A.2) to obtain

$$\int_0^\infty y^{10}\,\text{Ai}'^2(by^2)\,dy + b\int_0^\infty y^{12}\,\text{Ai}^2(by^2)\,dy = \frac{5\cdot 7^2}{2^{10}\cdot 3\cdot b^{11/2}}.$$

With the second integral (A.19), which we solved before, we obtain

$$\int_0^\infty y^{10}\,\text{Ai}'^2(by^2)\,dy = \frac{5\cdot 7\cdot 23}{2^{13}\cdot 3\cdot b^{11/2}}$$

and, with the substitution (A.20),

$$\int_0^\infty p^{8/3}\,\text{Ai}'^2\left(bp^{2/3}\right)dp = \frac{5\cdot 7\cdot 23}{2^{13}b^{11/2}}.\tag{A.25}$$

- For the integration of the angular spectral power density over frequency, we use this integral (A.24) and (A.16) in a different form, obtained by substituting $y = p^{1/3}$, $dy = 1/(3p^{2/3})\,dp$:

$$\int_0^\infty p^{4/3}\,\text{Ai}^2\left(bp^{2/3}\right)dp = \frac{5}{128b^{7/2}}, \qquad \int_0^\infty p^{2/3}\text{Ai}'^2\left(bp^{2/3}\right)dp = \frac{7}{128b^{5/2}}.\tag{A.26}$$

- For the integrated power spectrum we need two expressions that are obtained by partial integration:

$$\int_0^z z'\,\text{Ai}'(z')\,dz' = z\,\text{Ai}(z) - \frac{1}{3} + \int_z^\infty \text{Ai}(z)\,dz\tag{A.27}$$

and

$$\int_0^z z^2 \int_z^\infty \mathrm{Ai}(z') \, dz' \, dz = \frac{z^3}{3} \int_z^\infty \mathrm{Ai}(z) \, dz + \int_0^z \frac{z^3}{3} \mathrm{Ai}(z) \, dz$$

$$= \frac{z^3}{3} \int_z^\infty \mathrm{Ai}(z) \, dz + \frac{1}{3} \int_0^z z^2 \, \mathrm{Ai}''(z) \, dz$$

$$= \frac{1}{9} \left(2 + 3z^2 \, \mathrm{Ai}'(z) - 6z \mathrm{Ai}(z) - 3(2 - z^3) \int_z^\infty \mathrm{Ai}(z) \, dz \right). \quad \text{(A.28)}$$

- For the integral over the photon spectrum we need

$$\int_0^\infty \frac{\mathrm{Ai}'(z)}{\sqrt{z}} \, dz = -\frac{1}{\pi} \int_0^\infty \frac{u \sin(u^3/3 + uz)}{\sqrt{z}} \, du$$

$$= -\frac{1}{\pi} \int_0^\infty du \int_0^\infty \left(u \sin(u^3/3) \frac{\cos(uz)}{\sqrt{z}} + u \cos(u^3/3) \frac{\sin(uz)}{\sqrt{z}} \right) dz.$$

The integral over z can be found in Section 3.757 of [95],

$$\int_0^\infty \frac{\cos(uz)}{\sqrt{z}} \, dz = \int_0^\infty \frac{\sin(uz)}{\sqrt{z}} \, dz = \sqrt{\frac{\pi}{2u}}, \quad \text{(A.29)}$$

which gives

$$\int_0^\infty \frac{\mathrm{Ai}'(z)}{\sqrt{z}} \, dz = -\frac{1}{\sqrt{2\pi}} \int_0^\infty \sqrt{u} [\sin(u^3/3) + \cos(u^3/3)] \, du.$$

With the substitution $u^3 = 3v$ we bring this expression into the form (A.29):

$$\int_0^\infty \frac{\mathrm{Ai}'(z)}{\sqrt{z}} \, dz = -\frac{1}{\sqrt{6\pi}} \int_0^\infty \left(\frac{\cos v}{\sqrt{v}} + \frac{\sin v}{\sqrt{v}} \right) dv = -\frac{1}{\sqrt{3}}. \quad \text{(A.30)}$$

- The next integral can be reduced to the previous one with two integrations by parts:

$$\int_0^\infty \sqrt{z} \int_z^\infty \mathrm{Ai}(z') \, dz' \, dz = -\frac{2}{3} \int_0^\infty z^{3/2} \, \mathrm{Ai}(z) \, dz$$

$$= \frac{2}{3} \int_0^\infty \sqrt{z} \, \mathrm{Ai}''(z) \, dz = -\frac{1}{3} \int_0^\infty \frac{\mathrm{Ai}'(z)}{\sqrt{z}} \, dz = \frac{1}{3\sqrt{3}}. \quad \text{(A.31)}$$

- The next integrals are needed for the variance of the photon energy. Again, integrating twice in parts brings them into previously presented forms:

$$\int_0^\infty z^{5/2} \, \mathrm{Ai}'(z) \, dz = -\frac{5}{4\sqrt{3}} \quad \text{(A.32)}$$

and

$$\int_0^\infty z^{7/2} \, dz \int_z^\infty \mathrm{Ai}(z') \, dz' = \frac{35}{36\sqrt{3}}. \quad \text{(A.33)}$$

- To calculate the average circular polarization we need an integral of the form

$$\int_0^\infty y^5 \, \mathrm{Ai}(by^2) \, \mathrm{Ai}'(by^2) \, dy = \frac{d}{db} \left(\frac{1}{2} \int_0^\infty y^3 \, \mathrm{Ai}^2(by^2) \, dy \right).$$

We use the expression (A.8) for the square of the Airy function and the substitutions applied for the integral (A.11) to obtain

$$\frac{1}{2} \int_0^\infty y^3 \, \text{Ai}^2(by^2) \, dy = \frac{1}{12\pi b^2} \int_0^\infty z^{3/2} \, \text{Ai}(z) \, dz.$$

This integral is obtained by generalizing (A.30),

$$\int_0^\infty \frac{\text{Ai}'(cz)}{\sqrt{z}} \, dz = -\frac{1}{\sqrt{3}\sqrt{c}},$$

differentiating it with respect to c,

$$\int_0^\infty \sqrt{z} \, \text{Ai}''(cz) \, dc = \int_0^\infty cz^{3/2} \, \text{Ai}(cz) \, dz = \frac{1}{2\sqrt{3}c^{3/2}},$$

and setting $c = 1$:

$$\frac{1}{2} \int_0^\infty y^3 \, \text{Ai}^2(by^2) \, dy = \frac{1}{24\sqrt{3}\pi b^2}.$$

Differentiating this with respect to b gives our original integral:

$$\int_0^\infty y^5 \, \text{Ai}(by^2) \, \text{Ai}'(by^2) \, dy = -\frac{1}{12\sqrt{3}\pi b^3}.$$

We adapt it to the form we need in (5.38) with the substitution $y = p^{1/3}$, $dy = 1/(3p^{2/3}) \, dp$ and obtain

$$\int_0^\infty p \, \text{Ai}(bp^{2/3}) \, \text{Ai}'(bp^{2/3}) \, dp = -\frac{\sqrt{3}}{12\pi b^3}. \tag{A.34}$$

Appendix B

Bessel functions

B.1 General relations

Bessel functions are widely used and documented in many books such as [41, 95] and, in particular, [96]. Like for the Airy function we summarize some relevant properties, give an approximation for large order, and calculate two sums that are difficult to find.

We need here only Bessel functions of integer order n, which can be defined by the integral

$$J_n(z) = \frac{1}{\pi} \int_0^\pi \cos(z \sin \xi - n\xi) \, d\xi = \frac{1}{2\pi} \int_{-\pi}^\pi \cos(z \sin \xi - n\xi) \, d\xi. \qquad (B.1)$$

From this we can easily obtain the relation between different orders and the derivative,

$$(J_{n-1}(z) + J_{n+1}(z)) = \frac{2n}{z} J_n(z)$$

$$(J_{n-1}(z) - J_{n+1}(z)) = 2 \frac{dJ_n(z)}{dz} = 2J_n'(z), \qquad (B.2)$$

and, by taking combinations of these equations, also the two expressions for the derivative:

$$J_n'(z) = J_{n-1}(z) - \frac{n}{z} J_n(z) = -J_{n+1}(n) + \frac{n}{z} J_n(z). \qquad (B.3)$$

The Bessel function of non-negative integer order $n \geq 0$ can be expanded into a power series,

$$J_n(z) = \frac{z^n}{2^n n!} \left(1 - \frac{z^2}{2(2n+2)} + \frac{z^4}{2 \cdot 4(2n+2)(2n+4)} - \cdots \right), \qquad (B.4)$$

which can be extended to negative orders with the relation

$$J_{-n}(z) = (-1)^n J_n(z). \qquad (B.5)$$

From the above expression we also obtain the Bessel function for negative arguments:

$$J_n(-z) = (-1)^n J_n(z). \qquad (B.6)$$

With the relation (B.2) we obtain also the symmetry relations for the derivative of the Bessel function:

$$J_{-n}'(z) = (-1)^n J_n'(z), \qquad J_n'(-z) = -(-1)^n J_n(z). \qquad (B.7)$$

The product of two Bessel functions of the same argument but different orders with respect to n can be obtained by multiplying two integral representations (B.1) and is to be found in (6.681) of [95],

$$J_n(z)J_m(z) = \frac{2}{\pi} \int_0^{\pi/2} J_{n+m}(2z\cos\xi)\cos((n-m)\xi)\,d\xi, \tag{B.8}$$

which becomes, for the special case $m = n$,

$$J_n^2(z) = \frac{2}{\pi} \int_0^{\pi/2} J_{2n}(2z\cos\xi)\,d\xi. \tag{B.9}$$

B.2 The approximation for large order and arguments

To compare the synchrotron radiation emitted on a closed circle with the single-traversal case, we need an expression for the Bessel function and its derivative valid for large arguments and order. Such approximation are presented by equations (9.3.35) and (9.3.43) in [41] but their application for our need is rather lengthy. We give here an approach that is not as general and rigorous but shorter and more transparent and applies only to our special condition.

The vertical field component at the nth harmonic is given by an expression (4.26) of the form

$$nJ_n(n\beta\cos\psi) = \frac{n}{2\pi} \int_{-\pi}^{\pi} \cos(n\beta\cos\psi\sin\xi - n\xi)\,d\xi. \tag{B.10}$$

The ultra-relativistic approximation gives $\beta \approx 1$, $\gamma \gg 1$, $\psi \ll 1$, and

$$\beta\cos\psi \approx 1 - \frac{1+\gamma^2\psi^2}{2\gamma^2} < 1. \tag{B.11}$$

We obtain

$$nJ_n(n\beta\cos\psi) = \frac{n}{2\pi} \int_{-\pi}^{\pi} \cos\left[\left(1 - \frac{1+\gamma^2\psi^2}{2\gamma^2}\right)n\sin\xi - n\xi\right]d\xi.$$

The second term inside the cosine function is always larger than the first one and will, for $n \gg 1$, lead to a fast oscillation, producing nearly perfect cancelation. Therefore, only relatively small values $\xi < 1$ will be important for the integral and we can develop $\sin\xi$, giving for the argument of the cosine function

$$n\beta\cos\psi\sin\xi - n\xi \approx n\beta\cos\psi\left(\xi - \frac{\xi^3}{6}\right) - n\xi$$

$$\approx n\left(1 - \frac{1+\gamma^2\psi^2}{2\gamma^2}\right)\left(\xi - \frac{\xi^3}{6}\right) - n\xi$$

$$= -n\frac{\xi^3}{6} - n\frac{1+\gamma^2\psi^2}{2\gamma^2}\xi,$$

where we neglected the term $n\xi^3(1 + \gamma^2\theta^2)/(12\gamma^2)$. We make the change of variable

$$\xi = \left(\frac{2}{n}\right)^{1/3} \zeta$$

and obtain the limit of the integral for $n \gg 1$:

$$\zeta_m = \pm\left(\frac{n}{2}\right)^{1/3} \pi \approx \pm\infty.$$

Our original expression for the Bessel function (B.10) of very large order and argument is now

$$n J_n(n\beta \cos\psi) = \frac{2}{2\pi} \left(\frac{n}{2}\right)^{2/3} \int_{-\infty}^{\infty} \cos\left[\left(\frac{n}{2\gamma^3}\right)^{2/3} (1 + \gamma^2\psi^2)\zeta + \frac{\zeta^3}{3}\right] d\zeta.$$

On comparing this with the integral representation of the Airy function (A.1) we find

$$n J_n(n\beta \cos\psi) = 2\left(\frac{n}{2}\right)^{2/3} \text{Ai}\left[\left(\frac{n}{2\gamma^3}\right)^{2/3} (1 + \gamma^2\psi^2)\right]. \tag{B.12}$$

Next we seek a similar expression for the derivative of the Bessel function of large order and argument. We take the derivative with respect to the argument $z = n\beta \cos\pi$ of (B.10) and make the same approximation $\xi < 1$ as before:

$$J_n'(z) = J_n'(n\beta \cos\psi) = \frac{d J_n(n\beta \cos\psi)}{d(n\beta \cos\psi)}$$

$$= -\frac{1}{2\pi} \int_{-\pi}^{\pi} \sin\xi \sin(n\beta \cos\psi \sin\xi - n\xi) \sin\xi \, d\xi$$

$$\approx -\frac{1}{2\pi} \int_{-\pi}^{\pi} \left(\xi - \frac{\xi^3}{6}\right) \sin\left[n\left(1 - \frac{1 + \gamma^2\psi^2}{2\gamma^2}\right)\left(\xi - \frac{\xi^3}{6}\right) - n\xi\right] d\xi$$

$$\approx -\frac{1}{2\pi} \int_{-\pi}^{\pi} \xi \sin\left(-n\frac{\xi^3}{6} - n\frac{1 + \gamma^2\psi^2}{2\gamma^2}\xi\right) d\xi.$$

We transform the variable $\xi = \sqrt[3]{n/2}\zeta$, approximate the limits as before, and obtain for the derivative of the Bessel function multiplied by n for large order and arguments

$$n J_n'(n\beta \cos\psi) = -\frac{2}{2\pi} \left(\frac{n}{2}\right)^{1/3} \int_{-\infty}^{\infty} \zeta \sin\left[\left(\frac{n}{2\gamma^3}\right)^{2/3} (1 + \gamma^2\psi^2)\zeta + \frac{\zeta^3}{3}\right] d\zeta.$$

On comparing this with the expression for the derivative of the Airy function (A.1) we find

$$n J_n'(n\beta \cos\psi) = -2\left(\frac{n}{2}\right)^{1/3} \text{Ai}'\left[\left(\frac{n}{2\gamma^3}\right)^{2/3} (1 + \gamma^2\psi^2)\right]. \tag{B.13}$$

B.3 Sums over squares of Bessel functions

In calculating the total power radiated by a charge moving on a closed circle we need two sums over squares of Bessel functions, which were first given in [3] and will be explained here:

$$S_a = \sum_{n=1}^{\infty} n^2 J_n^2(nz), \qquad S_b = \sum_{n=1}^{\infty} n^2 J_n'^2(nz). \tag{B.14}$$

For our application we have $z = \beta \cos \psi < 1$. It is important that this quantity is smaller than unity in order for some operations to be valid.

We start with the first sum and express the square of the Bessel function by (B.9),

$$S_a = \frac{2}{\pi} \int_0^{\pi/2} d\xi \sum_{n=1}^{\infty} n^2 J_{2n}(2nz \cos \xi),$$

which contains a sum given by equation (17.22(8)) in [96],

$$\sum_{n=1}^{\infty} n^2 J_{2n}(nz') = \frac{8z'^2(4 + z'^2)}{(4 - z'^2)^4}. \tag{B.15}$$

With $z' = 2z \cos \xi$ we obtain the sum

$$S_a = \frac{2}{\pi} \int_0^{\pi/2} \frac{z^2 \cos^2 \xi \, (1 + z^2 \cos^2 \xi)}{2(1 - z^2 \cos^2 \xi)^4} \, d\xi, \tag{B.16}$$

which is integrated over ξ:

$$S_a = \sum_{n=1}^{\infty} n^2 J_n^2(nz) = \frac{z^2(4 + z^2)}{16(1 - z^2)^{7/2}}. \tag{B.17}$$

For the second sum, S_b, we express the derivative $J_n'(nz)$ once with the first and once with the second expression of (B.3),

$$J_n'^2(nz) = \frac{1}{z} J_n(nz)(J_{n-1}(nz) + J_{n+1}(nz)) - J_{n-1}(nz)J_{n+1}(nz) - \frac{1}{z^2} J_n^2(nz),$$

and collect Bessel functions of different orders, giving for the sum

$$\sum_{n=1}^{\infty} n^2 J_n'^2(nz) = \frac{1}{z^2} \sum_{n=1}^{\infty} n^2 J_n^2(nz) - \sum_{n=1}^{\infty} n^2 J_{n-1}(nz)J_{n+1}(nz). \tag{B.18}$$

The first sum has already been calculated as (B.17) and for the second sum we express the product of two Bessel functions by using the integral (B.8):

$$\sum_{n=1}^{\infty} n^2 J_{n-1}(nz)J_{n+1}(nz) = \frac{2}{\pi} \int_0^{\pi/2} d\xi \cos(2\xi) \sum_{n=1}^{\infty} n^2 J_{2n}(2z \cos \xi).$$

This latest sum is given by (B.15) for $z' = 2z \cos \xi$, resulting in

$$\sum_{n=1}^{\infty} n^2 J_{n-1}(nz)J_{n+1}(nz) = \frac{2}{\pi} \int_0^{\pi/2} \frac{z^2 \cos^2 \xi \, (1 + z^2 \cos^2 \xi) \cos(2\xi)}{2(1 - z^2 \cos^2 \xi)^4} \, d\xi$$

$$= \frac{z^2(2 + 3z^2)}{16(1 - z^2)^{7/2}}.$$

We obtain for the total expression (B.18) of the second sum

$$S_b = \sum_{n=1}^{\infty} n^2 J_n'^2(nz) = \frac{4 + 3z^2}{16(1 - z^2)^{5/2}}. \tag{B.19}$$

B.4 Series of Bessel functions

We rewrite the integral presentation (B.1) of the Bessel function as

$$J_n(z) = \frac{1}{\pi} \int_0^\pi [\cos(z \sin \xi) \cos(n\xi) + \sin(z \sin \xi) \sin(n\xi)] \, d\xi. \tag{B.20}$$

The function $\cos(z \sin \xi)$ is an even and periodic function with period π, whereas $\sin(z \sin \xi)$ is odd and has the same periodic properties. The two integrals represent the coefficients of the Fourier series of these functions which are given by the Bessel functions $J_n(z)$. The symmetry properties around the angle $\xi = \pi/2$ for the two functions are

$$\cos(z \sin \xi) = \cos(z \sin(\pi - \xi)), \qquad \sin(z \sin \xi) = \sin(z \sin(\pi - \xi)),$$

with, for $\cos(n\xi)$,

$$\cos(n(\pi - \xi)) = \cos(n\xi) \qquad \text{if } n \text{ is even}$$
$$\cos(n(\pi - \xi)) = -\cos(n\xi) \qquad \text{if } n \text{ is odd},$$

whereas for $\sin(n\xi)$ the symmetry is opposite,

$$\sin(n(\pi - \xi)) = -\sin(n\xi) \qquad \text{if } n \text{ is even}$$
$$\sin(n(\pi - \xi)) = \sin(n\xi) \qquad \text{if } n \text{ is odd}.$$

Therefore the first integral in (B.20) vanishes for odd values of n and the second integral vanishes for even values. With this we obtain for the Fourier series of the two functions

$$\cos(z \sin \xi) = J_0(z) + 2 \sum_{n=1}^{\infty} J_{2n}(z) \cos(2n\xi)$$

$$\sin(z \sin \xi) = 2 \sum_{n=1}^{\infty} J_{2n-1}(z) \sin((2n - 1)\xi),$$

where we used the symmetry relations of the Bessel functions with respect to order, (B.6), and of the trigonometric function with respect to argument.

We multiply the second equation by i, add to it the first equation, and thereby obtain a single complex Fourier series,

$$\cos(z \sin \xi) + i \sin(z \sin \xi) = e^{iz \sin \xi} = \sum_{n=-\infty}^{\infty} J_n(z) e^{in\xi}, \tag{B.21}$$

which we will use to express the radiation field of a strong undulator.

Appendix C

Developments of strong-undulator radiation

C.1 The plane-undulator radiation

In Section 8.3.6 the radiation from a plane undulator has been developed with respect to powers of the parameter K_u^*. The results for the angular and spectral power distributions are given here. Each harmonic is given only to the lowest power of K_u^*.

The angular power distributions for the two modes of polarization are

$$
\frac{dP_{1\sigma}}{d\Omega} \approx P_u \frac{3\gamma^{*2}}{\pi} \frac{(1 - \gamma^{*2}\theta^2 \cos(2\phi))^2)^2}{\left(1 + K_u^2/2\right)^2 (1 + \gamma^{*2}\theta^2)^5}
$$

$$
\frac{dP_{1\pi}}{d\Omega} \approx P_u \frac{3\gamma^{*2}}{\pi} \frac{(\gamma^{*2}\theta^2 \sin(2\phi))^2}{\left(1 + K_u^2/2\right)^2 (1 + \gamma^{*2}\theta^2)^5}
$$

$$
\frac{dP_{2\sigma}}{d\Omega} \approx P_u \frac{3\gamma^{*2}}{\pi} K_u^{*2} \frac{(\gamma^*\theta \cos\phi)^2(5(1 + \gamma^{*2}\theta^2) - 8(\gamma^*\theta \cos\phi)^2)^2}{(1 + K_u^2/2)^2(1 + \gamma^{*2}\theta^2)^7}
$$

$$
\frac{dP_{2\pi}}{d\Omega} \approx P_u \frac{3\gamma^{*2}}{\pi} K_u^{*2} \frac{(\gamma^*\theta \sin\phi)^2(1 + \gamma^{*2}\theta^2 - 8(\gamma^*\theta \cos\phi)^2)^2}{\left(1 + K_u^2/2\right)^2 (1 + \gamma^{*2}\theta^2)^7}
$$
(C.1)

$$
\frac{dP_{3\sigma}}{d\Omega} \approx P_u \frac{3\gamma^{*2}}{\pi} K_u^{*4} \frac{81}{64} \frac{((1 + \gamma^{*2}\theta^2)^2 - 18(1 + \gamma^{*2}\theta^2)(\gamma^*\theta \cos\phi)^2 + 24(\gamma^*\theta \cos\phi)^4)^2}{\left(1 + K_u^2/2\right)^2 (1 + \gamma^{*2}\theta^2)^9}
$$

$$
\frac{dP_{3\pi}}{d\Omega} \approx P_u \frac{3\gamma^{*2}}{\pi} K_u^{*4} \frac{729}{64} \frac{(\gamma^{*2}\theta^2 \sin(2\phi))^2(1 + \gamma^{*2}\theta^2 - 4(\gamma^*\theta \cos\phi)^2)^2}{\pi \left(1 + K_u^2/2\right)^2 (1 + \gamma^{*2}\theta^2)^9}.
$$

The spectral power distributions for the two modes of polarization are

$$
\frac{dP_{1\sigma}}{d\omega_1} \approx \frac{P_u}{\omega_{10}} \frac{3}{2\left(1 + K_u^2/2\right)^2} \left(\frac{\omega_1}{\omega_{10}}\right) \left(1 - 2\left(\frac{\omega_1}{\omega_{10}}\right) + 3\left(\frac{\omega_1}{\omega_{10}}\right)^2\right)
$$

$$
\frac{dP_{1\pi}}{d\omega_1} \approx \frac{P_u}{\omega_{10}} \frac{3}{2\left(1 + K_u^2/2\right)^2} \left(\frac{\omega_1}{\omega_{10}}\right) \left(1 - 2\left(\frac{\omega_1}{\omega_{10}}\right) + \left(\frac{\omega_1}{\omega_{10}}\right)^2\right)
$$

$$
\frac{dP_{2\sigma}}{d\omega_2} \approx \frac{P_u}{\omega_{10}} \frac{3K_u^{*2}}{32\left(1 + K_u^2/2\right)^2} \left(\frac{\omega_2}{\omega_{10}}\right)^2 \left(10 - 25\left(\frac{\omega_2}{\omega_{10}}\right) + 30\left(\frac{\omega_2}{\omega_{10}}\right)^2 - 10\left(\frac{\omega_2}{\omega_{10}}\right)^3\right)
$$

$$\frac{dP_{2\pi}}{d\omega_2} \approx \frac{P_u}{\omega_{10}} \frac{3K_u^{*2}}{32\left(1 + K_u^2/2\right)^2}\left(\frac{\omega_2}{\omega_{10}}\right)^2\left(10 - 17\left(\frac{\omega_2}{\omega_{10}}\right) + 10\left(\frac{\omega_2}{\omega_{10}}\right)^2 - 2\left(\frac{\omega_2}{\omega_{10}}\right)^3\right)$$

$$\frac{dP_{3\sigma}}{d\omega_3} \approx \frac{P_u}{\omega_{10}} \frac{3K_u^{*4}}{1152\left(1 + K_u^2/2\right)^2}\left(\frac{\omega_3}{\omega_{10}}\right)^3 \qquad\qquad\text{(C.2)}$$

$$\times \left(180 - 486\left(\frac{\omega_3}{\omega_{10}}\right) + 549\left(\frac{\omega_3}{\omega_{10}}\right)^2 - 240\left(\frac{\omega_3}{\omega_{10}}\right)^3 + 35\left(\frac{\omega_3}{\omega_{10}}\right)^4\right)$$

$$\frac{dP_{3\pi}}{d\omega_3} \approx \frac{P_u}{\omega_{10}} \frac{3K_u^{*4}}{1152\left(1 + K_u^2/2\right)^2}\left(\frac{\omega_3}{\omega_{10}}\right)^3$$

$$\times \left(162 - 270\left(\frac{\omega_3}{\omega_{10}}\right) + 171\left(\frac{\omega_3}{\omega_{10}}\right)^2 - 48\left(\frac{\omega_3}{\omega_{10}}\right)^3 + 5\left(\frac{\omega_3}{\omega_{10}}\right)^4\right).$$

C.2 The helical-undulator radiation

In Section 9.5.6 the radiation from a plane undulator has been developed with respect to powers of the parameter K_u^*. The results for the angular and spectral power distributions as well as the integrals giving the radiation in each harmonic are presented here. Each harmonic is given only to the lowest power of the helical-undulator parameter K_{hu}^*.

The development of the angular power distribution of helical-undulator radiation for the first three harmonics, each to the lowest order of K_{uh}^* is given below. The two terms F_{hx+} and F_{hx-} correspond to the two modes of helical polarization:

$$\frac{dP_1}{d\Omega} = P_h\gamma_h^{*2}(F_{h1+} + F_{h1-}) \approx P_h\gamma_h^{*2}\frac{3}{\pi}\frac{1 + \gamma_h^{*4}\theta^4}{\left(1 + \gamma^{*2}\theta^2\right)^5\left(1 + K_u^2\right)^2}$$

$$\frac{dP_2}{d\Omega} = P_h\gamma_h^{*2}(F_{h2+} + F_{h2-}) \approx P_h\gamma_h^{*2}\frac{3}{\pi}\frac{16K_{uh}^{*2}\gamma_h^{*2}\theta^2\left(1 + \gamma_h^{*4}\theta^4\right)}{\left(1 + \gamma^{*2}\theta^2\right)^7\left(1 + K_u^2\right)^2} \qquad\text{(C.3)}$$

$$\frac{dP_3}{d\Omega} = P_h\gamma_h^{*2}(F_{h3+} + F_{h3-}) \approx P_h\gamma_h^{*2}\frac{3}{\pi}\frac{729K_{uh}^{*4}\gamma_h^{*4}\theta^4\left(1 + \gamma_h^{*4}\theta^4\right)}{4(1 + \gamma^{*2}\theta^2)^9\left(1 + K_u^2\right)^2}.$$

The development of the spectral power distribution of helical-undulator radiation for the first three harmonics, each to the lowest order of K_{uh}^*, is given below. The two terms in the square brackets correspond to the two modes of helical polarization:

$$\frac{dP_1}{d\omega_1}\frac{\omega_{10}}{P_h} \approx \frac{3}{(1 + K^2)^2}\left(\frac{\omega_1}{\omega_{10}}\right)\left[\left(\frac{\omega_1}{\omega_{10}}\right)^2 + \left(1 - \frac{\omega_1}{\omega_{10}}\right)^2\right]$$

$$\frac{dP_{h2}}{d\omega_2}\frac{\omega_{10}}{P_h} \approx \frac{3K_{uh}^{*2}}{4\left(1 + K_u^2\right)^2}\left(\frac{\omega_2}{\omega_{10}}\right)^2\left(2 - \frac{\omega_2}{\omega_{10}}\right)\left[\left(\frac{\omega_2}{\omega_{10}}\right)^2 + \left(2 - \frac{\omega_2}{\omega_{10}}\right)^2\right] \qquad\text{(C.4)}$$

$$\frac{dP_{h3+}}{d\omega_3}\frac{\omega_{10}}{P_h} \approx \frac{3K_{uh}^{*4}}{36\left(1 + K_u^2\right)^2}\left(\frac{\omega_3}{\omega_{10}}\right)^3\left(3 - \frac{\omega_3}{\omega_{10}}\right)^2\left[\left(\frac{\omega_3}{\omega_{10}}\right)^2 + \left(3 - \frac{\omega_3}{\omega_{10}}\right)^2\right].$$

Powers for a helical undulator in each harmonic for the two modes of circular polarization and the total radiation are given below:

$$P_{1+} = P_h \frac{1}{\left(1 + K_u^2\right)^2} \left(\frac{3}{4} - \frac{5}{20} K_{uh}^{*2} + \frac{49}{1120} K_{uh}^{*4} + \cdots\right)$$

$$P_{1-} = P_h \frac{1}{\left(1 + K_u^2\right)^2} \left(\frac{1}{4} - \frac{3}{20} K_{uh}^{*2} + \frac{39}{1120} K_{uh}^{*4} + \cdots\right)$$

$$P_1 = P_h \frac{1}{\left(1 + K_u^2\right)^2} \left(1 - \frac{2}{5} K_{uh}^{*2} + \frac{11}{140} K_{uh}^{*4} + \cdots\right)$$

$$P_{2+} = P_h \frac{1}{\left(1 + K_u^2\right)^2} \left(\frac{8}{5} K_{uh}^{*2} - \frac{48}{35} K_{uh}^{*4} + \cdots\right)$$

$$P_{2-} = P_h \frac{1}{\left(1 + K_u^2\right)^2} \left(\frac{4}{5} K_{uh}^{*2} - \frac{32}{35} K_{uh}^{*4} + \cdots\right) \tag{C.5}$$

$$P_2 = P_h \frac{1}{\left(1 + K_u^2\right)^2} \left(\frac{12}{5} K_{uh}^{*2} - \frac{16}{7} K_{uh}^{*4} + \cdots\right)$$

$$P_{3+} = P_h \frac{1}{\left(1 + K_u^2\right)^2} \left(\frac{729}{224} K_{uh}^{*4} - \cdots\right)$$

$$P_{3-} = P_h \frac{1}{\left(1 + K_u^2\right)^2} \left(\frac{2187}{1120} K_{uh}^{*4} - \cdots\right)$$

$$P_2 = P_h \frac{1}{\left(1 + K_u^2\right)^2} \left(\frac{729}{140} K_{uh}^{*4} - \cdots\right).$$

References

[1] R. Y. Tsien, 'Pictures of Dynamic Electric Fields,' *Am. J. Phys.* **40** (1971) 46.

[2] T. Shintake, 'New 2D Real-time Radiation Field Simulator,' *Proceedings of the 8th European Particle Accelerator Conference, EPAC 2002, Paris 2002*, CERN, Geneva, 2002.

[3] G. A. Schott, *Electromagnetic Radiation*, Cambridge University Press, Cambridge, 1912.

[4] A. A. Sokolov and I. M. Ternov, *Synchrotron Radiation*, translated by M. August and H. R. Kissener, Pergamon Press, Rossendorf, 1966.

[5] V. N. Baier, V. M. Katkov, and V. S. Fadin, *Radiation from Relativistic Electrons*, Atomizdat, Moscow, 1973 (in Russian).

[6] A. A. Sokolov and I. M. Ternov, *Radiation from Relativistic Electrons*, translated by S. Chomet and edited by C.W. Kilmister, American Institute of Physics Translation Series, New York, 1986.

[7] V. A. Bordovitsyn, *Synchrotron Radiation Theory and its Development*, World Scientific, Singapore, 1999.

[8] H. Wiedemann, *Synchrotron Radiation*, Springer Verlag, Berlin, 2002.

[9] J. D. Jackson, *Classical Electrodynamics*, John Wiley, New York, 1962, 1974, and 1998.

[10] M. Schwartz, *Principles of Electrodynamics*, McGraw-Hill, New York, 1972.

[11] M. Zahn, *Electromagnetic Field Theory*, Wiley, New York, 1979.

[12] L. Eyges, *The Classical Electromagnetic Field*, Dover Publications, Inc., New York, 1980.

[13] J. Schwinger, L. L. DeRaad Jr, K. A. Milton, and Wu-yang Tsai, *Classical Electrodynamics*, Perseus Books/Westview Press, 1998.

[14] K. Wille, *Physics of Particle Accelerators and Synchrotron Radiation Sources*, B. G. Teubner, Stuttgart, 1992.

[15] H. Wiedemann, *Particle Accelerator Physics I and II*, Springer Verlag, Heidelberg, 1993.

[16] A. W. Chao and M. Tigner, *Handbook of Accelerator Physics and Engineering*, World Scientific, Singapore, 1998.

[17] H. Wiedemann, 'Synchrotron Radiation,' Chapter 3.1 in [16].

[18] M. Sands, 'The Physics of Electron Storage Rings,' SLAC Report 121, SLAC, Stanford, 1970.

[19] A. Hofmann, 'Theory of Synchrotron Radiation,' SLAC ACD-Note 38, SLAC, Stanford, 1986.

[20] Kwang-Je Kim, 'Characteristics of Synchrotron Radiation,' *AIP Conference Proceedings 184*, ed. M. Month and M. Dienes, AIP, New York, 1989, p. 567.

[21] H. Winick and S. Doniach, eds, *Synchrotron Radiation Research*, Plenum Press, New York, 1980.

[22] E. Koch, *Handbook of Synchrotron Radiation*, Volumes I and II, North-Holland Publishing Company, Amsterdam, 1983.

[23] G. N. Greaves and I. H. Munrow, eds, *Synchrotron Radiation, Sources and Applications*, 30th Scottish Universities Summer School in Physics, held in Aberdeen, 1985.

[24] H. Winick, ed., *Synchrotron Radiation Sources, a Primer*, World Scientific, Singapore, 1994.

[25] S. Krinsky, M. L. Perlman, and R. E. Watson, 'Characteristics of Synchrotron Radiation and of its Sources,' Chapter 2 of [22].

[26] J. P. Blewett, 'Synchrotron Radiation – 1873 to 1947,' *NIM* **A266** (1988) 1.

[27] I. M. Ternov, 'Synchrotron Radiation Research at Moscow State University,' *NIM* **152** (1978) 213.

[28] A. Liénard, 'Champ électrique et magnetique produit par une charge électrique concentrée en un point et animée d'un mouvement quelquonque' (Electric and magnetic field of a point charge undergoing a general motion), *L'Éclairage Électrique* **16** (1898) 5.

[29] E. Wiechert, *Archives Neerlandaises* (1900) 546.

[30] G. A. Schott, *Annalen Phys.* **24** (1907) 635.

[31] G. A. Schott, *Phil. Mag.* **13** (1907) 194.

[32] I. Ya. Pomeranchuk, *J. Phys. USSR* **2** (1940) 65.

[33] D. D. Iwanenko and I. Ya. Pomeranchuk, *C. R. Acad. Sci USSR* **44** (1944) 315.

[34] D. Iwanenko and K. Pomeranchuk, *Phys. Rev.* **65** (1944) 343.

[35] L. Arzimovich and I. Ya. Pomeranchuk, 'The Radiation of Fast Electrons in the Magnetic Field,' *J. Phys. USSR* **9** (1945) 267.

[36] J. Schwinger, 'On the Radiation by Electrons in a Betatron,' unpublished note (1945) and presentation at the American Physical Society meeting (1946) but edited by M. Furman and published as LBNL-39088, 1996.

[37] J. Schwinger, 'On the Classical Radiation of Accelerated Electrons,' *Phys. Rev.* 75 (1949) 1912.

[38] J. Blewett, *Phys. Rev.* **69** (1946) 87.

[39] E. G. Bessonov, 'Conventionally Strange Electromagnetic Waves,' *NIM* **A308** (1991) 135.

[40] B. M. Kincaid, 'Can an Electron Radiate DC?,' Lawrence Berkeley Laboratory report LSGN-244, 1996.

[41] M. Abramowitz and I. A. Stegun, *Handbook of Mathematical Functions*, Dover, New York, 1970.

[42] R. H. Helm, M. J. Lee, P. L. Morton, and M. Sands, 'Evaluation of Synchrotron Radiation Integrals,' *IEEE Trans. Nucl. Sci.* **NS-20** (1973) 900.

[43] J. K. Kim, 'Polarized Nature of Synchrotron Radiation,' Lawrence Berkeley Laboratory report LBL-31242, 1991.

[44] E. M. Purcell, 'Production of Synchrotron Radiation by Wiggler Magnets,' Cambridge Electron Accelerator Internal Note 1972.

[45] M. B. Moisev, M. M. Nikitin and N. I. Fedosov, *Sov. Phys. J.* **21** (1987) 332.

[46] J. K. Kim, 'A Synchrotron Radiation Source with Arbitrarily Adjustable Elliptic Polarization,' *NIM* **219** (1984) 425.

[47] V. L. Ginzburg, *Izv. Akad. Nauk SSSR, Ser. Fiz.* **11** (1947) 165.

[48] H. Motz, 'Applications of the Radiation from Fast Electron Beams,' *J. Appl. Phys.* **22** (1951) 527.

[49] H. Motz, W. Thon, and R. N. Whitehurst, 'Experiments on Radiation by Fast Electron Beams,' *J. Appl. Phys.* **24** (1953) 826.

[50] D. F. Alferov, Yu. A. Bashmakov, and E. G. Bessonov, *Synchrotron Radiation*, edited by N. G. Basov, Consultants Bureau, New York, 1976.

[51] V. N. Baier, V. M. Katkov, and V. M. Strakhovenko, 'Radiation Emitted by Relativistic Particles in Periodic Structures,' *Sov. Phys. JETP* **36** (1973) 1120.

[52] D. F. Aferov, Yu. A. Bashmakov, and E. G. Bessonov, 'Undulator Radiation,' *Zh. Tekh. Fiz.* **43** (1973) 2126.

[53] J. E. Spencer and H. Winick, 'Wiggler Systems as Sources of Electromagnetic Radiation,' Chapter 21 in [21].

[54] S. Krinsky, 'Undulators as Sources of Synchrotron Radiation,' *Proceedings of the 1983 Particle Accelerator Conference, IEEE Trans. Nucl. Sci.* **NS–30** (1983) 3078.

[55] V. Ya. Epp and G. K. Razina, 'Radiation in a Wiggler with Sinusoidal Magnetic Field,' *NIM* **A307** (1991) 562.

[56] V. Ya. Epp, 'Undulator Radiation,' Chapter 5 of [7].

[57] G. Brown, K. Halbach, J. Harris, and H. Winick, 'Wiggler and Undulator Magnets – a Review,' *NIM* **208** (1983) 65.

[58] R. P. Walker, 'Insertion Devices: Undulators and Wigglers,' contribution to the CERN Accelerator School Synchrotron Radiation and Free Electron Lasers, edited by S. Turner, CERN, Geneva, 1998.

[59] B. M. Kincaid, 'A Short-period Helical Wiggler as on Improved Source of Synchrotron Radiation,' *J. Appl. Phys.* **48** (1977) 2684.

[60] R. Coïsson, 'On Synchrotron Radiation in Non-uniform Magnetic Fields,' *Optics Commun.* **22** (1977) 135.

[61] A. I. Chechin and N. V. Smolyakov, 'Irregular Structure Undulator,' *NIM* **A308** (1991) 86.

[62] P. Chaix, D. Iracane, and F. Desrayaud, 'The Two Frequency Wiggler: Efficiency Enhancement and Sideband Inhibition,' *Nucl. Instrum. Methods. Phys. Res.* A **341** (1994) 215.

[63] A. Hofmann, 'Diagnostics with Undulator Radiation having a Gaussian Angular Distribution,' edited by S. Machida and K. Hirata, *KEK Proceedings* **95-7** (1995) 231.

[64] C. Bamer, S. C. Berridge, S. J. Boege, W. M. Bugg, C. Bula, D. L. Burke, F. C. Field, G. Horton-Schmid, T. Koffas, T. Kotseroglou, K. T. McDonald, A. C. Melissionos, D. D. Meyerhofer, E. J. Prebys, W. Ragg, D. A. Reis, K. Shmakov, J. E. Spencer, D. Walz, and A.W. Weidemann, 'Studies of Nonlinear QED in Collisions of 46.6 GeV Electrons with Intense Laser Pulses,' SLAC-PUB-8063, SLAC, Stanford, 1999.

[65] G. R. Fowles, *Introduction to Modern Optics*, Dover Publications, Inc., New York, 1989.

[66] A. Hofmann and K. W. Robinson, 'Measurement of the Cross Section of a High-Energy Electron Beam by means of the X-Ray Portion of the Synchrotron Radiation,' Proceedings of the 1971 Particle Accelerator Conference, *IEEE Trans. Nucl. Sci.* **NS-18** (1971) 973.

[67] A. Hofmann and F. Méot, 'Optical Resolution of Beam Cross-Section Measurements by means of Synchrotron Radiation,' *Nucl. Instrum. Methods Phys. Res.* 203 (1982) 483.

[68] A. Ogata, 'Focusing of Synchrotron Radiation,' *NIM* **A259** (1987) 566.

[69] O. Chubar; 'Resolution Improvement in Beam Profile Measurements with Synchrotron Light,' *Proceedings of the IEEE Particle Accelerator Conference PAC-93* (1993) p. 2510.

[70] Å. Andersson, M. Ericsson, and O. Chubar; 'Beam Profile Measurements with Visible Synchrotron Light on Max-II,' *Proceedings EPAC 1996*, Sitges, Barcelona, 1996, p. 1689.

[71] O. Chubar, P. Elleaume, and A. Snigirev, 'Phase Analysis and Focusing of Synchrotron Radiation,' *NIM* **A435** (1999) 495.

[72] A. P. Prudnikov, Yu. A. Brychkov, and O. I. Marichev, *Integrals and Series*, Vol. 2, Gordon and Breach Science Publishers, New York, 1986, p. 198.

[73] K. W. Robinson, 'Radiation Effects in Circular Electron Accelerators,' *Phys. Rev.* **111** (1958) 373.

[74] S. O. Rice, 'Mathematical Analysis of Random Noise,' *Selected Papers on Noise and Stochastic Processes*, edited by. N. Wax, Dover Publications, Inc., New York, 1954, p. 147.

[75] M. Sands, 'Synchrotron Oscillations Induced by Radiation Fluctuations,' *Phys. Rev.* **97** (1955) 470.

[76] J. M. Paterson, J. R. Rees, and H. Wiedemann, 'Beam Size Control in PEP,' SLAC PEP 125, SLAC, Stanford, 1975.

[77] M. Placidi, private communication, 1986.

[78] R. Coisson, 'Spatial Coherence of Synchrotron Radiation,' *Appl. Optics* **34** (1995) 904.

[79] J. Murphy, 'Synchrotron Radiation Source Data Book,' Brookhaven National Laboratory report BNL 423 33, 1996.

[80] L. I. Schiff, 'Production of Particle Energies Beyond 200 MeV,' *Rev. Sci. Instrum.* **17**.

[81] K. W. Robinson, 'Storage Ring for Obtaining Synchrotron Radiation with Line Spectra,' Cambridge Electron Accelerator Report CEAL-1032, 1966.

[82] L. V. Iogansen and M. S. Rabinovich, 'Coherent Electron Radiation in a Synchrotron I,' *Soviet Phys. JETP* **35** (1959) 708.

[83] L. V. Iogansen and M. S. Rabinovich, 'Coherent Electron Radiation in a Synchrotron II,' *Soviet Phys. JETP* **37** (1960) 83.

[84] S. A. Kheifets and B. Zotter, 'Coherent Synchrotron Radiation, Wake Field and Impedance,' CERN SL Report 85-43, CERN, Geneva, 1995.

[85] J. S. Nodvick and D. S. Saxon, 'Suppression of Coherent Radiation by Electrons in a Synchrotron,' *Phys. Rev.* **96** (1954) 180.

[86] S. A. Kheifets and B. Zotter, 'Shielding Effects on Coherent Synchrotron Radiation,' *Micro Bunches Workshop, Upton, N.Y. 1995*, edited by. E. B. Blum, M. Dienes, and J. B. Murphy, AIP, Woodbury, New York, 1995.

[87] T. Nakazato, M. Oyamada, N. Niimura, S. Urasawa, O. Konno, A. Kagaya, R. Kato, T. Kamiyama, Y. Torizuka, T. Nanba, Y. Kondo, Y. Shibata, K. Ishi, T. Obsaka, and M. Ikezawa, 'Observation of Coherent Synchrotron Radiation,' *Phys. Rev. Lett.* **63** (1989) 1245.

[88] R. Kato, T. Nakazato, M. Oyamada, S. Urasawa, T. Yamakawa, M. Yoshioka, M. Ikezawa, K. Ishi, T. Kanai, Y. Shiabata, and T. Takahashi, 'Suppression of Coherent Synchrotron Radiation in Conducting Boundaries,' *Proceedings of the IEEE Particle Accelerator Conference PAC-93* (1993) p. 1617.

[89] M. Abo-Bakr, J. Feikes, K. Holldack, G. Wüstefeld, and H. W. Hübers, 'Powerful Steady State Coherent Synchrotron Radiation at BESSY II,' *Proceedings of the 8th European Particle Accelerator Conference, EPAC-2002*, Paris, published by CERN, Geneva, 2002, p. 778.

[90] K. J. Kim, 'Brightness, Coherence and Propagation Characteristic of Synchrotron Radiation,' Lawrence Berkeley Laboratory report LBL-20181, 1983.

[91] R. Bossart, J. Bosser, L. Burnod, R. Coisson, E. D'Amico, A. Hofmann, and J. Mann, 'Observation of Visible Synchrotron Radiation Emitted by a High Energy Proton Beam at the Edge of the Magnetic Field,' *Nucl. Instrum. Methods Phys. Res.* **164** (1979) 375.

[92] R. Coisson, 'Monitoring High Energy Proton Beams by Narrow Band Synchrotron Radiation,' *1977 Particle Accelerator Conference, IEEE Trans. Nucl. Sci.* **NS-24** (1977) 1681.

[93] J. Bosser, L. Burnod, R. Coisson, G. Feroli, J. Mann, and F. Méot, 'Characteristics of the Radiation Emitted by Protons and Antiprotons in an Undulator,' CERN-SPS/83-5, CERN, Geneva, 1983.

[94] J. Arnold. T. Bohl, H. Burkhardt, R. Cornali, K. Cornelis, G. Engelmann, R. Giachino, A. Hofmann, M. Jonker, T. Linnecar, M. Meddahi, L. Normann, E. Shaposhnikova, A. Wagner, and B. Zotter, 'Energy Loss of Proton and Lead Beams in the CERN-SPS,' *Proceedings of the 1997 Particle Accelerator Conference, PAC97*, Vancouver, Canada, June 1997, p. 1813.

[95] I. S. Gradshteyn and I. M. Ryzhik, *Table of Integrals, Series and Products*, edited by A. Jeffrey, 5th edition, Academic Press, Boston, 1994.

[96] G. N. Watson, *A Treatise on the Theory of Bessel Functions*, Cambridge University Press, London, 1962.

Index